Renewable Energy Sources & Energy Storage

Series editor

Xiangjun Li, China Electric Power Research Institute, Beijing, China

The book series Renewable Energy Sources & Energy Storage publishes monographs, professional books and textbooks on the latest advances and developments in the field of renewable energy sources (RES; incl. wind, solar, hydrogen, biomass, etc.), energy storage (ES), conversion and applications from micro-grid to macro-grid level. Topics covered in the series are the key technologies and practices involved in RES and ES systems; solving problems in equipment manufacture; system integration; installation & debugging; SCADA; control; power prediction; status estimation; safety assessment; energy management; economic evaluation; experimental methods; establishment of standards and grid application.

More information about this series at http://www.springer.com/series/15433

Qiuye Sun

Energy Internet and We-Energy

Springer

Qiuye Sun
College of Information Science
and Engineering
Northeastern University
Shenyang, China

ISSN 2509-9698 ISSN 2509-9701 (electronic)
Renewable Energy Sources & Energy Storage
ISBN 978-981-13-0522-1 ISBN 978-981-13-0523-8 (eBook)
https://doi.org/10.1007/978-981-13-0523-8

Library of Congress Control Number: 2018942157

Printed on acid-free paper

This Springer imprint is published by the registered company Springer Nature Singapore Pte Ltd.
The registered company address is: 152 Beach Road, #21-01/04 Gateway East, Singapore 189721, Singapore

Preface

Background of This Book Energy is the driving force of social and economic development, and it is the important material basis for human survival. The use of energy varied over time, from the age of firewood to the coal age of fossil energy, the age of the oil and gas, the electrical age, and by now the era of clean energy represented by wind, solar energy, water energy, and biomass energy. The change of utilization of energy is accompanied by great progress of human civilization and a great leap in the productive forces of society and economy. Social development and scientific and technological progress increase human's dependence on energy. Thus taking the third industrial revolution as an opportunity, establishing a safe, efficient, economic and environmental new energy supply model has become a great challenge in the process of sustainable development of human society.

On the basis of the idea of free transmission and open sharing of Internet information, the practitioners of the energy industry propose to build a new environmental-friendly energy network, with the characteristics of open interconnection, interactive sharing, and economic-information-energy integration, to achieve low carbon production and consumption of energy and ensure the sustainable development of energy. Therefore, the Energy Internet comes into being and attracts widespread attention from energy and correlated industries worldwide within a short time. People try to interpret it from many aspects such as society, environment, economy, technology. However, what is the definition of Energy Internet? Is the energy network with various types of energy sources connected the Energy Internet? Is it a combination of energy and the Internet? Or is it a network just so-called smart grid 2.0? Perhaps all of opinions mentioned above depict features of Energy Internet partially, but not all characteristics of Energy Internet. The authors believe that the Energy Internet is a network with high complexity where information and energy are integrated in depth and share equal access to multiple types of energy resources through a variety of energy transmission media. What's more, it is a novel energy production, transmission, and consumption network which can realize the open sharing of information and energy internally and achieve efficient and environmental-friendly utilization of energy.

Since the Internet and the Energy Internet are inextricably linked, does the Energy Internet inherit all the characteristics of the Internet? Can information at any point of Energy Internet be obtained by the whole network? Can information be duplicated and restored infinitely in Energy Internet? Can any user in Energy Internet be the publisher and receiver of information at the same time? In order to answer these questions better, this book presents the concept of We-Energy, a novel energy interaction mode based on a cyber-physical-economy-energy model. On the basis of this concept, we will interpret the Energy Internet and its energy conversion process.

We-Energy is a combination of energy producers, energy storage devices, and consumers, it is capable to transform various types of energy such as electricity, district heat, and natural gas into desired energy types. It can exchange with others using advanced communication, electronic conversion, and automatic control technology. We-Energy is located at the bottom of the information network and energy network; it can absorb energy from the energy bus and can also provide energy for it. Unlike traditional energy suppliers, We-Energy adopts a structure which allows bottom-to-top power interaction from users. What's more, it holds a concept of point-to-point energy transmission. We-Energy has some important characteristics such as source and load coordination, multi-energy complementarity, peer-to-peer access, energy-information-economic coupling, and plug-and-play. We-Energy promotes the efficient use of energy and achieves the transformation from traditional vertical power dispatch pattern to a distributed and coordinated power dispatch pattern. Furthermore, it completes the transition from fossil energy to renewable energy and ultimately brings human society into the era of zero marginal energy costs.

Unwittingly, Energy Internet has attracted great attention worldwide; from being rarely known, various related technologies have sprung up. However, it still calls for great effort of all aspects from technology and economy to national policy to achieve the landing of Energy Internet down to earth. What is worth of mentioning, how to make more researchers and energy users to know Energy Internet and realize their positions in Energy Internet, and how to promote the generation and connection to grid of We-Energies have become the key factors during the smooth landing of Energy Internet. Over the past few years, the authors have been promoting related academic work through Energy Internet forums, reports, and conferences. Therefore, in this book, rather than present intricate formula deduction and complex system analysis, the authors strive to present Energy Internet to readers in an intuitive way by using straightaway sentences and diagrams.

This book is divided into two parts. The first part including Chaps. 1–4 mainly makes an overall introduction of Energy Internet and its integrated applications. In this part, the authors briefly introduce the origin of the concept of We-Energy and depict the structural features of the Energy Internet. Analysis on the network characteristics and security of information physics technologies have been done as well. The second part including Chaps. 5–10 focuses on specific research on Energy Internet. Coordinated control strategies for power management of hybrid micro-grid and distributed coordination control of multi-agent have been proposed. Research on energy router has also been done including its control strategy and stability analysis. What's more, the modeling and dual control strategies of energy hubs have been presented in this book. The energy flow calculation of the multi-energy system has been carried out including power flow calculation under steady state and reinforcement learning-based distributed energy flow calculation. Additionally, issues on optimal operation of Energy Internet have been done and consist of distributed cooperative management and reinforcement learning-based multi-WE-energy management.

As a complex network with energy, device, information, and economic coupling, Energy Internet needs to be viewed from a new perspective. The book tries to depict the characteristics of Energy Internet from multiple aspects, such as electrical, thermal, and information, for the first time. The authors hope to provide a helpful reference for readers.

The authors would like to acknowledge all of the help and encouragement received in the development of this book. Many of our graduate students and colleagues have contributed to the materials of this book. They are Jianguo Zhou, Fei Teng, Yushuai Li, Bingyu Wang, Yuyang Li, Rui Wang, Jingwei Hu, Lingxiao Yang, Ning Zhang, Danlu Wang, Yi Zhang, Xiaoting Yu, Qianyu Dong, Qian Sun, Dehao Qin. Finally, Qiuye Sun has checked the whole book carefully. At the same time, I want to thank the researchers for their efforts on Energy Internet so that I can successfully complete the book.

Shenyang, China Qiuye Sun

Contents

Chapter 1
Energy and Energy Internet

Abstract Energy is the foundation of the development of the world economy. The human society cannot live without energy. Throughout human history, the advancement of the human society is closely related to the exploitation and utilization of energy. The reformation of the energy construction cannot live without the development of science, while the connection between development of the economy and the support of energy cannot be separated. Energy, science and economy are mutually reinforcing and restriction. From the view of the sustainable development of society, the exploitation and utilization of the renewable energy and taking place of the traditional fossil fuel are important direction for the human to reform the construction of the energy. However, the existing grid construction is impossible to meet the demand of promoting energy conversion between different kinds of energy and improving the efficiency of the energy utilization. In order to solve this imminent problem, people begin to do research on a novel energy network construction which is known as Energy Internet. Constructed with the concept of the Internet, Energy Internet is a wide area network which combines economics, information and energy. Through building a large power grid as the backbone network and forming an open, equal economic, information and energy integrated framework, Energy Internet realizes the bidirectional energy transmission and dynamic balance utilization to maximize the adaptation to the access of the new energy. Moreover, in Energy Internet, energy is able to be conversed among electric energy, chemical energy and heat energy, meanwhile, as the hub of the energy conversion, power system should undertake the core energy conversion. This chapter introduces a novel energy network construction and makes a detailed discussion on the concept, characteristics, construction and the contained energy types of Energy Internet.

1.1 The Situation of World Energy

In recent years, the structure of energy consumption has changed dramatically, the United States, China and India have become the largest energy consumption countries, and made a big difference to the world.

Q. Sun, *Energy Internet and We-Energy*, Renewable Energy Sources & Energy Storage, https://doi.org/10.1007/978-981-13-0523-8_1

1.1.1 Energy Reserves and Distribution

At present, half of the world's ten largest oil producing countries are non-OPEC countries. Until 2013, the world's oil reserves are 1.6879 trillion barrels, which can satisfy 53.3 years global need. Over the past ten years, the world's oil proved reserves have increased by 27%. Moreover, at the end of 2013, the world's proved coal reserves amounted to 8600 billion tons. The United States, Russia, China, Australia, India and Germany were the world's top six coal resources countries. The six countries accounted for 80% of the world's coal resources. According to the latest data, by the end of 2013, the world's proved reserves of natural gas are 185.7 trillion cubic meters, enough to guarantee the need of 54.8 years. Iran has replaced Russia and become the world's largest natural gas resource country. In the field of production, the United States has replaced Russia and become the world's largest producer of natural gas.

In non-fossil energy sources, the total installed capacity of nuclear power generation increased year by year. By the end of 2013, with a total installed capacity of 372 million kilowatts, there are 436 nuclear power reactors could operate all over the world. In terms of regional distribution, North America, Europe and the Far East are still the main areas of nuclear power use. Meanwhile, hydropower development varies from place to place. The focal point of development of developed countries in North America and Europe has been transferred to the renovation and reformation of existing hydropower stations. Most developing countries in Asia, South America and other regions have made development plans. However, in the less-developed countries of Africa and other regions, the hydropower development is still facing many difficulties due to the constraints of capital and technology. Moreover, the development pace of wind power has slowed in Europe and America, but it is rapidly advancing in Asian countries. In 2013, the EU's wind power installed capacity reached a milestone of 100 million KW. But because of the European sovereign debt crisis and the global economic recession, the demand for fans in Europe will remain at a stable level in the coming years. In Asia, in 2013, China's installed capacity increased by 16088.7 MW and the increase was 21.4% year-on-year. The data of new installed and accumulative installation rank first in the world.

1.1.2 Energy Consumption

Emerging economic entities are still dominant in global energy demand. In 2013, the growth of energy demand in emerging economic entities accounted for 80%. Energy consumption in the EU has continued to decrease, reaching its lowest level since 1995, and energy consumption in Japan has declined to the lowest level since 1993.

(1) The consumption of fossil energy

At present, the world's energy consumption is still dominated by emerging economic entities. However, in EU, along with its economic development to a certain extent, some national energy demand amount decreased gradually from the beginning of 1995, in 2013 the demand for energy has reached the lowest energy since 1995. Moreover, the total energy demand in Japan in 2013 has reached a minimum value since 1993. In 2013, the world's demand for oil energy increased by 1.4%, but the world's total oil extraction only increased 0.8%, growth in demand exceeds supply growth. In terms of coal consumption, the Asia Pacific region, the Americas and Europe are the largest coal demand regions. In the field of natural gas consumption, the global gas consumption increased by 1.4%, and natural gas accounted for 23.7% of the primary energy consumption. China and the United States have the most significant increase in natural gas consumption, 10.8 and 2.4%, accounting for 81% of the global growth in natural gas consumption. The EU's gas consumption reached the lowest level since 1999.

(2) The consumption of non-fossil energy

In recent years, with the depletion of fossil energy and the increasing pollution, most of the countries in the world have paid more and more attention to the development of renewable energy. The development of renewable energy has been strongly supported worldwide. In 2013, the global consumption of renewable energy accounted for 2.7% of total energy consumption, increased by 0.8% since ten years ago. China has the largest growth rate of renewable energy consumption, followed by the US, while the major energy consumers in Europe like Germany, Spain and Italy are below average growth rate. On the global scale, wind energy consumption increased by 20.7% over the previous year, once again occupying more than half of the growth of renewable energy generation. The growth of solar power generation in the first half of the year has increased by 33% faster than the previous year; nuclear power generation accounts for 4.4% of global energy consumption, and it is the lowest since 1984. But nuclear power has increased by 0.9%, and the first growth is from 2010. The growth of nuclear power in the United States, China and Canada has compensated the decline in nuclear power generation in South Korea, Ukraine, Spain and Russia. Nuclear power accounts for 4.4% of the world's energy consumption, and hydropower accounts for 6.7% of the world's energy consumption. Although the proportion of non-fossil energy consumption is increasing, it still cannot significantly alleviate the world's energy and environmental difficulties. At present, electric vehicle and solar charging pile are still in research and popularization stage, while other energy conversion technologies are still in a small pilot stage, and their capacity is not enough to support people's daily life. Due to some large wind farm output increasing at night, it contradicts to the traditional power grid generation and there is no good solution. Therefore, to motivate the non-fossil energy to really enter the human energy development history, we also need to build a broader application platform.

1.2 Energy Conversion Technology

Energy is able to be divided into two categories: primary energy and secondary energy. As the energy and source which directly acquired from nature without any conversion, primary energy includes raw coal, crude oil, natural gas, oil shale, nuclear energy, solar energy, hydraulic power, wind power, wave energy, tidal energy, geothermal energy, biomass energy, ocean temperature difference energy and other energy. Secondary energy is the energy product which converted from primary energy, which includes electricity, steam, gas, gasoline, diesel, medium oil, liquefied petroleum gas, alcohol, biogas, hydrogen, coke and other energy. Moreover, primary energy is able to be further divided into renewable energy and non-renewable energy. As the energy which is recyclable and renewable in nature, renewable energy includes solar energy, hydraulic power, wind power, wave energy, tidal energy, geothermal energy, biomass energy, ocean temperature difference energy and other energy. Meanwhile, non-renewable energy which includes raw coal, crude oil, natural gas, oil shale, nuclear energy is impossible to renew in nature.

The utilization of primary energy and secondary energy reflects the development of a country. According to the analysis of the energy supply structure of the main countries, the developed countries have almost already accomplished the conversion from primary energy to secondary energy. Moreover, in developed countries, the traditional biomass energy utilization disappeared, the ratio of the coal utilization decreased sharply, the ratio of oil and natural gas utilization increased evidently, and the utilization of nuclear power, hydro power and biomass energy have already become the main body of the non-fossil energy. However, with the gradual exhaustion of resources, the utilization of wind power and solar energy should become the main direction of the future energy utilization. Compared with the developed country, the developing country has not accomplished the conversion from primary energy to secondary energy, the ratio of oil and natural gas is still low in primary energy structure, and the exploitation of nuclear power, hydropower and biomass energy still have great potentialities. From the energy consumption, taking the situation of China which is a typical developing country as a case, the consumption of coal, oil and natural gas still account for a large part, and it is of vital significance for China to motivate the conversion from primary energy to secondary energy. However, the energy development strategy of the developed country and the developing country are different, the former should mainly develop wind and solar energy, and the latter should develop nuclear energy, hydropower and biomass energy.

1.3 The Introduction of Energy Internet

Energy is the material basis for contemporary social and economic development. Considering that the traditional fossil fuels are gradually depleted and the global environmental pollution is worsening by the minute, it is high time what we should stimulate the development of the energy transition. The emergence of Energy Internet, which is the key to the third industrial revolution, is able to encourage the energy transition. Nowadays, the encouragement of Energy Internet development is an effective method to gradually replace fossil fuels with renewable energy sources, shift the structure of energy consumption to a low-carbon economy, and establish a green, sustainable energy utilization system. Because of the advancement of Energy Internet, Energy Internet has aroused great concern among governments and scholars from all over the world.

Energy Internet, as a new generation of more intensive, more environmentally friendly and more sustainable energy utilization pattern, plays a constructive role in the development of human life, society, economics and other aspects. To some extents, Energy Internet is able to be regarded as an integrated product which contains the field of energy, automation, information processing, network communication and others. Moreover, Energy Internet is able to be understood as a novel open energy ecosystem which regards power network, thermal network, natural gas network, traffic network and other complex networks as physical entities, regards energy technology, advanced control technology, intelligent optimization technology information processing technology and other technologies as the implementation methods. With the development of Energy Internet, the efficient utilization of renewable energy and the increase of the proportion of renewable energy in the production and consumption of primary energy are able to be achieved.

Figure 1.1 shows the development course of Energy Internet. In the 1930s, R. Buckminster Fuller (American thinker and non-specialist, 1895–1983) realized that if we interconnected the world through an electrical energy grid, the energy demand of population centers was able to be supplied by renewable resources from distant locations [1]. To the World Game seminar of 1969, R. Buckminster Fuller presented integrated, world-around, high-voltage electrical energy network concept [2]. Moreover, R. Buckminster Fuller presented the concept of global energy grid and concluded that "the global energy grid is the World Game's highest priority objective" in his book "*Critical Path (1983)*". Then in 1986, Global Energy Network Institute (GENI) was founded in 1986 by Peter Meisen to investigate the global

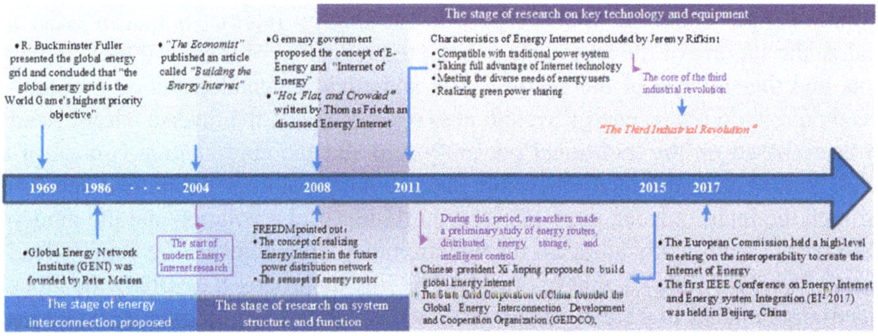

Fig. 1.1 The development course of energy internet

electric energy grid which proposed by R. Buckminster Fuller. The GENI conducts research and education related to interconnect electrical power grids between countries and continents, thereby creating an interconnected global energy grid, with an emphasis on linking local and remote renewable energy resources (wind, solar, hydro, geothermal, tidal and biomass) [3]. However, with the immaturity of the energy technology and the lack of the specific implementation steps, these are just quite a few future prospects for the future power system. The researchers from all over the world did not engage in the research of the contemporary Energy Internet until *"The Economist"* published an article called *"Building the Energy Internet"* in 2004 [4]. Then in 2008, many countries began to encourage the development of Energy Internet successively. In the United States, funded by the National Science Foundation, the future renewable electric energy delivery and management (FREEDN) systems center of North Carolina State University pointed out that in order to realize the plug and play of large-scale distributed renewable energy and distributed energy storage equipment, the power electronics technology and the information technology should be introduced into the power system, and the future power distribution network Energy Internet should be realized through the energy router and other power conversion or intelligent control equipment of the Energy Internet [5]. At the same time, Federal Ministry of Economics and Technology (BMWi) in Germany proposed the concept of E-Energy and with E-Energy, an "Internet of Energy" is developed. E-Energy is an energy system of future based on novel information and communication technologies (ICT). "The primary goal of E-Energy is to create E-Energy model regions that demonstrate how the tremendous potential for optimization presented by ICT can best be tapped to achieve greater efficiency, supply security and environmental compatibility in power supply, and how, in turn, new jobs and markets can be developed". Besides, BMW appropriated up to €140 million which be mobilized for the development of six E-Energy model regions [6]. Similarly, in September 2008, the book "*Hot, Flat, and Crowded: Why We Need a Green Revolution—And How It Can Renew America*", which is written by *New York Times* Foreign Affairs columnist Thomas Friedman, discussed detailly the Energy Internet [7]. In 2011, Jeremy Rifkin in his book *"The Third Industrial Revolution: How Lateral Power Is Transforming Energy, the Economy, and the World: Library Edition"* pointed out that Energy Internet is an important symbol of the third industrial revolution. Jeremy Rifkin also presented the main purpose of the Energy Internet is that in order to realize the sharing of the renewable energy, the reduction of the fossil fuels utilization, and the solution of the energy crisis and environmental pollution problems, we should build a new energy system network that meets the diverse energy needs of users based on the traditional power system and the coordination control of a large number of distributed renewable energy, energy storage equipment and load through the Internet technology [8]. The third industrial revolution and the modern Energy Internet quickly attracted the attention of a large number of country's leaders. In September 2015, Chinese President Xi Jinping proposed to satisfy the global electricity demand in a clean and green way through Energy Internet in the United Nations Development Summit. Then in 2016, the Global Energy Interconnection Development and Cooperation Organization (GEIDCO), with its permanent office domiciled in Beijing, China, is founded to promote the sustainable development of energy worldwide. The GEIDCO is launched by the State Grid Corporation of China and it is also the first cooperation and coordination organization of the global Energy Internet [9]. In May 2017, the European Commission held a high-level meeting on the interoperability to create the Internet of Energy [10]. In November 2017, the first IEEE Conference on Energy Internet and Energy system Integration (EI^2 2017) was held in Beijing, China. The conference aims to promote the interconnection, openness, sharing and coordination of various energy resources and shaping a green, low-carbon, efficient and low-cost energy ecosystem [11].

1.4 The Basic Definition of Energy Internet

As the next generation of the energy system, Energy Internet attracts a large amount of attention of the researchers from the United States, the European Union, China and other countries. Researchers have studied the Energy Internet from different layers, such as system design, physical entity, core technology, key equipment, resource configuration and other aspects. However, considering the compatibility and complexity of Energy Internet, the development of Energy Internet involves the different technologies from quite a few disciplines. Therefore, it is arduous to make a comprehensive and accurate definition of Energy Internet. Moreover, researchers from different fields have a different understanding of Energy Internet. Therefore, it is not likely to make a consensus definition of Energy Internet in a short term. Nowadays, there are four main definitions of Energy Internet which receive a wide range of recognition.

From the aspect of the purpose of building Energy Internet and the system design, Energy Internet is mainly based on the use of Internet technology to realize the coordination of distributed power, energy storage equipment and load in the wide area and realize the transformation from centralized fossil energy utilization to distributed renewable energy utilization. Therefore, Energy Internet should possess four principal characteristics. (1) The renewable energy is the main primary energy source. (2) Support the grid-connection of the large-scale distributed generation system and distributed energy storage system. (3) Realize the energy sharing in a wide area based on the Internet technology. (4) Support the electrification of the traffic system.

From the aspect of the physical entity of the complex network in Energy Internet, Energy Internet is a complex multi-energy flow system which couples tightly with the natural gas network, traffic network and other systems. Meanwhile, the power system is the core of Energy Internet, the Internet and other advanced information technology are the main means and the renewable source is the main primary energy.

From the aspect of the core technology and the key equipment of Energy Internet, some researchers point out that energy router is the key equipment of Energy Internet. Meanwhile, through the application of intelligent terminal, information acquisition and processing, prediction analysis, collaborative control, cloud computing, Internet of things, large data and power transformation control and other related technologies,

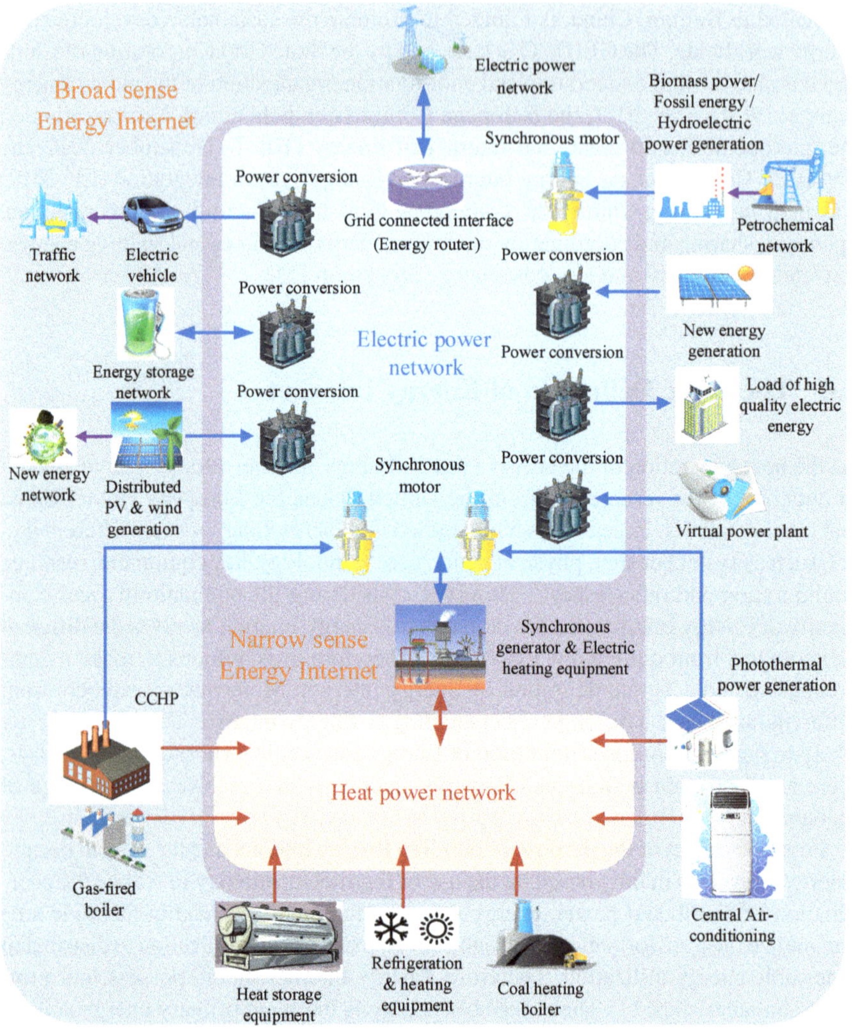

Fig. 1.2 Broad sense energy internet and narrow sense energy internet

Energy Internet realize the interconnection of the network nodes which composed of a large number of distributed energy acquisition and storage devices and various kinds of loads. Moreover, Energy Internet is able to realize energy transaction and sharing network of the bidirectional energy and information flow.

From the aspect of global energy resource distribution, Energy Internet can be regarded as the advanced stage of the development of strong smart grid. With the clean energy as the predominance and Ultra-High Voltage (UHV) as backbone network, various functions can be realized such as the power grid interconnections between different countries and continents, the global energy distribution of energy resources, the coordinated development of power grids at different levels, and the flexible access to the strong smart grid of various power sources and users.

From our perspective, according to the research status of Energy Internet, considering the structure and function of Energy Internet, Energy Internet can be defined as narrow sense Energy Internet and broad sense Energy Internet. As Fig. 1.2 shows, broad sense Energy Internet accumulates the scattered energy fragment in a huge area and form a super energy body with a flat structure. It deeply combines all kinds of energy resources and information resources which are utilized by human beings in different regions, such as energy-based network of current society (traffic, electric power, petroleum and other fields) and the information network which based on advanced information technology, to realize the interconnection and sharing of the production and consumption of the wide area energy resources and information resources. Compared with broad sense Energy Internet, narrow sense Energy Internet is a network which composes of certain topologies of energy production, consumption, and storage devices in a certain region, and realize coordination and complementation of the traditional energy networks through energy router and other energy conversion equipment. The interconnection of the different narrow sense Energy Internet combines all the energy-based networks and form broad sense Energy Internet.

1.5 The Characteristic of Energy Internet

Energy Internet is now a new concept, which is lack of entities equipment that can be used for research. The characteristics of Energy Internet appear in the differences between Energy Internet and smart grid/microgrid/ubiquitous energy network. This section will first list the distinctions between Energy Internet and the mainstream energy networks, based on which the characteristics of Energy Internet can be summarized.

(1) The Difference Between Energy Internet and Smart Grid

The smart grid is a modernized electrical grid with digital data streams, information technology, and advanced control methods to more efficiently produce, transmit and distribute electricity, which in turns to a reliable, secure, economic and environmentally friendly electrical grid [12]. The smart grid is designed to satisfy the electrical demand in the 21st century, which allows multiple kinds of the power supply (especially the renewable energy resources), optimizing the electricity market to the economic operation [13].

The main differences between Energy Internet and smart grid are as follows:

1. The power supply. The smart grid is a modernized electrical grid, while Energy Internet consists of an electrical grid, natural gas network, and heating network. Energy Internet is not just an integration of multiple power networks, but a unified power system in measurement, modeling, optimization and dispatch.
2. The access control. In smart grid, the distributed control of energy supply or dispatch is still a local control method, which needs a centralized control center

to collect information and route commands. In this way, each generator has to be "permitted" by the control center before it accesses the power system. Differently, Energy Internet emphasizes the distributed control, in which each energy generator can have "plug and play" function.

3. The information process. Compared with the traditional electricity grid, the smart grid is more "informationalized" because of its comprehensive database. But its information process is no better than that in a traditional grid, in which information system and physical system are two independent systems without integration. Differently, the cyber-physical system is an essential part of Energy Internet, which shows the interaction between information system and physical system.

(2) The Difference Between Energy Internet and Microgrid

Compared with the traditional power grid, the microgrid is a relatively small electricity network with multiple distributed generators and their loads. It can operate in parallel to the grid through static switch [14] or operate in island mode. When a microgrid is in island mode, it can be seen as an autonomous system.

The differences between Energy Internet and microgrid are as follows:

1. The function establishment. A microgrid is a supplement of the traditional power grid, and the relation between them is subordinate and principal. A microgrid consists of a locally distributed generator, energy storage and electricity loads, and it can minimize the disturbance towards the power grid in turns to improve reliability. Differently, Energy Internet aims at comprehensive energy optimization, which is to realize the autonomous access of multiple generators. The relation between Energy Internet and the traditional power grid is paratactic.
2. The network structure. The microgrid is based on master-slave structure, generally with a centralized control center. In a microgrid, the users follow orders from the control center, which means they have little autonomy. Differently, in the Energy Internet, each energy user has no authority to take orders. The energy dispatch is optimized with a distributed control to realize different targets such as cost minimization. Users of Energy Internet have more autonomy than those of the microgrid.
3. In a microgrid, the kinds of energy supply and the information process are similar to those of a smart grid. The differences in these aspects are discussed above.

(3) The Difference Between the Energy Internet and the Ubiquitous Energy Internet

The concept of a ubiquitous Energy Internet is similar to the E-Energy in Germany, which is an advanced network consisting of the current electricity grid and heating system and other energy systems. The ubiquitous Energy Internet aims at high energy efficiency, taking coordinated control in each part of the network such as energy supply, storage, usage and recycle, resulting in a synergy between energy input and output. In a ubiquitous Energy Internet, efficiency control system can give orders to realize cascade utilization in each energy flow, so that the network is of the highest efficiency and the structure is highly ordered.

The differences between Energy Internet and ubiquitous Energy Internet are as follows:

1. The hub of energy transportation and energy transformation. The ubiquitous Energy Internet generally has a ubiquitous energy hub (the energy hub in E-energy), to control the energy transportation and energy transformation. That is to say, there is still a control center, and the ubiquitous Energy Internet is star structure. Differently, in Energy Internet, each energy supply is equal, and the structure of Energy Internet is similar to the busbar construction in the information field. The same type of energy can be transported and transformed through the bus line, and different types of energy can be transformed through the energy conversion devices.
2. The design target. A ubiquitous Energy Internet is an energy network, which emphasizes the efficient comprehensive utilization of different types of energy. The key problem of energy network is energy transform. Differently, Energy Internet considers more than just the energy utilization. With the combination of physical entity and information network, the Energy Internet aims to realize the energy optimal utilization on the basis of safe and stable operation.
3. The main energy participant. The ubiquitous Energy Internet emphasizes the utilization of different kinds of energy, and the key problem is to improve the efficiency of energy transform. The main participants are energy enterprises. Differently, the main participants of Energy Internet are the users of the traditional power system. In Energy Internet, the users can be energy providers because of the bilateral energy network. In this way, the energy can be balanced on the spot, which reduces the cost of energy transport and in turns improves the energy utilization ratio.

(4) Conclusion

Based on the analysis above, the Energy Internet is an advanced complex energy network with multiple energy supplies, a high degree of integration of information and energy, a flat structural network, and a real-time energy trade-off. Its characteristics are as follows:

1. Energy Internet has multiple energy supplies and different energy access control. Because of the space-time feature of these different kinds of energy, the influences resulting from the intermittency and volatility of the renewable energy resources can be reduced, in order to guarantee the stability of the energy network. With multiple energy supplies, the efficiency of renewable energy resources is greatly improved, and the coordination between energy system and the environment is realized.
2. In Energy Internet, the energy network is highly integrated with an information network. With advanced methods of information collection, transmission and processing, combining with corresponding strategy of information calculation, estimation and perception, the Energy Internet has global observability and controllability. These control methods guarantee the safe and stable operation of

Energy Internet in every part such as energy production, energy transportation, energy storage and energy consumption.
3. Energy Internet has a flat structural network, which guarantees every eligible energy provider can be distributed accessed and can realize "plug and play" function. This horizontal structure promotes local produce and local consumption of the renewable energy, reduces the cost of energy transportation, and ensures the common sharing of multiple energy supplies in the Energy Internet.
4. Energy Internet has an equal, free and real-time energy trade-off, which prune the redundant link to realize a straight trade between energy producers and energy consumers. It provides a convenient transaction of low cost for all participants, promotes energy revolution.

1.6 The System Structure of Energy Internet

Energy Internet is an advanced complex network integrated energy system with information technology, control technology, and communication technology. Based on the current study, the analysis of the system structure of Energy Internet is as follows.

As the next energy system, the development of Energy Internet draws continuous attention in academia. Energy Internet has high compatibility and complexity, whose development comprises many technologies such as energy, information, control, communication, etc. Each research area has unique emphasis, so in the short run, the consensus on the framework of Energy Internet is difficult to be found. Based on existing research results, this section will analyze the framework of Energy Internet.

(1) Energy Internet of Energy Configuration in Large Regions
Energy Internet of energy configuration in large regions focuses on the configuration range and the regulation ability of energy resources, and it emphasizes the resource sharing among wild-area energy and energy configuration between nations and continents. Based on the strong smart grid, State Grid came up with the theory that the ultra-high voltage is the skeleton of the global Energy Internet. By means of wide interconnection of grids between nations and continents, a better global energy configuration can be achieved, hence a new energy system with wide service range, high configuration ability, high reliability and low-carbon characteristics can be constructed. The target of global Energy Internet is to connect power station in the North Pole and the equator, large energy bases in countries and continents and all kinds of distributed generation together, so as to transform clean energy such as solar energy, wind energy, hydro energy and ocean energy into electric energy used in industry and civilian by breaking through the lack of resources, the strong environmental constraints and the restrictions of space and time.

(2) Energy Internet of Multiple Resources Connected

Energy Internet of multiple resources connected is a structure of future energy system, which is a new energy supply system highly coupling from electrical grid, gas network, oil network, and transportation network.

Energy Internet formed by the interconnection between two or more typical energy networks is the prototype of future energy supply system. By means of various equipment for energy productivity, energy usage, and energy storage, the complex energy networks such as electricity network, thermal network, traffic network and petrochemical network can be connected by the "plug and play" facility, it turns to form a low-carbon combined energy supply system with renewable energy resources being the primary energy.

This kind of Energy Internet is a complex energy network in the broad sense, which connects distributed energy suppliers together in order to dispatch and manage them in a united way. It transforms primary energy which mainly is renewable energy to secondary energy which is easy to transport electricity and heat, so as to realize energy interconnection and sharing in large regions. Also, this Energy Internet can be applied to a small energy supply system which is comprised of energy production, transformation, storage, and usage. With primary energy being renewable energy resources, by means of distributed generation technology, small power stations connected with photovoltaic/thermal systems can power residential areas, hospitals, and factories, contributing to a low-carbon city.

(3) Energy Internet Emphasizing on Optimal Use of Energy

As the name implies, this kind of Energy Internet focuses on the "plug and play", the ability of access device for large-scale smart terminals such as distributed energy resources and electric vehicles, as well as bidirectional energy flow in the network. There are profound changes in facilities for energy management/transition, frameworks and operation modes of the energy system. When energy balances locally, optimal utilization of energy resources can be realized. According to the types of key equipment, this kind of energy has two categories.

One is considered as a multiple energy carrier networks, whose core is an interface —energy hub—between energy supplier and user. Based on energy need in the network, energy hub can achieve the transformation from the primary energy such as renewable energy, clean energy and fossil energy to the secondary energy such as electricity, heat and cold energy, in turns to allocate energy resources properly and energy sharing environmentally when the renewable energy is the majority of the primary energy.

The other one is considered as an energy supply system mainly based on the power grid. When building the Energy Internet, the major consideration is to design a smart energy transformation/regulation equipment—energy router, which can achieve the electricity transition between high/low voltage and dc/ac power in different kinds of generation facilities based on energy need in the network. At the same time, it can achieve voltage compensation, failure isolation and harmonic isolation to provide electric energy of high quality. By using advanced control theory, energy storage technology, and information technology, it can achieve intelligent management for the large-scale distributed generation, energy storage and energy user in the network. The smart control scheme of energy router is based on the bidirectional energy-information flow in Energy Internet and the tendency of integration in energy and information. And the target of the scheme is to empower the Energy Internet to achieve multiple tasks such as energy dispatch, allocation, quality test and load switching autonomously and in parallel, in order to guarantee the operation stability of Energy Internet.

(4) Energy Internet Based on Transformation of Existing Energy Network
This kind of Energy Internet takes internet concept for reference, mixing information technology with energy supply technology to form a clean energy network centering information and using renewable energy as the primary energy, making existing energy network more intelligent, environmental and low-carbon. It can ensure the coverage for all the physical entity of equipment from information internet to Energy Internet, which can promote efficiency, flexibility, and openness of the existing energy network. It can empower the distributed generation with the "plug and play" ability, achieve the equipment integration and configuration for energy supply, storage, and usage, and manage energy resources with high efficiency. This kind of Energy Internet is regarded as the future energy system.

In the network, the energy producer can communicate with the consumer through the information network and typical energy network to realize their connection in both informational and physical way. Energy flows by data package bilaterally between energy producer and consumer, which is the same with how information flows on the internet. Based on the cooperation between energy producers and information sharing with the grid, the Energy Internet can realize the intelligent energy dispatch, reliable energy transportation, and safe energy supply.

The four kinds of Energy Internet above have different emphasises but the same target which is to contribute to a low-carbon and sustainable future energy usage.

As the picture (graph) shows, the four kinds of Energy Internet explain the future energy usage in "solid-plane-line-dot" levels. Energy Internet of energy configuration in large regions is led by international and regional energy strategy, whose complexity is far more than a simple energy trading by involving international politics, diplomacy, and security;

Energy Internet of multiple resources connected mainly considers the conception of interconnection between multiple energy, which has massive challenges doing with society, institution, environment, and economics;

Energy Internet emphasizes on the optimal use of energy and Energy Internet based on transformation on existing energy network emphasize the inner framework of the Energy Internet and its implementation of the expected functions, which require more specific technology development.

This paper puts the former two kinds of Energy Internet into a category which promotes a low-carbon energy utilization with interconnection with multiple energy resources across regions. And the latter two are more specific, focusing on the key equipment in the network and its operational control problems, which is more interested in academia.

1.7 The Energy Systems in Energy Internet

Based on the current study, the Energy Internet can be classified into the global Energy Internet, the generalized Energy Internet, and the narrow-sense Energy Internet. Among them, the global Energy Internet has studied the major strategic issues and solutions of the world's energy sustainable development from a global perspective, and proposed a global energy concept that addresses energy issues in a global, historic, differentiated and open position. The generalized Energy Internet aggregates fragmented energy resources in a wide area into a "super energy" in which primary energy inputs are converted to secondary energy consumption and energy flows mainly in the network as electricity or heat, which integrates all the energy dispersed in different regions. And all kinds of energy are distributed, managed and scheduled in a unified manner. With the communication network, the optimal distribution of energy can be achieved, and the currently concentrated structure can be converted into a flat structure. In this highly decentralized and highly centralized energy system, the most notable feature is energy interconnection, sharing and user interaction. It contains all the energy-based networks (such as transport, electricity, petrochemicals, etc.) and advanced information networks. In the generalized Energy Internet, the electricity network, transportation network, petrochemical network, heat network, renewable energy network can be connected through a variety of electrical, cooling and heating equipment. And as a whole energy system, it can be connected to the narrow-sense Energy Internet through certain interface devices. In this super-energy, each current energy-based network is connected with the narrow-sense Energy Internet through the transmission devices of electric energy or heat energy, and the narrow-sense Energy Internet is divided into two major types of networks: electric energy network and thermal energy network. As the electricity being the most efficient in transmission and transformation, and thermal energy being easy-to-transform, these two kinds of energy become the core energy of the narrow-sense Energy Internet. This is also one of the important differences between the Energy Internet and the smart grid (with the power network as the core) (Fig. 1.3).

1.7.1 The Electricity Network

With the rising oil prices, the world's energy supply continued to be tight in 21 century. Proper development and utilization of green energy is the main solution to the energy problems in the future. At present, several common new energy sources include solar energy, wind energy and biomass energy, all of which are distributed energy sources.

Although DERs has the advantages of low investment, good environmental friendliness and high flexibility, its impact on large power grids is indeed an important issuc to be considered. IEEE P1547 provides a separate grid-connected standard for distributed energy: when a power system fails, the distributed energy must be taken

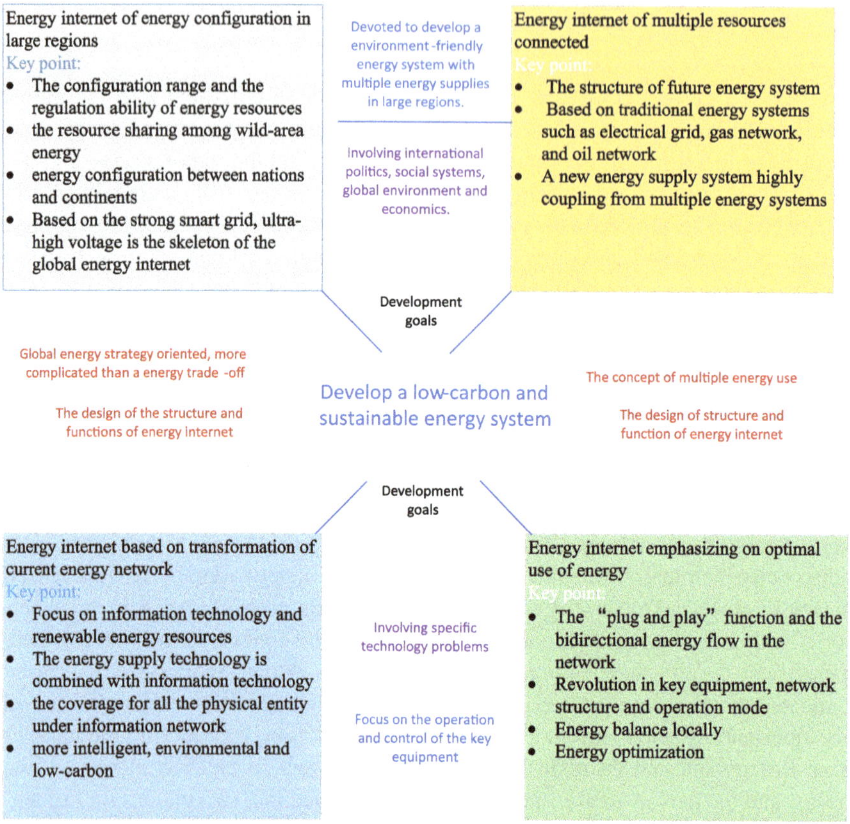

Fig. 1.3 Differences and relations between four kinds of energy internet

out of operation immediately. This greatly limits the full use of distributed energy, but also indirectly limits the use of new energy sources. In order to make most of the economic benefits and reliability improvements of distributed generation and to minimize its impact on the main network, the concept of microgrids has been proposed.

In 1999, for the first time CERTS conducted a study on the reliability, economy and environmental impact of microgrids. In 2002, a complete concept of microgrids was proposed. The definition of microgrids as given by CERTS is: Microgrids are a system of load and micro-power supplies that provide both electrical energy and heat; the conversion of power sources within the microgrid is mainly by the power electronics and is provided with the necessary control; micro-grid is a single controlled unit of the connected power grid, and can meet the user requirements for power quality and power supply security. DERs accessed by CETRS microgrids are all small units with a peak value of 2 MW or less, which avoids the use of fast but expensive controls and makes the system robust. The structure of the microgrid, shown in Fig. 1.4, shows

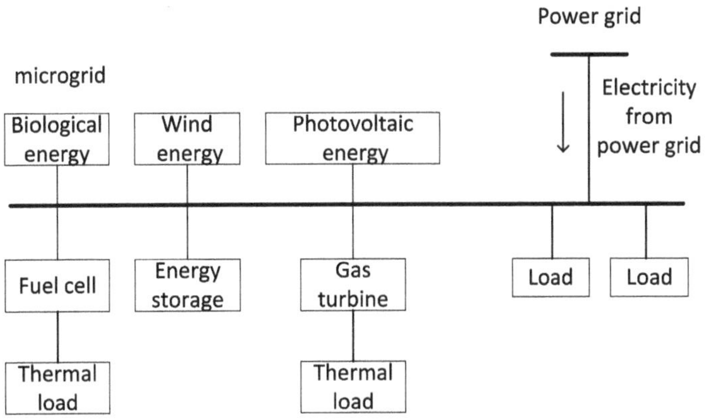

Fig. 1.4 The sketch of a microgrid

the micro-power forms of photovoltaic power generation, wind energy, fuel cells, micro gas turbines, some of which are close to the thermal load and provide local users with heat sources, thus improving the utilization of energy.

The current US CERTS microgrid research is mainly based on the "plug and play" and "peer" control ideas and design concepts. The preliminary theoretical research results of the US CERTS microgrid have been successfully tested on the laboratory microgrid platform. Microgrid demonstration projects such as the Mad River microgrid constructed in the northern United States power system have been tested on microgrid modeling and simulation methods, protection and control strategies and economic benefits, and the initial micro-grid management policies and regulations are formulated. From the perspective of the United States grid modernization, the focus of development of the US microgrid is to improve the reliability of power supply of important loads, meet the needs of users for a variety of power quality requirements, reduce costs and realize intelligence. The use and control of power electronic devices and many new energy sources in the CERTS microgrid provide a new idea for solving the problems of stabilizing and controlling the potential of renewable energy.

Japan has also started its research on microgrids based on the growing domestic shortage of energy resources and its growing load. To this end, Japan has also set up a special New Energy and Industrial Technology Development Organization (NEDO) to unify coordination of domestic universities, enterprises and national key laboratories on new energy and its applications. Japan's broadening of the definition of microgrid and the research on control and energy utilization based on the definition has provided a platform for the development of small-scale distribution systems and of large-scale independent systems based on traditional power sources. However, its development objective is mainly to diversify the energy supply, reduce pollution and meet the individual power needs of users.

Europe put forward the "Smart Grid" plan in 2005 and introduced its technical implementation plan in 2006 as the European 2020 and subsequent power development goals. At present, the theories of operation, control, protection, security and communication of microgrid have been initially formed in Europe. These theories are validated on the laboratory micro-grid platform. Its mission is to focus on researching advanced control strategies, developing corresponding standards, setting up demonstration projects, and preparing for the large-scale access of distributed and renewable sources of energy and the initial transition from traditional grid to smart grid. China's "*11th Five-Year Plan*" put forward the goal of building 5GW wind power and plans to build distributed power sources such as wind power and photovoltaic power to access the power grid. During the 12th Five-Year Plan period, China will set its goal in the solar energy and wind energy-dominated areas It will also promote the construction of 100 new energy demonstration cities. In the "*13th Five-Year Plan*", it pointed out that it is necessary to promote the energy revolution, speed up energy technology innovation, and build a clean, low-carbon, safe and efficient modern energy system.

1.7.2 The Transportation Network

The transportation network mainly refers to the location and capacity of charging piles, charging stations and other forms of energy supplement stations for the new energy vehicle under its popularization. As an effective way to solve the problems of energy shortage of traditional fossils, urban environmental pollution and global warming, electric vehicles have drawn worldwide attention. At present, electric vehicles have gradually become the focus of governments, automobile manufacturers and consumers. Electric vehicle charging pile, charging station planning is the focus of traffic network research, including site selection and fixed capacity. The planning is reasonable or not, will not only affect the convenience of electric vehicle users to travel so as to affect the promotion of the use of electric vehicles, but also affect power quality of distribution system. The State Council's *Energy-saving and New Energy Vehicle Industry Development Plan (2012-2020)* clearly put forward "by 2015, pure electric vehicles and plug-in hybrid vehicles are to reach the total production and sales to reach 500,000; By the 2020 Year, pure electric vehicles and plug-in hybrid vehicles production capacity is to reach 2000000, total production and sales to reach more than 5 million "industrial development goals. "2013 State Grid Corporation of Social Responsibility Report pointed out: The company will fully carry out the work of alternative energy and energy services to promote "substituting electricity for coal and oil, electricity from afar." The policy of "replacing oil with electricity" related to electric vehicles, and it may become one of the powerful measures to control urban haze. On February 25, 2014, Xinhua News Agency released the news that "Beijing will start large-scale construction of new energy vehicle charging pile for the first time, which would complete the construction of 1000 public quick-fill piles in 2014. And the report also pointed out that it would build services in the downtown area Radius of 5 km on average charging area. " The head of relevant departments introduced a set of data in the report: Since 2013, with the intensive introduction of various supportive policies in the country, the output of new energy vehicles in our country has jumped from 17,000 to 379,000 at an average annual increase rate of nearly 400%.

The biggest difference between the planning of electric vehicle charging station with that of the general power systems lies in the fact that it is necessary to consider the cost caused by network construction and land acquisition, as well as the market conditions after the charging station is built up and sales efficiency. Taking into account the electric vehicle ownership will be quite large in the future Energy Internet, dispatching agencies directly to each electric vehicle is unrealistic, electric vehicle mileage has been the focus of consumers. A well-equipped smart charging network helps alleviate the concerns of potential electric car customers and is crucial to the development of the entire electric car industry.

The development plan for electric vehicles and charging facilities is restricted to many factors. According to different development levels of power battery technologies, the quantity of the electric vehicle, the type of users, the coverage of vehicles, the differences of charging demand characteristics and the corresponding stages of charging facilities planning are different, Electric vehicle promotion can be divided into demonstration, public service, commercial operation:

(1) Demonstration

At this stage, the technology of electric vehicles has not yet been fully developed. The market for effectively and sustainably promoting the development of electric vehicles has not yet been established. The total amount and proportion of electric vehicles are very low. The majority of vehicles are mainly government-supported vehicles such as electric construction vehicles and rubbish trucks, etc. In addition, the driving range of electric vehicles are generally designated a smaller area or the specified route.

(2) Public service

This stage is the rapid development stage of electric vehicle technology, but still at a low level, and there are hidden constraints such as safety factors. At this stage the total amount and proportion of electric vehicles are still relatively low, the economic size is not high, and the development relies on government subsidies, leading publicity. The electric vehicles can be extended to electric buses, large enterprises and public vehicles, a few social vehicles.

(3) Commercial operation

At this stage, electric vehicle technology has basically matured. The total amount reached to a certain scale with more abundant species besides government and enterprises car, garbage truck, there are a large number of taxis and private cars, with larger charging needs. At this stage, EVs have the same or even higher economic efficiency than fuel trucks. The development mode is market-driven.

At present, electric vehicles are in the demonstration and promotion period. The convenience of charging electric vehicles will directly affect the purchase behavior of consumers for electric vehicles. Electric vehicle charging facilities are generally charging stations and charging piles. Charging modes are charging for the whole vehicle (fast or slow) and battery replacement. From the service object, charging stations can be divided into public charging stations for private cars and taxis, as well as dedicated charging stations for special vehicles such as electric buses. From the charging facility operation and practical point of view: public charging station service should first meet the needs of rapid charging of electric vehicles, only the rapid charger is used; charging pile for the long-time parking of electric vehicles to provide slow-charge service; the mode of replacing the battery need to solve the battery property rights, circulation, standards and other issues, which is currently more suitable for electric buses and other special vehicles.

Charging piles and dedicated charging station planning is easier to achieve, which can be set in the public parking or bus terminal stations and other private parking. Public charging station planning is different, not only to consider the scope and scale of services, but also consider the charge convenience. Public charging station layout planning model and calculation method are not yet mature. For the electric private car, taking into account the differences in the driving habits of the owners, the purpose of travel and travel lines, there are lots of random factors to provide charging services to the public charging station, and the determination process of corresponding charging load is also more complicated. In general, the layout planning of e public charging station for electric vehicles is a multi-objective and multivariable nonlinear optimization problem. The rationality and scientific reason of the scheme are very important to the sustainable development of smart grid and intelligent transportation network in the future significance.

1.7.3 The Heat Network

The thermal network, also known as heat pipe network, that is, a network starting from the boiler room, direct combustion engine room and heating center, etc., through the heat source to the thermal entrance of the building heating pipe, consisted of many heating pipes.

Hot water, as the heating medium, its design pressure is no more than 2.5 MPa, its design temperature is no more than 200 °C; steam, as the heating medium, its design pressure is no more than 1.6 MPa, its design temperature is no more than 350 °C. According to the different ways of energy heating, the thermal network can be divided into coal heating, oil heating, natural gas heating, electric heating, nuclear heating, geothermal heating, solar heating and so on. According to the different heating medium, the thermal network can be divided into hot water network and steam network, in which the steam heat network can also be divided into high pressure, medium pressure and low-pressure steam heat network according to the pressure level. According to the different water temperature, the heat energy network can be divided into the high-temperature hot water heat network ($T \geq 100$ Â °C) and low-temperature hot water heating network ($T \leq 95$ °C). According to their different status, the thermal energy network can be divided into first-grade pipe network and second-grade pipe network. First-grade pipe network is the network from the thermal core to the heat station. Second-grade pipe network consists of the pipes from the heat station to the user. According to different laying ways, the thermal energy network can be divided into a ditch laying, overhead laying and directly into the ground. According to the different forms of the system, the thermal network can be divided into closed systems and open systems. According to the classification of the supply and return the thermal network can be divided into the water supply networks and return networks.

As early as 1887, the United States established the first regional heating system in the world. France in 1893 and the former Soviet Union in 1903 had a district heating project. After the First World War, the heating industry in the developed countries as Canada, the Netherlands, Denmark, Hungary and Czechoslovakia, as well as other developing countries such as Poland, had also made some progress. After the Second World War, in the restoration and construction of the former Soviet Union, Eastern European countries and several Western European countries there were a large number of regional heating projects, and heating projects developed rapidly. At the same time due to the low fuel prices, the United States and many countries in western Europe developed heating system relatively slow. Since the world energy crisis in the 1970s, the district heating technology was re-emphasized and rapidly developed. Nowadays, the development of foreign heating industry has entered a period of steady development. Not only in its optimal management within the heating industry, technological innovation, but in government intervention in price regulation, access regulation and incentive regulation for the heating industry, heating industry reform has gained rich experience.

At present, the common heating methods in China are as follows: central heating, district heating, household heating and terminal heating. Centralized heating mainly refers to the centralized heat source as the main source of heat supply, the heat sources including coal-fired cogeneration, cogeneration gas and electricity, large coal-fired boiler room, large gas boiler room and large oil-fired boiler room, the main components are concentrated heat source, urban heat pipe network, heat exchange station, outdoor pipe network, indoor pipe network and the terminal cooling equipment. The terminal cooling equipment is mainly the radiator. The main features of urban central heating are the huge scale of heating, with the city as the object of heating, and with large-scale centralized heat source and two-stage heat pipe network. Regional heating is mainly based on the regional heat source heating method. The regional heat sources include coal, gas, oil-fired boiler room, electric boiler room, heat pump and cold-heat-electricity supply (CCHP). The main components are regional heat source, outdoor pipe network, indoor pipe network and end cooling equipment. The main features of regional heating are to have a certain scale of heating, the districts as heating objects, with outdoor heat sources and outdoor pipe network. Household heating mainly refers to the heat source placed in the home. Indoor heat sources mainly include coal-fired stoves, gas stoves, electric stoves (thermal storage and non-storage), water source heat pumps, air source heat pumps and ground source heat pumps. The main heat sources are the indoor heat source and the terminal cooling equipment. The terminal cooling equipment is mainly radiators, hot water radiant floor, radiators with thermostat valve and fan coil. The main features of household heating are the limited size of heating. With households as the heating object, the heat source located indoor. Although there may be indoor pipes exist, the pipes are heat radiators in the room, there is no heat loss, and no indoor heating indoor pipe network. The terminal heating system mainly refers to the system with heat source directly on the interior, in which the heat source is the terminal cooling equipment. The heat source mainly includes electric film, electric cable, electric heater and phase change energy storage heater. The heating is simple, the heat source is the terminal. The main feature of the terminal heating is limited to indoor heat supply, energy is generally electricity.

To achieve energy saving and environmental protection with minimum cost, the planning and optimization of the heat supply system must be realized first, including the selection of heat sources, the layout optimization of heat supply network, location and capacity. As the primary network of the central heating system can account for about 50% of the total investment of the system, so in the planning and optimization of the heating system, the network layout optimization is particularly important. Pipe network layout optimization commonly used simple tree algorithm. It is the optimal topology in all possible topologies, by respectively compared with the calculation, according to the weighted minimum. Pipe network layout optimization must first determine the pipe network cabling and heat exchange station capacity. Secondly, on the basis of planning and optimization, as well as the heat source, network topology, and heat transfer station have been determined on the basis of the pipe network design optimization, mainly the pipe diameter optimization. Because the actual standard pipe diameter is not a continuous variable, but a discrete variable, it must be considered in order to make the optimization results have practical guidance for engineering application.

1.7.4 The New Energy Network

When a lot of new energy is connected to Energy Internet, the Energy Internet will transform from the single energy allocation to the new energy interactive system with energy collection, energy transmission, energy storage and energy distribution. Reasonably planned and designed new energy network can effectively improve the utilization efficiency of new energy and improve the safety, economy and reliability of Energy Internet operation. However, if the type, installation location and capacity of the new energy are not compatible with the internet structure, not only the positive effects of new energy supply will be weakened, but also negative impacts will appear, such as increasing energy consumption and limited energy quality, and substantial increase in the risk of network failure, and so on. On the other hand, the planning of Energy Internet with new energy sources will be very complicated. The different investors will directly lead to the uncertainty of new energy's type, capacity, installation location and investment time, and due to intermittent characteristics of some renewable energies, the difficulty of load forecasting is significantly higher. As a large number of users install new energy sources, it is even more difficult for

Energy Internet planners to accurately predict the growth of the load, thus affecting subsequent planning.

1.7.5 The Petrochemical Network

With the development of fossil fuels, the natural gas-based petrochemical network has become the main way for the petrochemical transmission in the city. In the city's overall project of utilizing natural gas, the gas project is one of the important contents. The guiding role of the construction of gas projects is based on the city's overall planning and design of the overall gas-specific master plan, of which the natural gas pipeline network planning and design is an important part. The investment in urban natural gas system construction is huge. After the project is completed, it should not be easily rebuilt or expanded. Otherwise, it will affect the urban construction and affect the living conditions of urban residents, resulting in a huge waste of manpower, material resources and financial resources. Therefore, to ensure the healthy development of urban gas industry, natural gas pipeline network planning and design must be scientific. The goal of urban natural gas system engineering planning and design is not only to meet the requirements of users and process design but also to minimize the investment cost of urban natural gas system engineering and to ensure the operation of the entire natural gas network economically, safely, flexibly and reliably.

Due to the huge natural gas system, complicated structure and many restrictive conditions, it is difficult to achieve scientific and rational purposes by traditional manual design. The design of system optimization project must use advanced computer technology to solve the problems of low accuracy, low calculation efficiency and difficult to optimize the planning and design of traditional methods. It will guide the urban natural gas system engineering science and orderly construction and provide the urban managers with Decision-making basis, to achieve safe, economical, flexible and reliable gas supply, so as to save investment, improve the technical level and improve economic efficiency.

1.8 Artificial Methods in Energy Internet

1.8.1 AI Method in Single Energy Unit

We define the single energy unit (SEU) as an energy system which without any interconnection with other energy system. In other words, the generation, transmission and consumption of the energy merely happen in the SEU. When apply the AI methods in the SEU, we are able to categorize the methods in three aspects which include perception, decision and implement of the SEU.

In the perception of the SEU, we are also able to categorize into three aspects which include system prediction, fault detection, system monitoring and other aspects. The prediction of the amount of the power generation and consumption are the main aspects of the system prediction. When it comes to the prediction of the power generation, the AI methods are usually applied in renewable power forecasting. Moreover, besides applying AI methods to the power generation and consumption forecasting, AI methods are also able to utilized in electric price forecasting and other aspects. The AI methods are also applied in the power system fault detection and the fault includes the potential fault and the existed fault. The SEU is able to occur lots of kinds of faults, and most of faults is able to detected by AI methods and AI methods often perform more sensitive, fast and accurate than the conventional methods. Usually, by the means of analyzing the system voltage or frequency, we are able to detect the events occurring in the system.

In the decision of the SEU, we are able to categorize into optimization or management and assessment. The energy management which realized by AI methods can be categorized into end users management and system management. On the aspect of the management of end users. When it comes to the management of the SEU, the AI methods were mainly employed in demand side management in smart grid. Usually, the SEU state is able to be assessed by the AI methods, such as voltage stability, frequency dynamics, harmonic distortion, power quality, dynamic security, state estimation, etc.

In the implement of the SEU, we are able to categorize into the control and plan of the IES and other aspects. When AI methods applied to the SEU control, the control is able to be categorized into system control and equipment control. There are also many power converts' control which include AI methods. Moreover, the plan of the IES is mainly focus on the equipment placement and size.

1.8.2 AI Method in Multi Energy Units

In this section, we define the multi energy units (MEU) as an energy system in which communication network is essential. In other words, cooperation and competition happen between energy units in MEU to realize a coordinated goal of the whole system. In this section, we mainly review the AI methods which are utilized in the MEU. When apply the AI methods in the MEU, we are able to categorize the methods in two aspects which are consensus control and application of game theory.

In the consensus control of the MEU, there are stability control and optimal control, in which the communication networks are essential. When it comes with AI methods, the stability control often includes the consensus of frequency and voltage, and the power dispatch among energy units. The optimal control of energy internet is to minimize energy cost or energy consumption by optimize power dispatch, where AI methods are suitable for big data analysis with the growth of energy network.

In the application of game theory in MEU, especially in energy market, there are three parts which are energy provider, energy allocation and energy customer. As for energy provider, AI methods can be applied to assist the system to operate with the most economic generation plan, which includes generation scheduling, and the generation proportion of each provider. As for energy allocation, AI methods can be effective in solving the problems such as price forecast and price calculation. As for energy customer, AI method can be utilized to design an optimal pricing strategy for energy customers to be more benefit to choose from the main power grid or retail electricity market from the renewable resources.

References

1. The potential impact and importance of R. Buckminster Fuller's vision of a Global Energy Grid (2000) https://www.animatedsoftware.com/geni/rh2000ge.htm
2. R. Buckminster Fuller, *Critical Path* (1981)
3. GENI History—A Credible Foundation (2016) http://www.geni.org/globalenergy/issues/overview/history.shtml
4. The Economist, Building the Energy Internet (2004), http://www.economist.com/node/2476988
5. A.Q. Huang, M.L. Crow, G.T. Heydt, J.P. Zheng, S.J. Dale, The future renewable electric energy delivery and management (FREEDM) system: the energy internet, in *Proceedings of the IEEE*, vol. 99, no. 1, pp. 133–148 (2011)
6. E-Energy, *ICT-based Energy System of the Future* (2008) http://www.bmwi.de/Redaktion/EN/Publikationen/e-energy-ict-based-energy-system-of-the-future.html
7. T. Friedman, *Hot, Flat, and Crowded: Why We Need a Green Revolution and How It Can Renew America* (Farrar, Straus and Giroux, New York, 2008)
8. J. Rifkin, *The Third Industrial Revolution: How Lateral Power is Transforming Energy, the Economy, and the World* (Palgrave MacMillan, New York, 2011)
9. Overview of GEIDCO (2015), http://www.geidco.org/html/qqnyhlwen/col2017080765/2017-09/23/20170923162727440712239_1.html
10. High-Level Meeting: Interoperability to create the Internet of Energy (2017), https://ec.europa.eu/energy/en/events/high-level-meeting-interoperability-create-internet-energy
11. st IEEE Conference on Energy Internet and Energy System Integration (2017) <http://www.ieee-ei.org/>
12. O. Elma, U.S. Selamo ullari, An overview of demand response applications under smart grid concept, in *2017 4th International Conference on Electrical and Electronic Engineering (ICEEE)*, Ankara, Turkey, 2017, pp. 104–107
13. S.-W. Mei, J.-Q. Zhu, Mathematical and control scientific issues of smart grid and its prospects. Acta Automatica Sinica, 119–131 (2013)
14. X.-F. Yang, J. Su, Z.-P. Lü, H.-T. Liu, R. Li, Overview on micro-grid technology,in *Proceedings of the CSEE*, vol. 34, no. 1, pp. 57–70 (2014)

Chapter 2
Cyber-Physical Energy Internet

Abstract With the development of information technology, there are many important technology innovations in plenty of industries. The more innovations are used in industry, the harder we research information technology and traditional industry technology separately. So there is a new concept: Cyber-Physical Systems which are integrations of computation and physical processes [1]. These systems always integrate computer, communication and control technologies. As mentioned before, Energy Internet borrows from features of the Internet [2]. Many key technology features of Energy Internet are realized or will be realized by using Information Technology and more and more embedded computers have been designed into Energy Internet [3]. So Energy Internet which integrates energy technology and information technology can be seen as a complicated cyber-physical system. For the complicated cyber-physical system, there are many research interests in the new field. In this chapter, we explore some cyber-physical characteristics of Energy Internet mainly including the structure of cyber-physical energy internet, the relationship of the cyber resources and physical resources and the cyber security and safety of Energy Internet. In these characteristics, the security and safety of Energy Internet has attracted the attention of researchers and engineers. As Energy Internet presents an increased dependency on cyber resources which may be vulnerable to attack [4], the cyber-physical security and safety have become a new hot issue and this issue may include but not limited to the analysis of the influence of physical system by the cyber calculated attacks and the influence of structure failures to the whole system. Here we mainly analyze one of the calculated attacks for the some sub-system of Energy Internet.

2.1 Cyber-Physical Characteristics of Energy Internet

Energy Internet is a typical cyber-physical system, so Energy Internet has the common characteristics of the cyber-physical system and there are also some unique characteristics because of the system complexity. This part mainly introduces the structure of Cyber Physical Energy Internet which improves our understanding on the system and these abstracted cyber and physical layers will provide us a topologi-

Q. Sun, *Energy Internet and We-Energy*, Renewable Energy Sources & Energy Storage, https://doi.org/10.1007/978-981-13-0523-8_2

cal technique which can be used to analyze the security and stability of the structure under some failures and attacks. Then one of the cyber-physical characteristics of Energy Internet has been analyzed: the cyber-physical resources co-regulation of Energy Internet.

2.1.1 The Structure of Energy Internet—Cyber Layer and Physical Layer

In Energy Internet, energy sources are important material basis of social development, while reliable power supply is a crucial support for modern civilization [5]. Once the primary energy is being transformed into secondary energy in the form of energy carriers, it will be delivered to various users through energy transmission and distribution networks as end-use energy. The main type of energy carriers are heat and electricity. So there are two main energy transmission networks in Energy Internet, power grid and thermal grid.

In the two traditional networks, there exit some network characteristics because of the feature of different energy carriers. And as the knowledge of these characteristics has been deepened, some mathematical models with the constraint conditions has been built. However, these models and conditions can only describe the widespread characteristic of these networks. The real-time network state cannot be showed, which may result in some system accidents with economic losses, casualties and so on. However, this situation has been changed by the development of information technology, sensor technology and so on. These technologies provide us the chance to obtain the network state of the whole physical system. And it can be found that as the existence of the constraint conditions of the physical system, only the needed node data can be used to describe the state of the whole system. Through these data and mathematical models, engineers can control and optimize the system by some control and optimization methods. Here we should note that for different control, such as optimization methods, the different data and models are used, so it is common that the redundancy data is always needed [6]. So in order to understand the structure of cyber-physical system more clearly, we can divide the system into two layers—cyber layer and physical layer, and analyze the two layers by the network abstraction of the system. As for Energy Internet, the cyber layer and physical layer can also be built.

In Energy Internet, a hierarchical distributed structure in which all computers, embedded computers and energy devices connect to each other by energy lines (such as power line, natural gas pipeline) and telecommunication line would be applied. So here physical devices are abstracted as the physical nodes, embedded computers and computers with communication function are abstracted as the cyber nodes, the energy lines are abstracted as the physical links and the telecommunication lines are abstracted as the cyber links. The cyber-physical structure of Energy Internet are shown as Fig. 2.1 in which the black part represents the physical system and the blue part represents the cyber system. As the physical system of Energy Internet has been described earlier in the book, here we mainly illustrate the cyber part of Energy Internet.

As shown in Fig. 2.1, the information model of Energy Internet can be divided into two parts: measurement information flow part and control information flow part. Both of the two parts, from the perspective of information flow, modules of an information can be divided into four categories: (1) data-transmission; (2) data-processing (3) data-bus; and (4) data-storage modules.

Data-transmission can be defined as data exchange between different cyber units without any changes of data. It is a one-to-one mapping of data between different units.

Data-processing can be defined as the data transform. This transformation can be the changes of data format and the translation of various processes, such as square.

During the operation of Energy Internet, some modules' output data will be assembled together and then be inputted to some other modules. This multi-input and multi-output unit can be modeled as data-bus, in such a way as to include traditional data gathering and command issuing. Usually the above three categories of modules are not independent.

Data storage takes place in every specific device, such as Energy Router. There are two functions in this module: (1) storing the data which is important to local control or is redundant for the next data processing in the upper or lower cyber units; (2) assembling the data from all the redundant data in this specific device nodes and abandoning them.

Through these cyber nodes and links, we can build the topology structure of cyber network by matrix functions. These cyber matrix functions can be integrated with physical matrix functions which are built by the abstraction of physical system (the related method will be represented in the following chapters). These integration matrix functions can analyze the influence by the failures and attacks of Energy Internet, one of the concrete methods can refer to [7, 8].

2.1.2 The Cyber-Physical Characteristics

In Energy Internet, there are many cyber-physical characteristics should be studied. Here we mainly talk about one important characteristic: the efficiency of both cyber resources and physical resources. With the development of information technology and computing technology, it has become imperative for any resource scheduler to be aware of if, and when, it can scale back resources devoted to reasoning about the physical system and still maintain good quality of control and optimization [9]. For the complicated Energy Internet, there exist not only the complicated embedded smart energy devices, but also the complex control and optimization algorithms depending on the communication technology. The former one is about one specific energy device which is a under controlled smart device in Energy Internet and the latter one is about a whole system regardless of the size in Energy Internet. However, when we talk about the co-regulation of cyber resources and physical resources, the

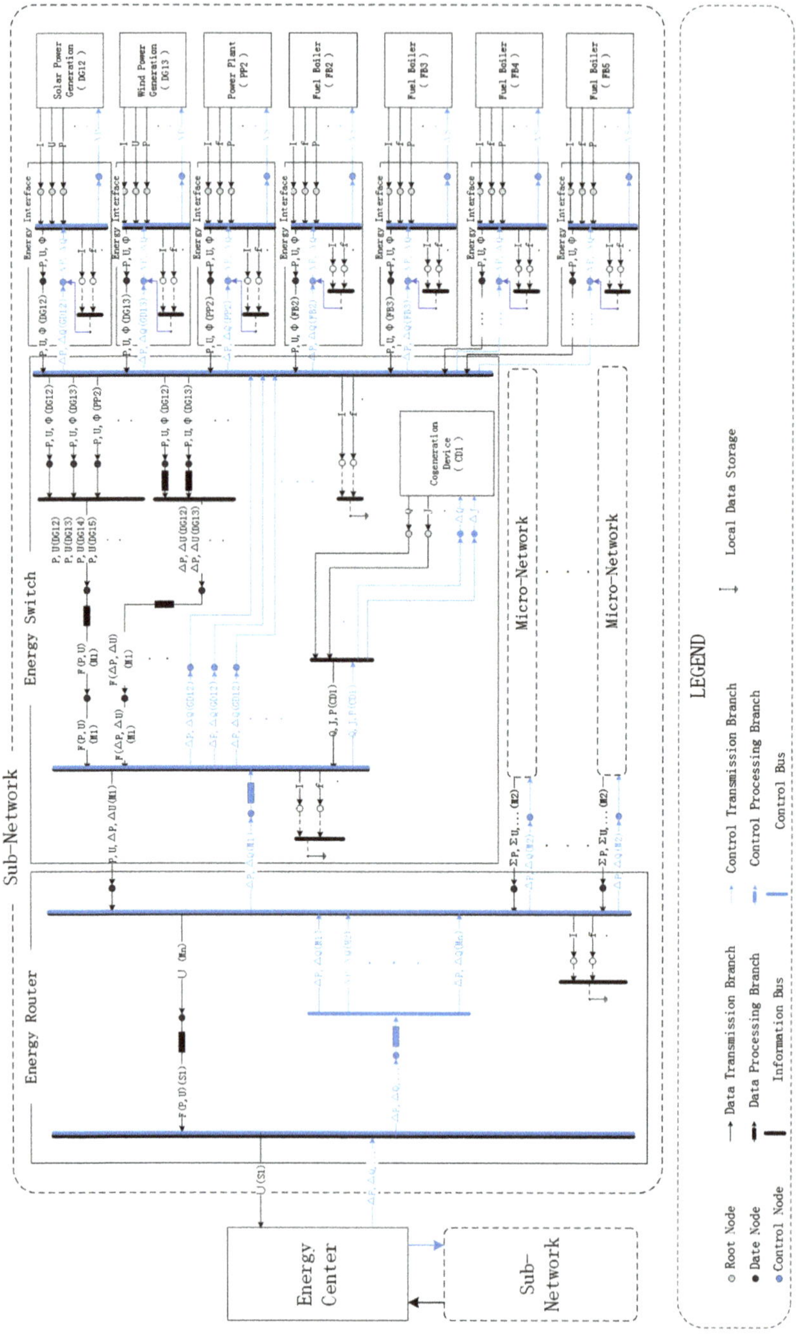

Fig. 2.1 The structure of the cyber-physical energy internet

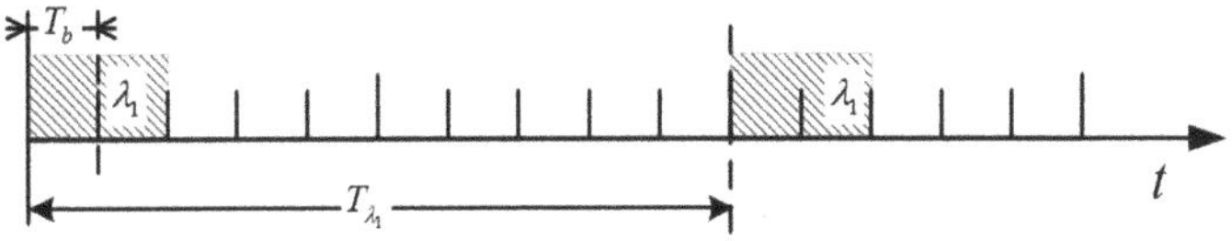

Fig. 2.2 Operation regulation

focus is always put on the cyber-physical resources co-regulation of the system and one of the common methods is event-triggered method. The cyber-physical resources co-regulation of the specific energy device receives little attention. In this part, we firstly talk about the cyber-physical resources co-regulation of the specific energy device before the event-triggered method is talked about.

A. The Cyber-Physical Resources Co-regulation

Here we talk about the abstraction of cyber resources of one device. For control system of the device, when a sample is completed, the control method will be asked and completed [10]. And computation and time complexity of a control math method is always same in the smart processors [11]. So the cyber resources utilization of a control system are related with the sampling frequency. And in the same time, the higher sampling frequency of one control function or task, the more utilization of cyber resources [12].

In a DG system, the cyber system becomes the gateway for virtually all aspects of the system. Control actuation inputs and control method computing, data collection, communication, voltage and frequency management, and island detection are potentially all being done simultaneously. Here the concept of task λ is used to explain how the control functions are operated in DSP. In Fig. 2.2, λ_1 is a cycle control task. From begin with a task until the next time it is been recalled again, it is held for T_{λ_1} seconds, which is the period of task λ_1 and T_{λ_1} is defined as the control task period. The basic time can be seen as the clock period, in which DSP just completed one of the most basic action. λ_1 always contains many clock periods when the DSP is operating. When DG system is being controlled, some control tasks are cyclical repeated, and there are subtasks in these control tasks. These tasks are always being departed into higher priority tasks and soft real-time tasks. In the reason that alternating current is the current whose direction changes periodically in the grid and the PWM modulating is normally used, the tasks in microchips in the grid-connected inverter are usually processed periodically and some of them usually share one cycle, which can be done successively. Meanwhile in control cycle, for preemptive scheduling,the only thing we know is that λ_1 is completed in T_{λ_1}. And we can hardly know precisely when the task is executed. For each of the microchip, it can only execute one task at a time. And since the computing resources in microchips are finite, it is important for computer engineers to provide hard timing guarantees particularly for safety-critical physical system controllers with best-effort execution of soft real-time tasks.

Here the cyber executing rate of a task is introduced from [12], which can be represented symbolically by λ.

$$x_c(\lambda) = f_\lambda t - n(t) \tag{2.1}$$

where $f_\lambda = 1/T_\lambda$ is the task executing frequency and T_λ the task period from the end time of this task to the begin of the next same task, $x_c(\lambda)$ is cyber executing rate ranging from 0 to 1. $n(t)$ is the cycle task count, $n(t)T_\lambda \leq t \leq (n(t)+1)T_\lambda$. And it implies that $\dot{x}_c(\lambda) = f_\lambda$.

Then the cyber resources model can be built $\dot{x}_c(\lambda) = [0]x_c(\lambda) + [1]f_\lambda$ and can be combined with the physical model (mainly be built under the control methods). Through the whole model and the method of optimal control, the cyber-physical resources co-regulation can be realized. Here a gird-connected inverter in a distributed generation system is used as an example.

In Energy Internet, for the reason that the voltage and frequency fluctuation of grid and the random output of distributed generation, synchronization control of inverter is always a challenge. There are three reasons for the inverter to adjust its output: to maximize energy harvesting, communication with other inverter, or scientific data acquisition in general. The inverter control strategy must be planned and carried out within narrow time constraints, and it is critical that controllers be capable of rejecting disturbances to achieve these goals. For inverter in Energy Internet, there are also many other tasks such as communication with other inverter and with the development of smart grid, there will be more other tasks for inverter to handle. All these tasks are completed in the one intelligent chip in inverter control system. In order to complete all the task in desired time, the main function of inverter control system can be accomplished by using smaller cyber resources under the same or better control performance is wanted. As mentioned before, in the reason that the power fluctuations and changes in weather, fixed sampling frequency maybe let this fluctuations to the reference of the output, this is not what we want. So we assume that the intelligent chip in inverter is running a real-time operating system capable of dynamically adjusting the period of the control tasks as long as the control tasks are not running or in an interrupted state. Before the cyber control method is talked to keep away from these fluctuations and changes, inverter under cyber-physical system should be modeled considering the control method of physical system. Though the study of inverter, the main control method can be seen as a change to the state equation, not the input variables but also the state coefficient matrix. It is easy known that the state equation remains the same during each sampling period, and after a sampling period, the state equation will update based on the control method and the sampling value of physical system. So the cyber-physical system is studied, it should be modeled considering the influence of sampling and control method. After the control system can dynamically adjust the period of the control tasks, considering that the computing complexity of each control task is fixed for the designed method, we can easy to understand that if the sampling frequency is reduced, the cyber resource utilization will be reduced.

For the gird-connected inverter, the two control parts are the model of inverter and DC/DC converter. The physical model with boost circuit and single phase inverter is analyzed as example.

(1) Single-phase Inverter Model

The conventional model of single-phase inverter is given

$$\begin{cases} L_1\dot{i}_{L_1} = u_{inv} - u_{C_1} - i_{C_1}R_{C_1} \\ L_2\dot{i}_{L_2} = u_{C_1} + i_{C_1}R - u_g \\ C_1\dot{u}_{C_1} = i_{C_1} \end{cases} \tag{2.2}$$

where $i_{C_1} = i_{L_1} - i_{L_2}$.

The PWM voltage inverter can be seen as an amplifying element, and if the amplitude of triangular carrier wave is U_{tri}, the transfer function of inverter bridge $K_{PWM} = U_{dc}/U_{tri}$. And formulation (2.2) can be rewritten as follow:

$$\begin{bmatrix} \dot{i}_{L_1} \\ \dot{i}_{L_2} \\ \dot{u}_{C_1} \end{bmatrix} = \begin{bmatrix} -\frac{R_{C_1}}{L_1} & \frac{R_{C_1}}{L_1} & -\frac{1}{L_1} \\ \frac{R_{C_1}}{L_2} & -\frac{R_{C_1}}{L_2} & \frac{1}{L_2} \\ \frac{1}{C_1} & -\frac{1}{C_2} & 0 \end{bmatrix} \begin{bmatrix} i_{L_1} \\ i_{L_2} \\ u_{C_1} \end{bmatrix} + \begin{bmatrix} \frac{K_{PWM}}{L_1} \\ 0 \\ 0 \end{bmatrix} u_m - \begin{bmatrix} 0 \\ \frac{1}{L_2} \\ 0 \end{bmatrix} u_g \tag{2.3}$$

Here let $\boldsymbol{x}_i = [i_{L_1}, i_{L_2}, u_{C_1}]$. Then formulation (2.3) can be rewritten

$$\dot{\boldsymbol{x}}_i = \boldsymbol{A}_i\boldsymbol{x}_i + \boldsymbol{B}_{i\_1}u_m + \boldsymbol{B}_{i\_2}u_g \tag{2.4}$$

In which, u_g is the perturbation vector. In current control of grid-inverter, the voltage distortion exciting in the grid maybe break the stability of the physical system, so the controller with feedforward grid voltage compensator is used to eliminate some bad influence. Considering capacitances and inductances in inverter, the gain of compensator is usually set as the reciprocal of K_{PWM}, i.e. $W_f = 1/K_{PWM}$. After feedforward voltage compensator, the model of single-phase inverter can be simplified as

$$\dot{\boldsymbol{x}}_i = \boldsymbol{A}_i\boldsymbol{x}_i + \boldsymbol{B}_{i-1}\boldsymbol{u}_i \tag{2.5}$$

For VSIs, the current output usually follows the grid voltage, which means the phase of grid voltage is the reference of the phase of i_{L_2-ref}. The input u_i is changed after one control period.

(2) DC/DC Converter Model

The boost current is used to analyze the physical model. For boost current, there are two operation modes: IGBT turn-on mode and IGBT turn off mode.

When IGBT turns on, the model of boost current can be shown as follow:

$$\begin{bmatrix} \dot{i}_{L_0} \\ \dot{u}_{C_{DG}} \\ \dot{u}_{C_{DC}} \end{bmatrix} = \begin{bmatrix} 0 & 1/L_0 & 0 \\ 1/C_{DG} & 1/R_0C_{DG} & 0 \\ 0 & 0 & -1/RC_{DC} \end{bmatrix} \begin{bmatrix} i_{L_0} \\ u_{C_{DG}} \\ u_{C_{DC}} \end{bmatrix} + \begin{bmatrix} 0 \\ -1/R_0C_{DG} \\ 0 \end{bmatrix} [u_{in}], \tag{2.6}$$

where R is the equivalent resistance of the after current.

When IGBT turns off, the model of boost current can be shown as follow:

$$\begin{bmatrix} \dot{i}_{L_0} \\ \dot{u}_{C_{DG}} \\ \dot{u}_{C_{DC}} \end{bmatrix} = \begin{bmatrix} 0 & 1/L_0 & -1/L_0 \\ -1/C_{DG} & -1/R_0C_{DG} & 0 \\ 1/C_{DC} & 0 & -1/RC_{DC} \end{bmatrix} \begin{bmatrix} i_{L_0} \\ u_{C_{DG}} \\ u_{C_{DC}} \end{bmatrix} + \begin{bmatrix} 0 \\ 1/R_0C_{DG} \\ 0 \end{bmatrix} [u_{in}] \tag{2.7}$$

As mentioned before, in gird-connected inverter, the first control part changes the duty ratio value of IGBT and then the voltage output of DC bus can be adjusted. If we assume the duty ratio of current control cycle is D_k,then the model of boost converter as follows:

$$\begin{bmatrix} \dot{i}_{L_0} \\ \dot{u}_{C_{DG}} \\ \dot{u}_{C_{DC}} \end{bmatrix} = \begin{bmatrix} 0 & \frac{2}{L_0} - \frac{D_k}{L_0} & -\frac{1}{L_0} + \frac{D_k}{L_0} \\ \frac{D_k}{C_{DG}} & \frac{D_k}{R_0C_{DG}} & 0 \\ \frac{1}{C_{DG}} - \frac{D_k}{C_{DC}} & 0 & -\frac{2}{RC_{DC}} + \frac{D_k}{RC_{DC}} \end{bmatrix} \begin{bmatrix} i_{L_0} \\ u_{C_{DG}} \\ u_{C_{DC}} \end{bmatrix} + \begin{bmatrix} 0 \\ \frac{D_k}{R_0C_{DG}} \\ 0 \end{bmatrix} u_{in} \tag{2.8}$$

Here let $\boldsymbol{x}_b = \left[i_{L_1}, u_{C_1}, u_{C_2}\right]$. Then formulation (2.8) can be rewritten

$$\dot{\boldsymbol{x}}_b = \boldsymbol{A}_b(D_k)\boldsymbol{x}_b + \boldsymbol{B}_b(D_k)u_b \tag{2.9}$$

In this part, the state coefficient matrix $\boldsymbol{A}_b(D_k)$ and $\boldsymbol{B}_b(D_k)$ is changed after one control period.

(3) Cyber-Physical Model of gird-connected inverter

In gird-connected inverter, there are two main physical models and for each of them, there is a control task period i.e. the cyber state variable. Considering one control period of gird-connected inverter, the continuous state equation can be shown as follows:

$$\begin{bmatrix} \dot{\boldsymbol{x}}_i \\ \dot{\boldsymbol{x}}_b \\ \dot{x}_{\lambda\_i} \\ \dot{x}_{\lambda\_b} \end{bmatrix} = \begin{bmatrix} \boldsymbol{A}_i & & & \\ & \boldsymbol{A}_b(D_k) & & \\ & & 0 & \\ & & & 0 \end{bmatrix} \begin{bmatrix} \boldsymbol{x}_i \\ \boldsymbol{x}_b \\ x_{\lambda\_i} \\ x_{\lambda\_b} \end{bmatrix} + \begin{bmatrix} \boldsymbol{B}_{i-1} & & & \\ & \boldsymbol{B}_b(D_k) & & \\ & & 1 & \\ & & & 1 \end{bmatrix} \begin{bmatrix} \boldsymbol{u}_i \\ u_b \\ u_{\lambda\_i} \\ u_{\lambda\_b} \end{bmatrix} \tag{2.10}$$

where x_{λ_i} represents the inverter control task and u_{λ_i} is the frequency of it, x_{λ_b} represents the MPPT control task and u_{λ_b} is the frequency of it.

The synchronization control with the grid of the inverters is a classic CPS challenge because of the power quality requirements. For better quality, there are more functions must be finished on time in gird-connected inverter control system, such as complex control algorithm, elaborate data processing, communication with other systems. Considering the main cyber resource is been used to finish the complex control algorithm for the MPPT and grid-connected control, whether some cyber resource utilizations can be reduced with same or keep the cyber resource utilization with better control quality by changing the control frequency in gird-connected inverter. As mentioned before, for the exiting grid voltage distortion and a random change in the weather, the two control frequencies can be adjusted dynamically based on the performance of each control quality to reduce or keep some cyber utilization. How to build a cyber control law in gird-connected inverter is analyzed. Before it, the discrete model of gird-connected inverter is first analyzed, for the real-time operating system capable of dynamically adjusting the period of the control tasks as long as the control tasks are not running or in an interrupted state.

(4) Discrete gird-connected inverter Model

With the development of information and embedded technology, an inverter under control is a hybrid system. It is to say that under a sampling or switching period, the state variables of the inverter are continuous dynamics and control commands stay the same. And at the point of next period arriving, the state variables are also continuous dynamics, but the control commands "jump", which make the state equations change. So after digital controlling, the physical model of an inverter become discrete. And for a given sampling period, the discrete inverter model of the inverter can be written. Here the discrete inverter model under a given period is given.

The model of gird-connected inverter is discrete. The two models of gird-connected inverter can be seen as linear time invariant system in each period. And the coefficient matrix for each period can be captured. So after a zero-order held, the physical system can be written as

$$\begin{bmatrix} \boldsymbol{x}_i(k+1) \\ \boldsymbol{x}_b(k+1) \\ x_{\lambda\_i}(k+1) \\ x_{\lambda\_b}(k+1) \end{bmatrix} = \begin{bmatrix} \boldsymbol{G}_i(k) & & & \\ & \boldsymbol{G}_b(k) & & \\ & & 1 & \\ & & & 1 \end{bmatrix} \begin{bmatrix} \boldsymbol{x}_i(k) \\ \boldsymbol{x}_b(k) \\ x_{\lambda\_i}(k) \\ x_{\lambda\_b}(k) \end{bmatrix} + \begin{bmatrix} \boldsymbol{H}_i(k) & & & \\ & \boldsymbol{H}_b(k) & & \\ & & T_{\lambda\_i}(k) & \\ & & & T_{\lambda\_b}(k) \end{bmatrix} \begin{bmatrix} \boldsymbol{u}_i(k) \\ \boldsymbol{u}_b(k) \\ u_{\lambda\_i}(k) \\ u_{\lambda\_b}(k) \end{bmatrix} \quad (2.11)$$

And $\boldsymbol{G}_b(k) = e^{A_b(D_k)T_{\lambda\_b}(k)}$, $\boldsymbol{G}_i(k) = e^{A_i T_{\lambda\_i}(k)}$, $\boldsymbol{H}_b(k) = \int_0^{T_{\lambda\_b}} e^{A_b(D_k)\eta} d\eta B_b(D_k)$ and $\boldsymbol{H}_i(k) = \int_0^{T_{\lambda\_i}} e^{A_i\eta} d\eta B_i$. From the equation the coefficient matrix is based on task period T. And $x_{\lambda\_i}(k) = 1/T_{\lambda\_i}(k)$,$x_{\lambda\_b}(k) = 1/T_{\lambda\_b}(k)$. After this, we can design the cyber control law $u_{\lambda\_i}(k)$ and $u_{\lambda\_b}(k)$ of gird-connected inverter.

(5) Cyber Control Law

In order to design a control law for the cyber inverter model, which consider the influence of the voltage distortion in grid, the dependence between the cyber and physical systems should be considered. The gird-connected inverter control system focuses on the MPPT and the flowing with the grid voltage. Therefore, in order to guarantee the performance of physical system, the cyber system control law is designed based on the physical states trajectory. As mentioned before, there are two cyber state variables in the inverter: the cyber state variable of MPPT x_{λ_b} and the cyber state variable of the grid-connected control strategy x_{λ_i}. And the two cyber state variables usually influence each other. So this influence should also be considered when the cyber system control law is designed. We assume the form of cyber control law as follows:

$$\mathbf{u}_\lambda(k) = \mathbf{G}_{\lambda p}\big(\mathbf{x}_p(k) - \mathbf{x}_{p,r}(k-1)\big) - \mathbf{G}_\lambda\big(\mathbf{x}_\lambda(k) - \mathbf{x}_{\lambda,r}\big), \quad (2.12)$$

where $\mathbf{G}_{\lambda p}$ is the physical-cyber gain matrix, $\mathbf{G}_\lambda$ is the cyber gain matrix, $\mathbf{x}_{\lambda,r}$ is the cyber system reference trajectory, $\mathbf{x}_\lambda(k)$ is the sampling frequency at k moment. For giving detail cyber control law of inverter, these parameters should be set up in (2.12).

Here we first analyze the situation that the low grid voltage distortion and the same weather. the cyber system reference trajectories are analyzed, i.e. the desired sampling rate for λ_i and λ_b. Before choosing the desired sampling, the bounds of the sampling rate for an inverter should be analyzed.

We should notice that in the common control method the sampling frequency is usually been set the same as the switching frequency of IGBT or higher. This sampling frequency satisfy the Rremann sampling and can hardly influence the control stability of inverter. And the arithmetic speed of CPU in DSP can be 1 GHz. So considering the cyber resources of DSP and the stability of physical system, there should be upper and lower bounds for sampling rate. For our inverter, we specify 10 times switching frequency as the upper bound, and switching frequency as the lower bound, both for λ_i and λ_b.

$$\begin{aligned} x_{c\_\lambda,\max} &= 10 * f_{sw} \\ x_{c\_\lambda,\min} &= f_{sw} \end{aligned} \tag{2.13}$$

Then we analyze the physical-cyber gain $\mathbf{G}_{\lambda p}$. In this cyber control law, the cyber system can adjust the sampling rate according the performance of physical system is wanted. For MPPT control part, the deviation value between the twice before and after sampling should be estimated. So when there exits the deviation, the sampling frequency should be increased. And for inverter control, if it exits the deviation between the output and reference, the sampling frequency should be also increased. So we can have that

$$G_{\lambda\_ip} = \begin{cases} g_{\lambda\_ip,i}, & if\ x_{p,i}(k) - x_{p,i,r} \geq 0 \\ -g_{\lambda\_ip,i}, & if\ x_{p,i}(k) - x_{p,i,r} < 0 \end{cases} \tag{2.14}$$

$$G_{\lambda\_bp} = \begin{cases} g_{\lambda\_bp,i}, & if\ P_{\lambda\_b}(k+1) - P_{\lambda\_b}(k) \geq 0 \\ -g_{\lambda\_bp,i}, & if\ P_{\lambda\_b}(k+1) - P_{\lambda\_b}(k) < 0 \end{cases} \tag{2.15}$$

Last, we analyze the cyber gain matrix $\mathbf{G}_{\lambda}$. In a task, when the sampling frequency is greater than the reference frequency, the sampling frequency should be decrease. When the sampling frequency is less than the reference frequency, the sampling frequency should be increased. So $g_{\lambda,ii} > 0$. In DG system, that the cyber resource utilization in MPPT and inverter control are confirmed to leave some resource to other functions is wanted. So the control precision of the two control parts is a reciprocal process. That is to say, when the sampling frequency of MPPT is higher than the reference sampling frequency, the sampling frequency of inverter control certain degree should be improved. When the sampling frequency of MPPT is lower than the reference sampling frequency, the sampling frequency of inverter control certain degree is decreased. On the other hand, the versa. So $g_{\lambda\_ij} < 0$. So the cyber control law can be written as follow:

$$\begin{bmatrix} u_{\lambda\_i}(k) \\ u_{\lambda\_b}(k) \end{bmatrix} = \begin{bmatrix} G_{\lambda\_ip} & \\ & G_{\lambda\_bp} \end{bmatrix} \begin{bmatrix} x_{p,i}(k) - x_{p,i,r} \\ P_{\lambda\_b}(k+1) - P_{\lambda\_b}(k) \end{bmatrix} - \mathbf{G}_{\lambda} \begin{bmatrix} x_{\lambda\_i}(k) - x_{\lambda\_i,r} \\ x_{\lambda\_b}(k) - x_{\lambda\_b,r} \end{bmatrix} \tag{2.16}$$

Now we consider the grid voltage distortion and the change weather. The grid voltage distortion usually appears and considers the influence to DG system, that if the sampling frequency can be decreased i.e. the control frequency of inverter when the grid voltage distortion is allowed within the scope of DG system to filter out part of the power grid voltage distortion is hoped. (Abnormal fluctuations trigger DG system to cut itself from grid.) When the grid normally operating, the sampling frequency should be increased to make the DG system follow the grid quickly. Meanwhile, the environment of DG system is unstable for quickly MPPT, whether the reference control frequency of MPPT control can be increased? When DG system is running at the MPP, decreasing the sampling frequency considering the stability of the system is being hoped.

As mentioned before, one of purposes in this section is to make the cyber resource utilization remain same for inverter and MPPT control parts. So when the inverter sampling reference frequency is decreased, the MPPT sampling reference frequency can be increased or kept. And on normal operation of grid, the inverter control is increased and the MPPT sampling reference frequency is decreased for interconnection with grid with accuracy and speed.

For this purpose, a judging criterion for information control law is added. According this, the sampling reference frequency is changed for both MPPT and inverter control of cyber control law to realize our purpose. The characteristics of the two levels of control should be considered for this criterion.

For inverter control, considering the voltage stabilization, there is an ideal voltage value for each bus in the grid. When we connect the inverter of DG system to the grid, we should also test the voltage value of buses. So a threshold value is chosen related to it as the judging criteria of reference frequency of inverter control.

$$\varepsilon = \alpha U_m * \frac{f}{f_{sw}}, \tag{2.17}$$

where $\alpha > 1$ is the coefficient of the threshold value and it can be chosen by the current grid operation condition. For cyber control law, the difference between twice sampling value of grid voltage is judged with the threshold value ε. If the difference is over ε, the reference frequency of inverter control is decreased. And on the other side, the reference frequency is increased or kept. Sometimes it can be decreased for special situation.

For MPPT control method, difference between twice computing output power of gird-connected inverter should be judged to give the reference value of output voltage of it. If the difference is big, we consider there is not the MPP of DG system and the environment outside is changing. Here a threshold value ε_1 is given of it, if the difference is bigger than ε_1, the reference frequency of MPPT should be increased. And if the difference is smaller than ε_1, the reference frequency of MPPT should be decreased or kept. Sometimes it can be decreased for special situation.

Considering there are two control parts for smart processors, when requiring cyber resources of the conflicting two control parts, it should be dealt with though strong conditions to judge. The strong conditions contain: (1) the difference between twice sampling value of grid voltage is over ε; (2) difference between twice computing output power of DG s bigger than ε_1.

After the analysis, the reference sampling frequency of the two control parts can be as follow (Table 2.1):

Table 2.1 Step of $\Delta x_{c\_\lambda_1}$, $\Delta x_{c\_\lambda_2}$

	$\lvert P(k) - P(k+1)\rvert > \varepsilon_1$	$\lvert P(k) - P(k+1)\rvert \le \varepsilon_1$
$\lvert U_k - U_{k+1}\rvert > \varepsilon_2$	$\Delta x_{c\_\lambda_1} < 0$, $\Delta x_{c\_\lambda_2} > 0$	$\Delta x_{c\_\lambda_1} < 0$, $\Delta x_{c\_\lambda_2} > 0$
$\lvert U_k - U_{k+1}\rvert \le \varepsilon_2$	$\Delta x_{c\_\lambda_1} < 0$, $\Delta x_{c\_\lambda_2} > 0$	$\Delta x_{c\_\lambda_1} \ge 0$, $\Delta x_{c\_\lambda_2} \le 0$

$$\begin{aligned} x_{c\_\lambda_1,rk} &= x_{c\_\lambda_1,rk} + \Delta x_{c\_\lambda_1} \\ x_{c\_\lambda_2,rk} &= x_{c\_\lambda_2,rk} + \Delta x_{c\_\lambda_2} \end{aligned} \tag{2.18}$$

B. Event-Triggered Update Strategy

Event-triggered update strategy has been proposed to balance the communication resources and control performance of a whole system. This strategy can determine the update moment based on the system status, and the unnecessary update moments can be avoided, which is a general phenomenon in periodic update strategy. Here we use a microgrid under distributed cooperative control methods which is common in Energy Internet to analyze the event-triggered update strategy.

For DGs in microgrids, in order to realize the reactive power sharing, the update conditions of event-triggered strategy should not only based on the status of itself, but also the status from its neighbors. When the event-triggered update strategy is designed, the stability and consensus of the microgrid can be realized. However, some kinds of event-triggered mechanisms which are not properly designed, the Zeno phenomenon may happen. That is to say, the system may have an infinite number of sampling instants in finite time. In this section, we will design a periodic event-triggered update strategy in which a properly designed time interval is added for the mircogrid motivated by [13]. In this event-triggered update strategy, the Zeno phenomenon can be avoided since the designed time interval exists. The distributed cooperative control methods have been introduced to solve the accurate power sharing, voltage and frequency restoration and so on. In this way, the performance of microgrids can be improved.

The structure of the microgrid system with accurate reactive power sharing is shown in Fig. 2.3. The strategy consists of two control levels. The primary level consists of the inner control loop and the droop control of inverters, which can regulate the output voltage of each DG at an acceptance range of the rated frequency and amplitude for small load variation. As the large loads variations are always expected when the system is under isolation and the impedance of each line from DGs to loads is always mismatched, existing Q-V droop controllers can hardly realize the accurate reactive sharing. This may cause a large-scale current circulation and the stability and reliability of the system may be broken. Here from reference [14, 15], we assume that the load is relatively close to the DG1. And when the load reactive power increases at some point, the output reactive power of DG1 increases sharply to satisfy the load change, and the reactive power of the other DGs increase marginally since they are located further away from the load. During this process, if the voltage of the DG1 drops down to the minimum value, the DG1 will take place

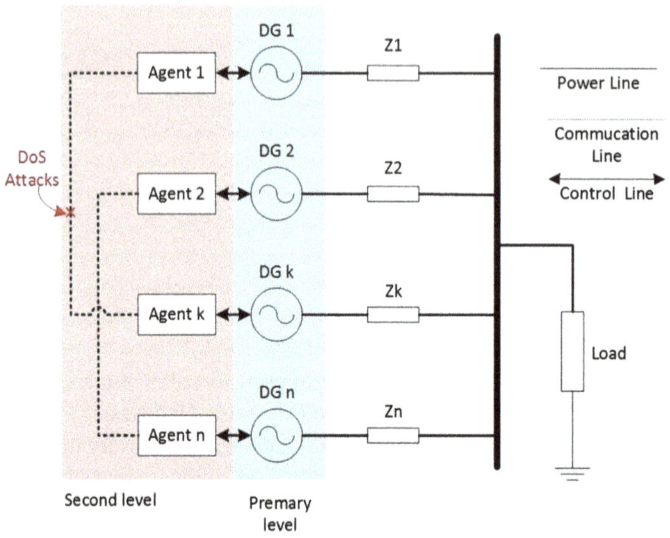

Fig. 2.3 The structure of the microgrid system

in overloading protection and activates the ac load shedding system. Thus, it is of great significance for multi-DGs to realize the reactive power sharing accurately in real time. So the secondary level is proposed to ensure the accurate reactive power sharing through the distributed controllers with the introduced virtual impedance.

A. Droop Controllers and Distributed Consensus Controllers

When two or more DGs are connected, droop controllers are used for the voltage regulation which can be expressed as

$$V_i = V_i^d - m_i\big(Q_i - Q_i^d\big), \tag{2.19}$$

where V_i and V_i^d represent the amplitude and the desired amplitude of DG_i output voltage, respectively; m_i is the Q-V droop coefficient of DG_i; Q_i and Q_i^d represent the output and desired reactive powers of DG_i, respectively. As mentioned before, though the Q-V droop control method is proposed to share the reactive power among DGs, it suffers from poor results if the line impedance is taken into consideration. It will cause the current circulation among DGs and limit the maximum output power of them.

So the secondary control level has been proposed based on adaptive virtual impedance and distributed consensus control method to accurate reactive power sharing [16, 17]. For the distributed consensus method in the secondary control level, digraph should be used to introduce the structure of the microgrid. The DGs and their inverters are usually numbered by a set of nodes, denoted by $V = \{1, 2, \ldots, N\}$, which means that there are N DGs in this microgrid.

For the microgrid, the communication network model of the microgrid is captured by a communication digraph $\mathsf{G} = \{\mathsf{V}, \mathsf{E}, \boldsymbol{A}\}$, which consists of the node set V, a communication edge set $\mathsf{E} \subset \mathsf{V} \times \mathsf{V}$ and the associated adjacency matrix $\boldsymbol{A} = \left(a_{ij}\right)_{N\times N}$. If there exists a communication link between node i and node j, denoted by $(i, j) \in \mathsf{E}$, nodes i and j are called as communication neighbors. The set of neighbors of agent i is denoted as $N_i = \{j \in \mathsf{V}|(i, j) \in \mathsf{E}\}$. For adjacency matrix $\boldsymbol{A}$, a_{ij} is the weight of edge $(i, j) \in \mathsf{E}$ with

$$a_{ij} = \begin{cases} 1, \; if \; j \in N_i \\ 0, \; otherwise \end{cases}. \tag{2.20}$$

The Laplacian matrix of the digraph G is defined as $\boldsymbol{L} = \boldsymbol{D} - \boldsymbol{A}$, where $\boldsymbol{D} = diag\{d_i\}$ with $d_i = \sum_{j\in N_i} a_{ij}$. We should note that if there is a root node with a direct path from that node to every other node, a digraph contains a directed spanning tree in the graph. Then the whole system of the secondary control level can be written as

$$\dot{\boldsymbol{Q}}^{\wedge}(t) = \boldsymbol{u}_Q(t), \tag{2.21}$$

$$\boldsymbol{u}_Q(t) = -C\boldsymbol{q}_Q(t), \tag{2.22}$$

$$\boldsymbol{q}_Q = \boldsymbol{L}\boldsymbol{Q}^{\wedge}, \tag{2.23}$$

where the variables are defined as $\boldsymbol{Q}^{\wedge} = \left[m_1 Q_1 \; m_2 Q_2 \cdots m_N Q_N\right]^T$, $\boldsymbol{u}_Q = \left[u_{Q1} \; u_{Q2} \cdots u_{Q_N}\right]^T$, $\boldsymbol{q}_Q = \left[q_{Q1} \; q_{Q2} \cdots q_{QN}\right]^T$, in which $q_{Qi}(t) = \sum\limits_{j\in N_i} a_{ij}\left(Q_i^{\wedge}(t) - Q_j^{\wedge}(t)\right)$ reflects the local neighbor reactive power sharing error, and C is the coupling gain. And to adapt the virtual impedance correction term δQ_i, a proportional integral (PI) controller $D_i(s)$ has been introduced with the reactive power mismatch shown as following

$$\delta Q_i = D_i(s)u_{Qi}, \tag{2.24}$$

where u_{Qi} is the reactive power controller at DG_i. To keep the system stable, we should establish the relationship between q_{Qi} and δQ_i though the coupling gain and the PI control gain, which can be designed and chosen by the theory on optimal and nonlinear control [19]. In the distributed control approach, a monitor is needed for each inverter to monitor the reactive power flow $Q_i(t)$ of itself and a data receiver is needed to receive its neighbors' information of the reactive power flow $Q_j(t)$.

When the method and other similar methods have been achieved, DGs are usually considered as agents. They can exchange state information with neighboring DGs through the communication network of microgrids.

As the communication over the information network is packet-based, the measurements of neighbors' reactive power flow are transmitted at discrete instants. As the input of control system must be continuous, the zero-order holder fashion is always used between updates of the information. Then the controller u_{Qi} is shown as

$$u_i(t) = -C \sum_{j\in N_i} a_{ij}\left(Q_i^{\wedge}\left(t_k^i\right) - Q_j^{\wedge}\left(t_k^j\right)\right), \quad t \in \left[t_k^i, t_{k+1}^i\right), \tag{2.25}$$

where t_k^i is the desired update time instant for DG_i with $k \in \mathbb{N}$.

Typically, the communication between DGs is packet-based. It means that the state information of DG-agents can only be transmitted at discrete instants. In the controller (2.25), the reactive power at each DG is measured and coupled with the Q-V droop coefficient and then transmitted to its neighboring DG at some time instants denoted by

$$0 \le t_0^i < t_1^i < \cdots < t_k^i, k \in \mathbb{N}. \tag{2.26}$$

Hence at each transmission instant t_k^i, the value of $Q_i^{\wedge}$ is updated until the next transmission instant t_{k+1}^i.

Here we consider the distributed consensus controllers to realize the accurate reactive power sharing:

$$\dot{Q}_i^{\wedge}(t) = u_i(t), \tag{2.27}$$

where $Q_i^{\wedge} = n_i Q_i$. It is easy to know that at steady state, DG_i and DG_j share their reactive powers proportionally if equation

$$\lim_{t\to\infty} \left\| Q_i^{\wedge}(t) - Q_j^{\wedge}(t) \right\| = 0, \quad i, j = 1, 2, \cdots, N \tag{2.28}$$

holds. From the distributed consensus control perspectives, this implies that all $Q_i^{\wedge}$ achieve consensus as time goes to infinity.

In order to realize the accurate reactive power sharing, the control law for each DG based on discrete states is proposed as

$$u_i(t) = -C \sum_{j\in N_i} a_{ij}\left(Q_i^{\wedge}\left(t_k^i\right) - Q_j^{\wedge}\left(t_k^i\right)\right), \quad t \in \left[t_k^i, t_{k+1}^i\right), \tag{2.29}$$

where C is the coupling gain, a_{ij} are the elements which reflect connected condition change of DGs in the adjacency matrix and t_k^i represents the latest event-triggered instant of DG_i. Here it should be noticed that we choose this control law rather than $u_i(t) = -C \sum_{j\in N_i} a_{ij}\left(Q_i^{\wedge}\left(t_k^i\right) - Q_j^{\wedge}\left(t_{k'(t)}^i\right)\right)$, where $k'(t) = \arg \min_{l\in\mathbb{N}: t \ge t_l^j} \left\{t - t_l^j\right\}$, because in this control law the update of controller and the triggering times of any DG are independent of triggering time instants of other DGs [18]. In order to design the update strategy, the measurement error $e_{Qi}(t) = Q_i^{\wedge}\left(t_k^i\right) - Q_i^{\wedge}(t), \quad i = 1, 2, \ldots, N$ of each DG should be used, which represents the error between the value of the measurement at the last successful control update and the value of the measurement at the current time. To determine the event triggering time instants $\{t_0^i, t_1^i, \ldots, t_k^i, \ldots\}$ for DG_i, the triggering condition can be developed, usually as the follow

$$t_{k+1}^{Qi} = \min\left\{t > t_k^{Qi} \middle| \left(\|e_{Qi}(t)\| - \sigma_i \|q_{Qi}(t)\| \le 0\right)\right\}. \tag{2.30}$$

In this strategy, the Zeno phenomenon may happen. To avoid it, here we introduce a periodic event-triggered mechanism

$$t_{k+1}^{Qi} = \min\left\{t > t_k^{Qi} + h_i \middle| \left(\|e_{Qi}(t)\| - \sigma_i \|q_{Qi}(t)\| \le 0\right)\right\}, \tag{2.31}$$

where $\sigma_i \in \mathbb{R}_{>0}$ is the suitable design parameters and $h_i \in \mathbb{R}_{>0}$ is a constant waiting time, where the inter-event times is not less than h_i. With the suitable parameter, the stability of each DG will not be destroyed and the consensus can be achieved.

Let $\phi_i(t) = Q_i^{\wedge}(t) - Q_1^{\wedge}(t), i = 1, 2, \ldots, N$, then for $i = 1$ $\phi_1(t) = Q_1^{\wedge}(t) - Q_1^{\wedge}(t) = 0$ which is used in the following proof o. Denote $\boldsymbol{\phi}(t) = \left(\phi_1^T(t), \phi_2^T(t), \ldots, \phi_N^T(t)\right)^T$. For $i = 2, 3, \ldots, N$, from (2.28) the system can realize consensus if $\lim_{t\to\infty} \|\phi_i(t)\| = 0$. Notice that

$$\dot{\phi}_i(t) = u_i(t) - u_1(t) \quad i = 2, 3, \ldots, N. \tag{2.32}$$

Then we have

$$\begin{aligned}
\dot{\phi}_i(t) = &-C \sum_{j\in N_i} a_{ij}\phi_i(t) + C\left(\sum_{j=2}^{N_i} a_{ij} - \sum_{j=2}^{N_i} a_{1j}\right)\phi_j(t) \\
&+ (1-\chi(t))\left(C\sum_{j\in N_i} a_{ij}\int_{t-\tau(t)}^{t}\dot{\phi}_i(t)dt - C\left(\sum_{j=2}^{N_i} a_{ij} - \sum_{j=2}^{N_i} a_{1j}\right)\int_{t-\tau(t)}^{t}\dot{\phi}_j(t)dt\right) \\
&+ \chi(t)\left(C\left(\sum_{j=2}^{N_i} a_{ij} - \sum_{j=2}^{N_i} a_{1j}\right)e_j(t) + C\sum_{j\in N_1} a_{1j}e_1(t) + Ca_{i1}e_1(t) - C\sum_{j\in N_i} a_{ij}e_i(t)\right).
\end{aligned} \tag{2.33}$$

where $\chi(t) = \begin{cases} 0, & t \in [s_k, s_k + h) \\ 1, & t \in [s_k + h, s_{k+1}) \end{cases}$, $\tau(t) = t - t_k^i \le h, \quad t \in \left[s_k^i, s_k^i + h\right), e_{Qi}(t) = Q_i^{\wedge}\left(t_k^i\right) - Q_i^{\wedge}(t), \quad t \in [s_k + h, s_{k+1})$. Denote $\bar{\boldsymbol{\phi}}(t) = \left(\phi_2^T(t), \phi_3^T(t), \ldots, \phi_N^T(t)\right)^T$, $\bar{\boldsymbol{e}}(t) = \left(e_2^T(t), e_3^T(t), \ldots, e_N^T(t)\right)^T$, and $\bar{\boldsymbol{e}}_1(t) = \left(e_1^T(t), e_1^T(t), \ldots, e_1^T(t)\right)^T$.

Thus, the compact form of (2.33) is

$$\dot{\bar{\boldsymbol{\phi}}}(t) = -H \otimes C\bar{\boldsymbol{\phi}}(t) + (1-\chi(t)) \otimes H \otimes C\int_{t-\tau(t)}^{t}\dot{\bar{\boldsymbol{\phi}}}(s)ds + \chi(t) \otimes (D_1 \otimes C\bar{\boldsymbol{e}}_1(t) - H \otimes C\bar{\boldsymbol{e}}(t)), \tag{2.34}$$

where

$$H = \begin{bmatrix} \sum\limits_{j\in N_2} a_{2j} + a_{12} & -a_{23} + a_{13} & \cdots & -a_{2N} + a_{1N} \\ -a_{32} + a_{12} & \sum\limits_{j\in N_3} a_{3j} + a_{13} & \cdots & -a_{3N} + a_{1N} \\ \vdots & \vdots & \ddots & \vdots \\ -a_{N2} + a_{12} & -a_{N3} + a_{12} & \cdots & \sum\limits_{j\in N_N} a_{Nj} + a_{1N} \end{bmatrix}, \text{ and } D_1$$

$$= diag\left\{d_1 + a_{21}, d_1 + a_{31}, \ldots, d_1 + a_{N1}\right\}.$$

Then the consensus analysis of the accurate reactive power sharing control system has been transferred to the asymptotic stability of system (2.34).

B. Control Update Policy and Consensus Analysis

For the consensus controller, we can choose the coupling gain C that makes all the eigenvalues of $-H \otimes C$ are in the open left half plane by the linear system theory [19]. Then there exist constants $\gamma_1 > 0$ such that

$$\left\| e^{-H\otimes C(t-t_0)} \right\| \leq e^{-\gamma_1(t-t_0)}, t \geq t_0. \tag{2.35}$$

And we denote that $k_1 := \|(-H \otimes C)\|$, $k_2 := \|D_1 \otimes C\|$ and $k_3 := \max\{\|(-H \otimes C)\|, \|D_1 \otimes C\|\}$, $\bar{N} = \max\{N_i\}$, $\sigma_{\max} = \max\{\sigma_i\}$, $b_1 = \sigma_{\max}\sqrt{2}\left(\bar{N} + \sqrt{N}\sqrt{\bar{N}}\right) \Big/ \left(1 - \sigma_{\max}\bar{N} - \sigma_{\max}\sqrt{N}\sqrt{\bar{N}}\right)$ in the following analysis.

Theorem 1 *Consider the reactive power sharing consensus controller system composed of (2.21) and (2.22) and control input (2.29), where all the eigenvalues of $-H \otimes C$ have negative real parts. For the periodic event-triggered mechanism (2.31), the consensus can be achieved for all $t \in \mathbb{R}_{\geq 0}$, with σ_i such that*

$$\sigma_{max} \leq \frac{\gamma_1}{\left(\sqrt{2}k_3 + \gamma_1\right)\left(\bar{N} + \sqrt{N}\sqrt{\bar{N}}\right)}. \tag{2.36}$$

And then it exits $\lambda \in (0, \gamma_1)$, which satisfies that $\frac{k_1^2 e^{\lambda h}\left(e^{\lambda h} - 1\right)}{\lambda(\gamma_1 - \lambda)} < 1$ and $\frac{k_3 b_1}{\gamma_1 - \lambda} < 1$.

Proof (1) When $\chi(t) = 0$ and $t \in \left[s_k^i + h_i, s_{k+1}^i\right)$, the system can be rewritten as

$$\dot{\bar{\boldsymbol{\phi}}}(t) = -H \otimes C\bar{\boldsymbol{\phi}}(t) + (D_1 \otimes C\bar{\boldsymbol{e}}_1(t) - H \otimes C\bar{\boldsymbol{e}}(t)). \tag{2.37}$$

Using the variation of parameter formula, (2.37) can be rewritten as

$$\bar{\boldsymbol{\phi}}(t) = e^{(-H\otimes C)(t-t_k^i)}\bar{\boldsymbol{\phi}}(t_k^i) + \int_{t_k^i}^{t} e^{(-H\otimes C)(t-s)}(-H \otimes C\bar{\boldsymbol{e}}(s) + +D_1 \otimes C\bar{\boldsymbol{e}}_1(s))ds. \tag{2.38}$$

Recalling (2.35), then

$$\left\|\bar{\boldsymbol{\phi}}(t)\right\| \le e^{-\gamma_1(t-t_k^i)}\left\|\bar{\boldsymbol{\phi}}(t_k^i)\right\| + \int_{t_k^i}^{t} e^{-\gamma_1(t-s)}(k_1\|\bar{\boldsymbol{e}}(s)\| + k_2\|\bar{\boldsymbol{e}}_1(s)\|)ds, \tag{2.39}$$

Notice that for DG_i, there exists that

$$h\big(e_{Qi}(t), q_{Qi}(t)\big) = \left\|e_{Qi}(t)\right\| - \sigma_i\left\|q_{Qi}(t)\right\| \le 0, \tag{2.40}$$

so we have that

$$\begin{aligned}\left\|e_{Qi}(t)\right\| &\le \sigma_i\left\|\sum_{j\in N_i} a_{ij}\big(Q_i^\wedge(t_k^i) - Q_j^\wedge(t_k^i)\big)\right\| \\ &\le \sigma_i\left\|\sum_{j\in N_i} a_{ij}\phi_i(t) + \sum_{j\in N_i} a_{ij}e_{Qi}(t) - \sum_{j=2}^{N_i} a_{ij}\phi_j(t) - \sum_{j\in N_i} a_{ij}e_{Qj}(t)\right\| \\ &\le \sigma_i N_i\|\phi_i(t)\| + \sigma_i N_i\left\|e_{Qi}(t)\right\| + \sigma_i\sqrt{N_i}\|\boldsymbol{\phi}(t)\| + \sigma_i\sqrt{N_i}\left\|\boldsymbol{e}_Q(t)\right\|.\end{aligned} \tag{2.41}$$

Accumulate (2.41) from $i = 1$ to N, then

$$\begin{aligned}&\sum_{i=1}^{N}\left\|e_{Qi}(t)\right\| \\ &\quad\le \sqrt{N}\left\|\boldsymbol{e}_Q(t)\right\| \\ &\quad\le \sigma_{\max}\left(\sqrt{N}\bar{N}\|\boldsymbol{\phi}(t)\| + \sqrt{N}\bar{N}\left\|\boldsymbol{e}_Q(t)\right\| + N\sqrt{\bar{N}}\|\boldsymbol{\phi}(t)\| + N\sqrt{\bar{N}}\left\|\boldsymbol{e}_Q(t)\right\|\right)\end{aligned} \tag{2.42}$$

As $\phi_1(t) = 0$ has held true for anytime, one can have that $\left\|\bar{\boldsymbol{\phi}}(t)\right\| = \|\boldsymbol{\phi}(t)\|$, then (2.42) can rewrite as follows

$$\left(1 - \sigma_{\max}\bar{N} - \sigma_{\max}\sqrt{N}\sqrt{\bar{N}}\right)\left\|\boldsymbol{e}_Q(t)\right\| \le \sigma_{\max}\left(\bar{N} + \sqrt{N}\sqrt{\bar{N}}\right)\left\|\bar{\boldsymbol{\phi}}(t)\right\|. \tag{2.43}$$

As $\bar{\boldsymbol{e}}(t) = \left(e_2^T(t), e_3^T(t), \ldots, e_N^T(t)\right)^T$ and $\bar{\boldsymbol{e}}_1(t) = \left(e_1^T(t), e_1^T(t), \ldots, e_1^T(t)\right)^T$, then

$$\|\bar{\boldsymbol{e}}(t)\| + \|\bar{\boldsymbol{e}}_1(t)\| \le \sqrt{2}\left\|\boldsymbol{e}_Q(t)\right\| \le b_1\left\|\bar{\boldsymbol{\phi}}(t)\right\| \tag{2.44}$$

Then we have that

$$k_1\|\bar{e}(s)\| + k_2\|\bar{e}_1(s)\| \le k_3(\|\bar{e}(s)\| + \|\bar{e}_1(s)\|) \le k_3\|e_Q(t)\| \tag{2.45}$$

Substituting (2.45) into (2.39)

$$\|\bar{\phi}(t)\| \le e^{-\gamma_1(t-t_k^i)}\|\bar{\phi}(t_k^i)\| + k_3 b_1 \int_{t_k^i}^{t} e^{-\gamma_1(t-s)}\|\bar{\phi}(s)\|ds. \tag{2.46}$$

Then we will prove that

$$\|\phi(t)\| < e^{-\lambda(t-t_k^i)}\|\phi(t_k^i)\| := v(t),\, t > t_k^i, \tag{2.47}$$

where $t > t_0$, $\lambda \in (0, \gamma_1)$, and satisfies that $\frac{k_3 b_1}{\gamma_1-\lambda} < 1\|\phi(t)\| \le e^{-\lambda(t-t_k^i)}\|\phi(t_k^i)\|$, $t \ge t_k^i$, $\dot{\bar{\phi}}(t) = -H \otimes C\bar{\phi}(t) + H \otimes C \int_{t-\tau(t)}^{t} \dot{\bar{\phi}}(s)ds$.

If (2.47) does not hold, there must exist a $t^* > t_k^i$ for $\|\phi^*(t)\| = v(t^*)$ and $\|\phi(t^*)\| < v(t^*)$ for $t < t^*$. Then by (2.46), we have

$$\begin{aligned} v(t^*) = \|\phi(t^*)\| &< e^{-\gamma_1(t*-t_k^i)}\|\phi(t_k^i)\| + k_3 b_1\|\phi(t_k^i)\| \int_{t_k^i}^{t} e^{-\gamma_1(t*-s)-\lambda(s-t_0)}ds \\ &= \|\phi(t_k^i)\|\left(e^{-\gamma_1(t*-t_k^i)} + \frac{k_3 b_1}{\gamma_1-\lambda}\left(e^{-\lambda(t*-t_k^i)} - e^{\gamma_1(t*-t_k^i)}\right)\right). \end{aligned} \tag{2.48}$$

As $\sigma\text{max} \le \frac{\gamma_1}{(\sqrt{2}k_3+\gamma_1)(\bar{N}+\sqrt{N}\sqrt{\bar{N}})}$, we have that $1 - \frac{k_3 b_1}{\gamma_1-\lambda} > 0$ and there exits that

$$v(t^*) < e^{-\lambda(t*-t_k^i)}\|\phi(t_k^i)\| = v(t^*). \tag{2.49}$$

The contradiction shows that (2.48) is valid, and when $t = t_k^i$, we have that $\|\phi(t_k^i)\| < e^{-\lambda(t_k^i-t_k^i)}\|\phi(t_k^i)\| = \|\phi(t_k^i)\|$, then one has

$$\|\phi(t)\| \le e^{-\lambda(t-t_k^i)}\|\phi(t_k^i)\|,\, t \ge t_k^i, \tag{2.50}$$

(2) When $\chi(t) = 1$ and $t \in [s_k^i, s_k^i + h_i)$, the system can be rewritten as

$$\dot{\bar{\phi}}(t) = -H \otimes C\bar{\phi}(t) + H \otimes C \int_{t-\tau(t)}^{t} \dot{\bar{\phi}}(s)ds, \tag{2.51}$$

By the Newton-Leibnitz formula, $\dot{\bar{\phi}}(t) = -H \otimes C\bar{\phi}(t-\tau(t))$, which is useful in the following.

Then by the variation of parameter formula, one can have that

$$\bar{\phi}(t) = e^{(-H\otimes C)(t-t_k^i)}\bar{\phi}(t_k^i) + \int_{t_k^i}^{t} e^{(-H\otimes C)(t-\theta)}\left(H \otimes C \int_{\theta-\tau(t)}^{\theta} \dot{\bar{\phi}}(s)ds\right)d\theta. \tag{2.52}$$

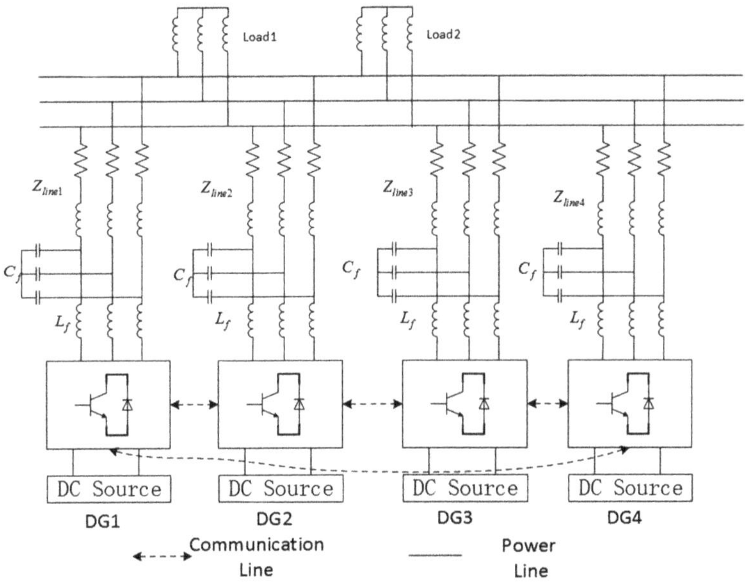

Fig. 2.4 The simulation test microgrid system

As $\dot{\bar{\boldsymbol{\phi}}}(t) = -H \otimes C\bar{\boldsymbol{\phi}}(t - \tau(t))$, then we have that

$$\left\|\bar{\boldsymbol{\phi}}(t)\right\| \leq e^{-\gamma_1(t-t_k^i)}\left\|\bar{\boldsymbol{\phi}}\left(t_k^i\right)\right\| + \int_{t_k^i}^{t} e^{-\gamma_1(t-\theta)}\left(k_1\int_{\theta-\tau(t)}^{\theta} k_1\left\|\bar{\boldsymbol{\phi}}(s-\tau(t))\right\|ds\right)d\theta \tag{2.53}$$

As $\left\|\bar{\boldsymbol{\phi}}(t)\right\| = \|\boldsymbol{\phi}(t)\|$, then we can prove that with the same method as above

$$\|\boldsymbol{\phi}(t)\| < e^{-\lambda(t-t_0)}\left\|\boldsymbol{\phi}\left(t_k^i\right)\right\| := \boldsymbol{v}_1(t), t > t_k^i, \tag{2.54}$$

and satisfies that $\frac{k_1^2 e^{\lambda h}\left(e^{\lambda h}-1\right)}{\lambda(\gamma_1-\lambda)} < 1$.

Then the consensus of can be achieved exponentially and the reactive power of microgrids sharing can be achieved.

To explore the influence for microgrid by DoS attacks, a system shown in Fig. 2.4 has been simulated in MATLAB/Simulation, where the configuration of DG units with same rated powers is considered. The parameters are presented in Table 2.2.

In this case, the performance of reactive power sharing control method under the proposed periodic event-triggered mechanism is compared with the conventional droop control method. And the change of the load is considered. The simulation process has been divided into three steps: first, conventional droop control without power sharing method is used; second, at $t = 2$ s, the reactive power sharing control method under the periodic event-triggered control is started; third, at $t = 6$ s, the

Table 2.2 System parameters

Parameter	Value	Frequency droop coefficients	
DC voltage	400 V	2.18e-5 rad/(w*s)	
MG voltage	208 V	Voltage droop coefficients	
MG frequency	60 Hz	1e-3 V/Var	
Line parameters		Load parameters	
Z_{line1}	0.3Ω + 0.5mH	P_{load1}	30 kW
Z_{line2}	0.2Ω+0.6mH	Q_{load1}	1.5kVar
Z_{line3}	0.175Ω+0.95mH	P_{load2}	5 kW
Z_{line4}	0.175Ω+1.55mH	Q_{load2}	0.25kVar
C	15	h	0.01 ms
σ	0.05		

load 2 is disconnected; finally, at $t = 8$ s, the load 2 is connected. The simulation results are given in Fig. 2.5. As seen in Fig. 2.5b, the reactive powers are shared without destroying the active power sharing all the times. The event time instants for each DG's controller are given in Fig. 2.5c.

2.2 Cyber Security and Safety of Energy Internet

Energy Internet should provide the users with stable and secure energy environment. So it is a strong need to analyze the safety and security of control system in energy internet in spite of the presence of malicious attacks [20]. In recent years, several security incidents have occurred which resulted in an energy internet crush through malicious attacks on its information system, such as the 2015 Ukraine Blackout [21]. In these reasons, it is of great importance to study the energy internet control strategies with cyber-securities despite the malicious attacks.

For micro-girds, there may exit many kinds of attacks, such as injection attacks, *Denial-of-Service* (DoS) attacks and so on. Nowadays the attention of researchers of microgrids has been focused on the injection attacks because of the Ukraine Blackout [22], and few attention is focused on DoS attacks, which aim to prevent the communication between system components and are common attacks on internet.

2.2.1 *DoS Attacks in Energy Internet*

For Energy Internet, there may exit many kinds of attacks, such as injection attacks, *Denial-of-Service* (DoS) attacks and so on. Nowadays the attention of researchers of microgrids has been focused on the injection attacks because of the Ukraine

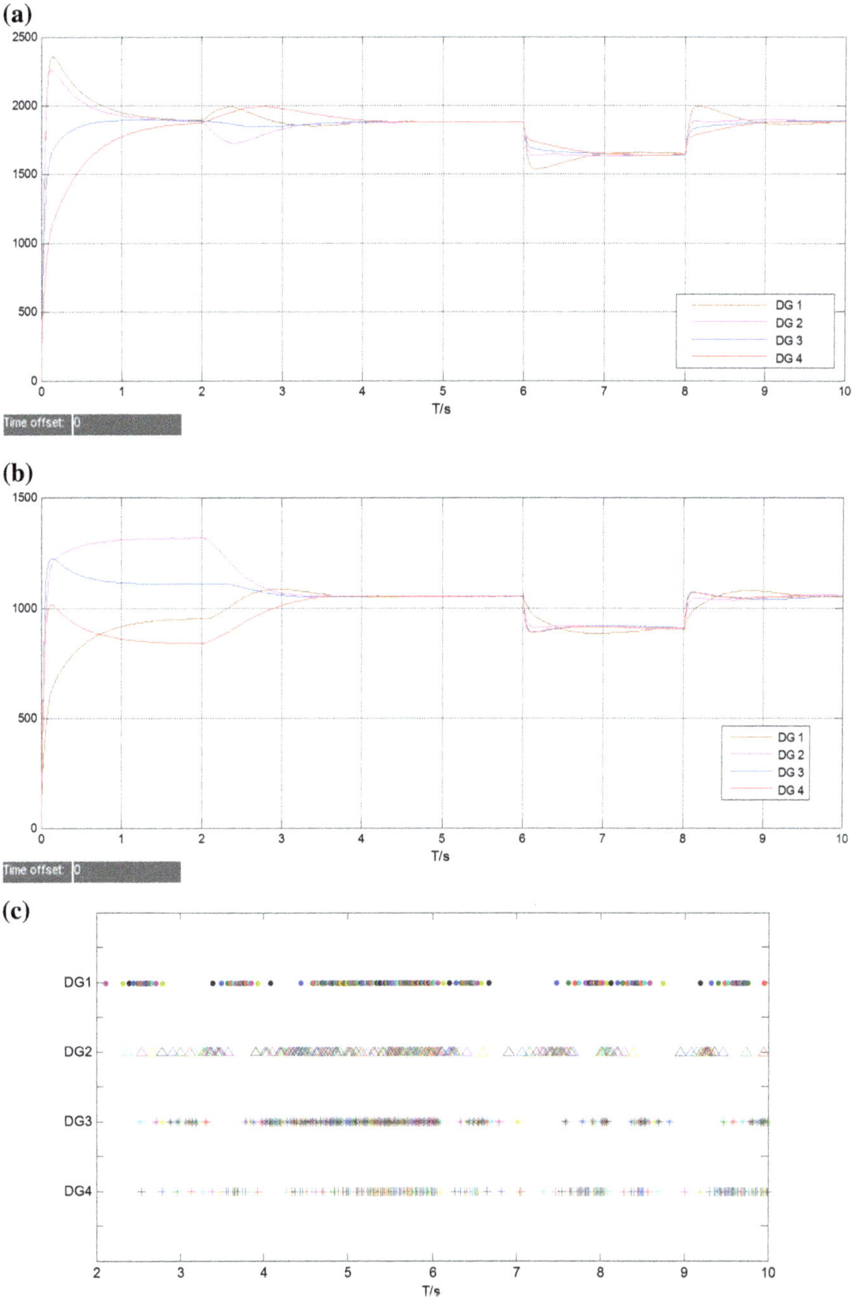

Fig. 2.5 The simulation results without DoS attacks: **a** Reactive power; **b** Active Power; **c** The Triggering Sequence of Each DG

Blackout [23], and few attention is focused on DoS attacks, which aim to prevent the communication between system components and are common attacks on internet. Here we notice that communication between some elements in microgrids must be keeping guaranteeing the systems' stability in most of control methods. And as the using of digital control technologies, micro-gids present hybrid dynamic properties. That is to say, the state of microgrids is piecewise continuous. The state information of equipment is sampled at one moment and is always used to control methods for a period of time until the next sampling instant. And mostly the stability of the system is related with the keeping period time. When DoS attacks happen, they may lengthen several or most of the keeping period time, which can influence or even destroy the stability of microgrids and we do not want to see this film. So studying the influence of DoS attacks into microgrids and designing the control methods with tolerance for DoS attacks will become one of the researching detections of the security of microgrids. Nowadays the research aspects of the security problem of control systems by DoS attacks are the detection of DoS attacks [24], the system tolerance for DoS attacks [25–27], optimal DoS attack scheduling [28] and so on. In [25], for a special kind of DoS attacks, named PWM jammers with the firmed attack frequency and during time, an event-triggered control update strategy has been studied under the Input-to-State Stability. A more general DoS attack model has been proposed in [26] with the consideration of attacker's strategy for the frequency and duration of attacks and the Input-to-State Stability of the control system has been learned despite the presence of the DoS attack model. Then in [27], authors talk about the stability of networked control systems under DoS attacks. The above literatures mainly study the control update strategy for a whole system and learn the tolerance of DoS attacks under one stability condition. These methods can provide references for study the tolerance for DoS attacks in microgrids. In microgrids, as distributed control methods have been always used for greater flexibility, such as multi-agent method, attackers may aim at one or some distributed subsystems, not the whole mirogrid, which can also influence the performance of microgrids. These should be considered when studying microgrids under DoS attacks.

2.2.2 DoS Attack and Model

Here we also use the microgrid to analyze the DoS attacks in Energy Internet. In the secondary control level of the microgrid, the information is transformed between the DG neighboring agents, then the accurate reactive power sharing can be achieved. It should and must be based on the reliability and security of the information network in the secondary control level. However, the information network of the micro-grid is usually vulnerable for DoS jammers because of the lack of network security awareness in industry system. As previously mentioned, DoS phenomenon can be expressed as the periodic denial of the updates of control strategy which can affect the information network of the secondary control level. It will prevent the controller u_{Qi} from being acted at the desired time instants. Then the reactive power can hardly be shared for serious attacks. In such case, when load changes, the reactive power sharing will fail and the stability and reliability of the microgrid may be broken as mentioned before. In micro-rids, there are many communication links, and each of them may be attacked by the jammers. Let $h^{i,j}_{k,k\in\mathbb{N}_0}$ denote the time of kth attack occurrence between DG_i and DG_j and $\tau^{i,j}_k \in \mathbb{R}_{\geq 0}$ denote the duration of kth attack between DG_i and DG_J. Then for kth DoS attack, we let

$$H^{i,j}_k := \left[h^{i,j}_k, h^{i,j}_k + \tau^{i,j}_k\right) \tag{2.55}$$

represent the kth DoS time interval of the communication link between i and j and $H^{i,j} = \bigcup\limits_{k\in\mathbb{N}\geq 0} H^{i,j}_k$. Here we introduce a new time sequence $H_{n\in\mathbb{N}_{\geq 0}}$ with a new sequence number n, which denotes the whole DoS attacks happening in the microgrid and $H = \bigcup\limits_{i,j\in N} H^{i,j}$. Let $\Xi(T_1, T_2)$ and $\Theta(T_1, T_2)$ represent the sets of time intervals in which communication is denied and allowed respectively as follow

$$\Xi(T_1, T_2) := \bigcup_{n\in\mathbb{N}_0} H_n \cap [T_1, T_2], \tag{2.56}$$

$$\Theta(T_1, T_2) := [T_1, T_2] \backslash \Xi(T_1, T_2), \tag{2.57}$$

For $T_1, T_2 \in \mathbb{R}_{\geq 0}$. Then the definition of DoS frequency and duration can be introduced which can represent the natural characterization of DoS attacks.

Definition 1 ([26, 27], *DoS frequency*). Given any $T_1, T_2 \in \mathbb{R}_{\geq 0}$ with $T_2 > T_1$, let $n(T_1, T_2)$ denote the number of DoS off/on transitions occurring in the interval $[T_1, T_2[$. For a given sequence of DoS attacks $\{H_n\}_{n\in\mathbb{N}}$ satisfies the DoS frequency constraint if there exist $\mu \in \mathbb{R}_{\geq 0}$ and $\xi_D \in \mathbb{R}_{\geq 0}$ that

$$n(T_1, T_2) \leq \mu + (T_2 - T_1)\xi_D \tag{2.58}$$

.

Definition 2 ([26, 27], *DoS duration*). Given any $T_1, T_2 \in \mathbb{R}_{\geq 0}$ with $T_2 > T_1$, let $|\Xi(T_1, T_2)|$ denote the length of the interval over which communication is disabled. For a given sequence of DoS attacks $\{H_n\}_{n\in\mathbb{N}}$ satisfies the DoS duration constraint if there exist $\rho \in \mathbb{R}_{\geq 0}$ and $T_D \in \mathbb{R}_{>0}s$ that

$$|\Xi(T_1, T_2)| \leq \rho + (T_2 - T_1)T_D \tag{2.59}$$

T_D is the upper bound on the average duration of DoS per unit time

Then from lemma 2 in [26], it is easy to know that for any $T_1, T_2 \subset \mathbb{R}_{\geq 0}$, with $0 \leq T_1 \leq T_2$, the interval $[T_1, T_2[$ is the disjoint union of $\overline{\Theta}(T_1, T_2)$ and, $\overline{\Xi}(T_1, T_2)$, where $\overline{\Theta}(T_1, T_2)$ is the union of sub-intervals of $[T_1, T_2[$ the condition of (2.31) holds and $\overline{\Xi}(T_1, T_2)$ is also the union of sub-intervals of $[T_1, T_2[$. Then there exist two sequences of non-negative and positive real numbers $\{\zeta_m\}_{m\in\mathbb{N}_0}$ and $\{\upsilon_m\}_{m\in\mathbb{N}_0}$ such that

$$\overline{\Theta}(T_1, T_2) := \bigcup_{m\in\mathbb{N}_0} Z_m \cap [T_1, T_2[, \tag{2.60}$$

$$\overline{\Xi}(T_1, T_2) := \bigcup_{m\in\mathbb{N}_0} W_m \cap [T_1, T_2[, \tag{2.61}$$

where

$$Z_m := \{\zeta_m\} \bigcup [\zeta_m, \zeta_m + \upsilon_m[, \tag{2.62}$$

$$W_m := \{\zeta_m + \upsilon_m\} \bigcup [\zeta_m + \upsilon_m, \zeta_{m+1}[. \tag{2.63}$$

And in fact, for each $m \in \mathbb{N}_0$, a successful control update occurs at ζ_m and one DoS attack occurs at υ_m, then there is no DoS over Z_m and the communication is denial over W_m. From this lemma, it can be seen that the influence by DoS attacks is not only depends on the attacks sequence but also the update strategy that the microgrid adopts. So when analyzing the stability and consistency of the microgrid under DoS attacks, the update strategies must be considered. As the event-based update control method has been studied to reduce the communication burden, the event-triggered update strategy can be widely used in many control systems, also in microgrids. So the event-triggered control update strategy is further introduced to analyze the stability and consistency of the microgrid under DoS attacks.

2.2.3 *Analysis of Influence from DoS Attacks*

Now we derive the main result that: *the reactive power sharing control system under the periodic event-triggered control update rule keep consensus for DoS signal that satisfies definition 1 and 2 with ξ_D and T_D sufficiently large*. The following results holds.

As mentioned earlier, we can have that for $\chi(t) = 0$, $\left\|\boldsymbol{e}_Q(t)\right\| \leq (\gamma_1 - \lambda)\|\boldsymbol{\phi}(\theta)\|$ and for $\chi(t) = 1$, $\left\|\boldsymbol{e}_Q(t)\right\| \leq \frac{\sigma_{\max}\left(\bar{N}+\sqrt{N}\sqrt{\bar{N}}\right)}{\left(1-\sigma_{\max}\bar{N}-\sigma_{\max}\sqrt{N}\sqrt{\bar{N}}\right)}\|\boldsymbol{\phi}(t)\|$. Here we denote $\tilde{\sigma} = \max\left\{\gamma_1 - \lambda, \frac{\sigma_{\max}\left(\bar{N}+\sqrt{N}\sqrt{\bar{N}}\right)}{\left(1-\sigma_{\max}\bar{N}-\sigma_{\max}\sqrt{N}\sqrt{\bar{N}}\right)}\right\}$, then

$$\left\|\boldsymbol{e}_Q(t)\right\| \leq \tilde{\sigma}\,\|\boldsymbol{\phi}(t)\| \tag{2.64}$$

exits when there are no DoS attacks. And it is easy to know that the (2.64) does not hold in the DoS intervals. Then the analysis of the influence of DoS attacks for the microgrid is to decompose the time axis into intervals where it is possible to satisfy (2.64) and intervals where, due to the occurrence of DoS, (2.64) need not hold.

We here analyze the norm inequality of system in the intervals Z_m and W_m respectively. Consider the intervals Z_m, from (2.54) we have

$$\|\phi(t)\| \le e^{-\lambda(t-\zeta_m)}\|\phi(\zeta_m)\| \tag{2.65}$$

for all $t \in Z_m$ and all $m \in \mathbb{N}_0$.

For one of DGs in the microgrid, if it has been under DoS attack, the triggering condition need not hold. Recall that $\|e_{Qi}(t)\| \le \|\phi_i(t_{k(\zeta_m+\upsilon_m)}) - \phi_i(t)\|$ for all $t \in Z_m$, where $t_{k(\zeta_m+\upsilon_m)}$ represents the last successful control update up to $\zeta_m + \upsilon_m$. Consider the continuity of $Q_i^{\wedge}(t)$, we have

$$\|\bar{\phi}(t_{k(\zeta_m+\upsilon_m)})\| \le (1+\tilde{\sigma})\|\bar{\phi}(\zeta_m+\upsilon_m)\| \tag{2.66}$$

for all $m \in \mathbb{N}_0$. Then we can have that

$$\|\bar{e}_Q(t)\| \le \|\bar{\phi}(t)\| + (1+\tilde{\sigma})\|\bar{\phi}(\zeta_m+\upsilon_m)\| \tag{2.67}$$

for all $m \in \mathbb{N}_0$.

Recall that

$$\|\phi(t)\| \le e^{-\gamma_1(t-(\zeta_m+\upsilon_m))}\|\phi(\zeta_m+\upsilon_m)\| + k_1\int_{(\zeta_m+\upsilon_m)}^{t} e^{-\gamma_1(t-s)}\|\bar{e}(s)\|ds \tag{2.68}$$

holds for all $t \ge (\zeta_m + \upsilon_m)$. Substituting (2.67) into (2.68), we get

$$\begin{aligned}\|\phi(t)\| &< e^{-\gamma_1(t-(\zeta_m+\upsilon_m))}\|\phi(\zeta_m+\upsilon_m)\| + k_1\int_{t_{k(\zeta_m+\upsilon_m)}}^{t} e^{-\gamma_1(t-s)}\|\phi(s)\|\,ds\\ &\quad + \left(1+\tilde{\sigma}\right)k_1\|\phi(\zeta_m+\upsilon_m)\|\int_{(\zeta_m+\upsilon_m)}^{t} e^{-\gamma_1(t-s)}ds.\end{aligned} \tag{2.69}$$

Then we will prove that

$$\|\phi(t)\| \le e^{\tilde{\lambda}(t-(\zeta_m+\upsilon_m))}\|\phi(\zeta_m+\upsilon_m)\|, t \ge (\zeta_m+\upsilon_m), \tag{2.70}$$

where $\gamma_1 + \tilde{\lambda} > k_1$. There are two parts in this proof:

(1) When $t = (\zeta_m + \upsilon_m)$, there exists $\|\phi(\zeta_m+\upsilon_m)\| = e^{\tilde{\lambda}\times 0}\|\phi(\zeta_m+\upsilon_m)\|$.

(2) When $t > (\zeta_m + \upsilon_m)$, $\|\phi(t)\| < e^{\tilde{\lambda}(t-(\zeta_m+\upsilon_m))}\|\phi(\zeta_m+\upsilon_m)\| := \tilde{\nu}_1(t)$ should hold. If it does not hold, it must exist a $t^* > t_0$ for $\|\phi^*(t)\| = \tilde{\nu}(t^*)$ and $\|\phi(t^*)\| < \tilde{\nu}(t^*)$ for $t < t^*$. Then by (2.69), we have

$$\tilde{v}(t^*) = \left\| \phi(t^*) \right\| < e^{-\gamma_1(t*-(\zeta_m+\upsilon_m))} \|\phi(\zeta_m+\upsilon_m)\| + k_1 \|\phi(\zeta_m+\upsilon_m)\| \int_{t_{k(\zeta_m+\upsilon_m)}}^{t*} e^{-\gamma_1(t*-s)} e^{\tilde{\lambda}(t*-(\zeta_m+\upsilon_m))} ds \, + (1+\tilde{\sigma}) k_1 \|\phi(\zeta_m+\upsilon_m)\| \int_{(\zeta_m+\upsilon_m)}^{t*} e^{-\gamma_1(t-s)} ds$$

$$= \|\phi(\zeta_m+\upsilon_m)\| \left(\begin{array}{l} \left(1 - \frac{k_1}{\gamma_1+\tilde{\lambda}}\right) e^{-\gamma_1(t-(\zeta_m+\upsilon_m))} + \frac{k_1}{\gamma_1+\tilde{\lambda}} e^{\tilde{\lambda}(t-(\zeta_m+\upsilon_m))} \\ + \frac{(1+\tilde{\sigma})k_1}{\gamma_1}\left(1 - e^{-\gamma_1(t-(\zeta_m+\upsilon_m))}\right) \end{array} \right) \tag{2.71}$$

$$\tilde{v}(t^*) < \|\phi(\zeta_m+\upsilon_m)\| \left(\begin{array}{l} \left(1 - \frac{k_1}{\gamma_1+\tilde{\lambda}} - \frac{(1+\sigma)k_1}{\gamma_1}\right) e^{-\gamma_1(t-(\zeta_m+\upsilon_m))} \\ + \frac{k_1}{\gamma_1+\tilde{\lambda}} e^{\tilde{\lambda}(t-(\zeta_m+\upsilon_m))} + \frac{(1+\sigma)k_1}{\gamma_1} \end{array} \right). \tag{2.72}$$

As $e^{-\gamma_1(t-(\zeta_m+\upsilon_m))} < 1$,then

$$\left(1 - \frac{k_1}{\gamma_1+\tilde{\lambda}} - \frac{(1+\sigma)k_1}{\gamma_1}\right) e^{-\gamma_1(t-(\zeta_m+\upsilon_m))} + \frac{k_1}{\gamma_1+\tilde{\lambda}} e^{\tilde{\lambda}(t-(\zeta_m+\upsilon_m))} + \frac{(1+\sigma)k_1}{\gamma_1} < \left(1 - \frac{k_1}{\gamma_1+\tilde{\lambda}}\right) e^{-\gamma_1(t-(\zeta_m+\upsilon_m))} + \frac{k_1}{\gamma_1+\tilde{\lambda}} e^{\tilde{\lambda}(t-(\zeta_m+\upsilon_m))} \tag{2.73}$$

With $\gamma_1 + \tilde{\lambda} > k_1$, it makes that

$$\tilde{v}(t^*) < e^{\tilde{\lambda}(t*-(\zeta_m+\upsilon_m))} \|\phi(\zeta_m+\upsilon_m)\| = \tilde{v}(t^*). \tag{2.74}$$

It shows that (2.71) is valid.

Then we will analyze the relationship between DoS frequency and duration and λ and $\tilde{\lambda}$.

Before it, first considering the interval $W_0 = [0, \zeta_1[$, during which there is no communication. It means that (2.64) does not hold. Then

$$\|\phi(t)\| \le e^{\tilde{\lambda}\zeta_1} \|\phi(0)\| = e^{\tilde{\lambda}|\Xi(0,\zeta_1)|} \|\phi(0)\| \tag{2.75}$$

holds. In the next interval $Z_1 = [\zeta_1, \zeta_1 + \upsilon_1[$, (2.64) holds, one has

$$\|\phi(\zeta_1+\upsilon_1)\| \le e^{-\lambda((\zeta_1+\upsilon_1)-\zeta_1)} \|\phi(\zeta_1)\|. \tag{2.76}$$

As $\left|\Xi(0, \zeta_1+\upsilon_1)\right| = \zeta_1$ and $\left|\Theta(0, \zeta_1+\upsilon_1)\right| = \upsilon_1$, (2.76) can be rewritten as

$$\|\phi(\zeta_1+\upsilon_1)\| \le e^{-\lambda|\Theta(0,\zeta_1+\upsilon_1)|} e^{\tilde{\lambda}|\Xi(0,\zeta_1+\upsilon_1)|} \|\phi(0)\|. \tag{2.77}$$

In the next interval $W_1 = [\zeta_1 + \upsilon_1, \zeta_2[$, (2.64) holds, one has

$$\|\phi(\zeta_2)\| \le e^{\tilde{\lambda}(\zeta_2-(\zeta_1+\upsilon_1))}\|\phi(\zeta_2)\|. \tag{2.78}$$

As $\left|\Xi(0, \zeta_2)\right| = \zeta_1 + \zeta_2 - (\zeta_1 + \upsilon_1)$ and $\left|\overline{\Theta}(0, \zeta_2)\right| = \upsilon_1$, we have

$$\|\phi(\zeta_2)\| \le e^{-\lambda\overline{\Theta}(0,\zeta_2)}e^{\tilde{\lambda}|\Xi(0,\zeta_2)|}\|\phi(0)\|\|\phi(\zeta_2)\|. \tag{2.79}$$

Then with the mathematical induction, we have

$$\|\phi(t)\| \le e^{-\lambda\left|\overline{\Theta}(0,t)\right|}e^{\tilde{\lambda}\left|\Xi(0,t)\right|}\|\phi(0)\| \tag{2.80}$$

holds for all $t \in \mathbb{R}_{\ge 0}$.

Theorem 2 *For the reactive power sharing consensus controller system composed of (2.21) and (2.22) and control input (2.29), with the periodic event-triggered mechanism (2.31), the consensus can be realized under any DoS sequence satisfying definition 1 and 2 with ξ_D and T_D such that*

$$\frac{1}{T_D} + \xi_D h < \frac{\lambda}{\lambda + \tilde{\lambda}}, \tag{2.81}$$

where $\gamma_1 + \tilde{\lambda} > k_1$.

Before recalling the definition 1 and 2, first we analyze the influence to the system from the DoS time intervals. As we want to find the tolerance of DoS attacks for the consensus control of the microgrid, during $[T_1, T_2[$, the denial time of DoS attacks can be upper bounded by the $|\Xi(T_1, T_2)|$, which means the total time length of DoS, plus the number $n(T_1, T_2)$ of DoS multiplied by minimum triggering-time h, which is related with the communication periods. And considering that DoS attacks may occur at the beginning of $[T_1, T_2[$, so we have

$$\left|\bar{\Xi}(T_1, T_2)\right| \le |\Xi(T_1, T_2)| + (1 + n(T_1, T_2))h \tag{2.82}$$

for all $T_1, T_2 \in \mathbb{R}_{\ge 0}$ with $0 \le T_1 \le T_2$. Then we have

$$\left|\bar{\Xi}(T_1, T_2)\right| \le \rho + \frac{T_2 - T_1}{T_D} + (1 + \mu + (T_2 - T_1)\xi_D)h. \tag{2.83}$$

Let $\rho_* := \rho + (1 + \mu)h$ and $T_* := \frac{T_D}{1+\xi_D\Delta_* T_D}$, then (2.83) can be rewritten as

$$\left|\bar{\Xi}(T_1, T_2)\right| \le \rho_* \, \frac{T_2 - T_1}{T_*}. \tag{2.84}$$

Then

$$\left|\overline{\Theta}(T_1, T_2)\right| = T_2 - T_1 - \left|\bar{\Xi}(T_1, T_2)\right| \geq T_2 - T_1 - \left(\rho_* + \frac{T_2 - T_1}{T_*}\right). \quad (2.85)$$

Substituting (2.84) and (2.85) into (2.80), one has

$$\|\boldsymbol{\phi}(t)\| \leq e^{-\lambda\left(t - \left(\rho_* + \frac{t}{T_*}\right)\right) + \tilde{\lambda}\left(\rho_* + \frac{t}{T_*}\right)} \|\boldsymbol{\phi}(0)\| = e^{\rho_*(\lambda + \tilde{\lambda})} e^{-t\left(\lambda - \frac{\lambda + \tilde{\lambda}}{T_*}\right)} \|\boldsymbol{\phi}(0)\|. \quad (2.86)$$

So with $\frac{1}{T_D} + \xi_D h < \frac{\lambda}{\lambda + \tilde{\lambda}}$, we can have that $t\left(\lambda - \frac{\lambda+\tilde{\lambda}}{T_*}\right) > 0$, then the consensus of the system can achieved.

Here the above example is used to explain the influence on the system by DoS attacks. When the DoS attacks happen, the communication channel is denied and the update policies can hardly work. In this case, we talk about the tolerant DoS attacks for our microgrid. As shown in the before, we can know that $\gamma_1 = 60$, $k_1 = 64.603$, $k_2 = 45$, $k_3 = 64.603$, $b_1 = 0.4501$, $\lambda = 30.9216$, $\tilde{\lambda} = 4.603$. Further we have that $\lambda \big/ \left(\lambda + \tilde{\lambda}\right) = 0.8704$, which means that the reactive power sharing of the microgrid can still works under a maximum of 87% of communication denial. In this part, we will discuss one of the tolerant DoS attacks to the microgrid. First the reactive power sharing control method under the periodic event-triggered control is used; at $t = 3$ s, we let the DoS attacks in; finally at $t = 6$, load2 is disconnected. Here we attack the acceptor of each DG by $H^1 = \{[3, 6)\}$, $H^2 = \{[4, 5), [9, 10)\}$, $H^3 = \{[4, 5)\}$ and $H^4 = \{[3, 4), [8, 10)\}$, so the characters of the DoS attacks can be written as $1 \big/ T_D + \xi_D h < 0.7134$. And the reactive power sharing cannot be influenced, shown in Fig. 2.6.

When the DoS attacks are serious, the performance of microgrid is influenced. And the voltage stability can be also being influenced. In this case, we choose that $H^1 = \{[3, 7), [9, 10)\}$, $H^2 = \{[4, 6), [7, 10)\}$, $H^3 = \{[4, 5), [6, 7), [8, 10)\}$ and $H^4 = \{[3, 6), [7, 10)\}$, which makes the character of the DoS attacks $H^4 = \{[3, 6), [7, 10)\}$ almost close to 1. So the consensus of the system is broken, shown as Fig. 2.7. Here we should notice that when system is consensus, the DoS attacks cannot influence the system performance, and the consensus can be broken under DoS attacks if the system has been changed, such as load change. And it is formal in the true microgrid system.

From the above simulations, we can find that the system performance can be influenced by DoS attacks.

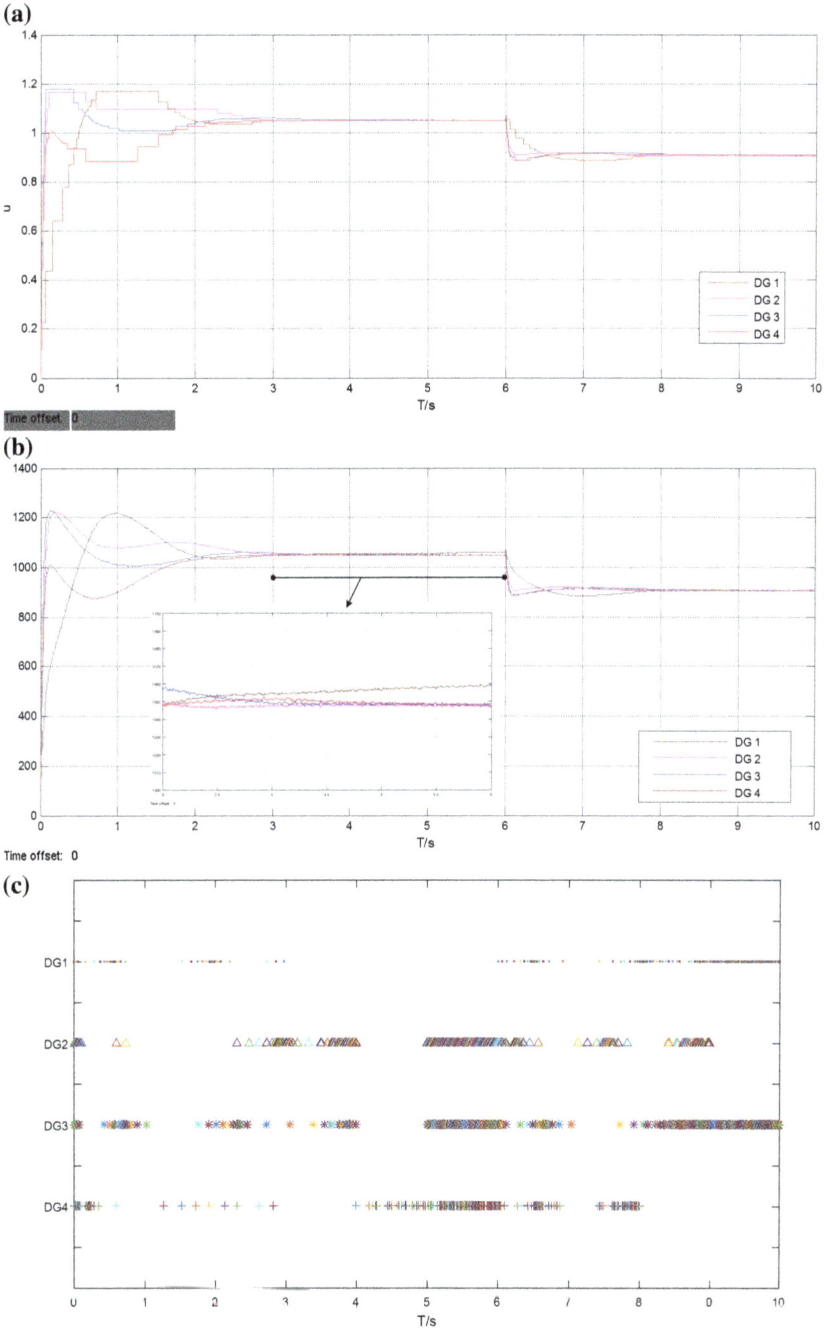

Fig. 2.6 The simulation results under tolerant DoS attacks: **a** the Control Value of the Consensus Controller; **b** the Reactive Power; **c** the Triggering Sequence

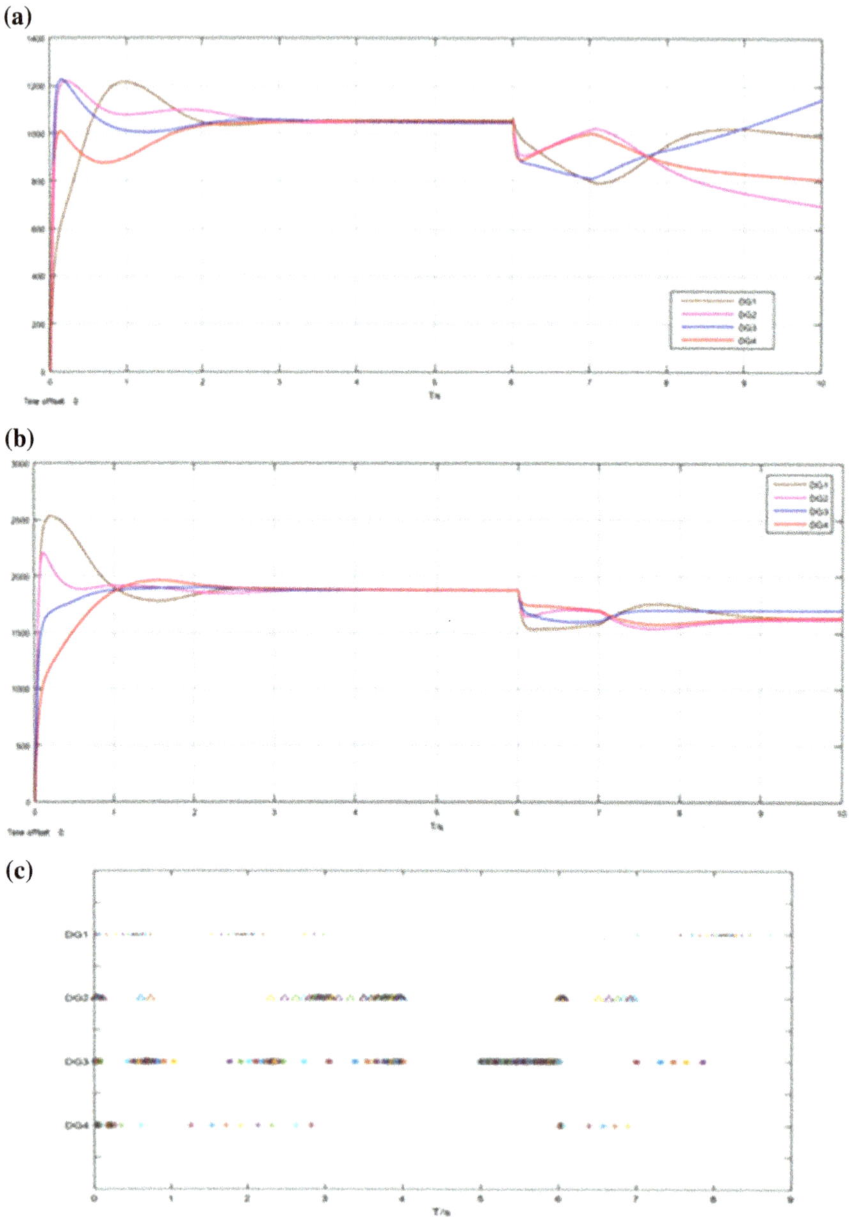

Fig. 2.7 The simulation results under serious DoS Attacks: **a** the Reactive Power; **b** the Active Power; **c** The Triggering Sequence

2.3 Conclusion

Energy Internet can be seen as a complicated cyber-physical system. For the complicated cyber-physical system, there are many research interests in the new field. In this chapter, we explore some cyber-physical characteristics of Energy Internet mainly including the structure of cyber-physical energy internet, the relationship of the cyber resources and physical resources and the cyber security and safety of Energy Internet. In these characteristics, the security and safety of Energy Internet has attracted the attention of researchers and engineers. As Energy Internet present an increased dependency on cyber resources which may be vulnerable to attack [29], the cyber-physical security and safety have become a new hot issue and this issue may include but not limited to the analysis of the influence of physical system by the cyber calculated attacks and the influence of structure failures to the whole system. Here we mainly analyze one of the calculated attacks for some sub-system of Energy Internet. However, these researches are the beginning and there are still problems in these researches.

References

1. P. Derler, E.A. Lee, A.S. Vincentelli, Modeling cyber–physical systems. Proc. IEEE **100**(1), 13–28 (2012)
2. A.Q. Huang, M.L. Crow, G.T. Heydt et al., The future renewable electric energy delivery and management (FREEDM) system: the energy internet. Proc. IEEE **99**(1), 133–148 (2011)
3. W. Wolf, Cyber-physical systems. Computer **42**(3), 88–89 (2009)
4. Lee A, Guidelines for smart grid cyber security. NIST Interagen-cy/Internal Report (NISTIR)-7628, 2010
5. X.U.E. Yusheng, Energy internet or comprehensive energy network? J. Modern Power Syst. Clean Energy **3**(3), 297–301 (2015)
6. S. Xin, Q. Guo, H. Sun et al., Cyber-physical modeling and cyber-contingency assessment of hierarchical control systems. IEEE Trans. Smart Grid **6**(5), 2375–2385 (2015)
7. J. Kim, L. Tong, On topology attack of a smart grid: undetectable attacks and countermeasures. IEEE J. Sel. Areas Commun. **31**(7), 1294–1305 (2013)
8. Z. Li, M. Shahidehpour, A. Alabdulwahab et al., Bilevel model for analyzing co-ordinated cyber-physical attacks on power systems. IEEE Trans. Smart Grid **7**(5), 2260–2272 (2016)
9. J.M. Bradley, EM Atkins, Computational-physical state co-regulation in cyber-physical systems, in *2011 IEEE/ACM International Conference on Cyber-Physical Systems* (*ICCPS*). IEEE (2011), pp. 119–128
10. R. Bhattacharya, G.J. Balas, Anytime control algorithm: model reduction approach. J. Guid. Control Dyn. **27**(5), 767–776 (2004)
11. JM Bradley, EM Atkins, Computational-physical state co-regulation in cyber-physical systems, in *2011 IEEE/ACM International Conference on Cyber-Physical Systems* (*ICCPS*). IEEE (2011), pp. 119–128
12. J.M. Bradley, E.M. Atkins, Coupled cyber-physical system modeling and coregulation of a cubesat. IEEE Trans. Rob. **31**(2), 443–456 (2015)
13. A. Selivanov, E. Fridman, Event-triggered H-∞ Control: a switching approach. IEEE Trans. Autom. Control **61**(10), 3221–3226 (2016)
14. Q.C. Zhong, Robust droop controller for accurate proportional load sharing among inverters operated in parallel. IEEE Trans. Ind. Electron. **60**(4), 1281–1290 (2013)

15. N. Eghtedarpour, E. Farjah, Power control and management in a hybrid AC/DC microgrid. IEEE Trans. Smart Grid **5**(3), 1494–1505 (2014)
16. H. Zhang, S. Kim, Q. Sun et al., Distributed adaptive virtual impedance control for accurate reactive power sharing based on consensus control in microgrids. IEEE Trans. Smart Grid (2016)
17. S. Kim, H. Zhang, Q. Sun et al., Consensus-based distributed control for accurate reactive, harmonic and imbalance power sharing in microgrids. IEEE Trans. Smart Grid (2016)
18. W. Hu, L. Liu, G. Feng, Consensus of linear multi-agent systems by distributed event-triggered strategy. IEEE Trans. Cybern. **46**(1), 148–157 (2016)
19. D.Z. Zheng, *Linear System Theory* (Press of Tsinghua University, Beijing, 2002)
20. F. Pasqualetti, F. Dorfler, F. Bullo, Control-theoretic methods for cyberphysical security: geometric principles for optimal cross-layer resilient control systems. Control Syst. IEEE **35**(1), 110–127 (2015)
21. G. Liang, S.R. Weller, J. Zhao et al., The 2015 ukraine blackout: implications for false data injection attacks. IEEE Trans. Power Syst. **PP**(99), 1–1 (2016)
22. Z. Li, M. Shahidehpour, A. Alabdulwahab, A. Abusorrah, Bilevel model for analyzing coordinated cyber-physical attacks on power systems. IEEE Trans. Smart Grid **7**(5), 2260–2272 (2016)
23. Z. Li, M. Shahidehpour, A. Alabdulwahab, A. Abusorrah, Bilevel model for analyzing coordinated cyber-physical attacks on power systems. IEEE Trans. Smart Grid **7**(5), 2260–2272 (2016)
24. F. Pasqualetti, F. Dörfler, F. Bullo, Attack Detection and identification in cyber-physical systems. IEEE Trans. Autom. Control **58**(11), 2715–2729 (2013)
25. H. Shisheh Foroush, S. Martínez, On event-triggered control of linear systems under periodic denial-of-service jamming attacks, in *2012 IEEE 51st IEEE Conference on Decision and Control (CDC), Maui, HI* (2012), pp. 2551–2556
26. C. De Persis, P. Tesi, Input-to-state stabilizing control under denial-of-service. IEEE Trans. Autom. Control **60**(11), 2930–2944 (2015)
27. V.S. Dolk, P. Tesi, C. De Persis, W.P.M.H. Heemels, Event-triggered control systems under denial-of-service attacks. IEEE Trans. Control Netw. Syst. **4**(1), 93–105 (2017)
28. H. Zhang, P. Cheng, L. Shi, J. Chen, Optimal DoS attack scheduling in wireless networked control system. IEEE Trans. Control Syst. Technol. **24**(3), 843–852 (2016)
29. A. Lee, Guidelines for smart grid cyber security. NIST Interagency/Internal Report (NISTIR)-7628 (2010)

Chapter 3
We-Energy Modelling

Abstract In this chapter, a mechanism model of We-Energy based on its structure is proposed, which embodies the distinguishing features of bi-directional power transformation and energy coupling. A quaternary model of WE is established under steady and transient state, which can be divided into normal state, alert state, emergency state and recovery state. And the interaction process of quaternary model is described as Cyber-Physics-Economy-Energy. Simulation results validate that the proposed model is of high identification accuracy and has better generalization performance, and can effectively fit the state variation of each node of the whole system under different operation modes.

3.1 Introduction

Nowadays, different carriers of energy, such as electricity, district heat and natural gas, are mostly produced, transmitted, and distributed on each framework in which the modellings of energy carriers are also independently. Compared to these conventional energy frameworks, an integrated framework with multi-energy coupled carriers appears more effective in allocating resources and can bring more economic benefits [1]. Simultaneously, global warming, shortage of fossil fuels and policy incentives have accelerated the employment of renewable energy. Driven by these factors, a variety of energy generation units, typically including combined heat and power (CHP), photovoltaic, wind power, geothermal resources and biomass energy, are displayed in the energy framework [2]. With the increasing penetration of these small and medium-sized energy generation units, some areas have energy output performance. A number of devices, such as bi-directional inverter, solid state transformer (SST) and energy router, also provide the possibility for energy infeed. However, since the output of renewable energy units is generally highly stochastic and intermittent, it is difficult to dispatch energy in centralized control [3]. Specifically, an individual area with energy generation units tends to maximize their own benefit. Thus, a distributed and plug-and-play framework with multi-energy coupled carriers, so called Energy Internet, was proposed.

Q. Sun, *Energy Internet and We-Energy*, Renewable Energy Sources & Energy Storage, https://doi.org/10.1007/978-981-13-0523-8_3

In the previous studies, a hybrid system combined gas and electricity network model which aims to minimize costs of the whole system was developed in [4–6]. A steady-state power flow analysis of hybrid system was researched in [7]. Combined heat and power dispatch models considering storage and transfer were studied in [8, 9]. The integrated systems with electricity, natural gas, and district heating were established in [10–12]. However, these models with different carriers of energy are considered under the steady state.

In order to establish a general model for Energy Internet under steady and transient state, a multi-energy coupled area for the Energy Internet, named "We-Energy (WE)", is proposed which is characterized by producer-consumer integration, coupling and complementarity, openness, and regionalization. WE is an aggregation of energy production devices, energy storage devices and user loads, proposed in this chapter which differs from traditional energy supply mode dominated by traditional energy supply companies.

3.2 We-Energy Concept

With the influence of the traditional energy suppliers weakening gradually, terminal users no longer receive energy from the unified energy suppliers passively, but have some right, such as energy production and sale. The openness, sharing, peer-to-peer and plug-and-play of the Energy Internet will provide ultimate users with an energy interaction platform. However, energy quality will be affected by the fluctuations caused by energy storage devices and energy production devices. And the total fluctuation will increase or offset when several fluctuations superpose together. For Energy Internet, if energy terminal users can accept energy fluctuations by themselves, it is enough to regulate the energy quality in points of common coupling (PCCs). In some small areas, if they can market their own products and supply energy for Energy Internet, energy quality that Energy Internet really concerns is the output part. Similarly, what Energy Internet needs to guarantee is the energy quality of PCCs or inputs of some small areas.

3.2.1 Definition of We-Energy

Inspired by the "We-Media" from the point of view of information transmission in the Internet, including blog, WeChat, post bar and Bulletin Board System (BBS) which are characterized by diversity, openness and popularization, "We-Energy", as a novel energy interacting area for Energy Internet [13], is proposed in this chapter. Compared with the traditional energy networks that different energies are supplied independently, WEs have the capacity to transform various type of energy into the desired energy. According to energy pricing, peak shaving can be realized by transforming energy under meeting demand response. The strong coupling and complementarity of the energy production and the energy demand contribute to achieving the energy balance among WEs, reducing the cost of energy transmission, and improving the utility efficiency of renewable energy.

In the Energy Internet, WE is not only energy consumer but also energy producer which trade energy with others by the advanced communication, electronic conversion and automatic control technology. For the scale, a personal energy entity, villa, enterprise and community which are provided with energy production devices or energy storage devices (e.g. distributed generations, electricity/heat storage devices, CCHP (Combined Cooling Heating and Power)) can be accessed by WE. Here, a note about this multi-energy coupled area is that the devices contained in a WE are as a unity to participate in energy trading. According to the definition, what makes WE different from traditional energy area is that WE is no long a passive consumer, but also a potential energy supplier in the energy interaction, which trades energy with others in the principle of 'peer-to-peer' instead of 'peer-to-plane'. The main features of the WE are given as follows:

(1) Producer-Consumer Integration: the traditional energy supply mode is broken by WEs which can not only market their own products but trade with Energy Internet. Common energy terminal can participate in energy trades and transmissions. And the energy transmission is changed as bi-directional transmission. Therefore, the ability of energy production, transmission, storage and consumption can be integrated by WEs in the energy internet;
(2) Energy Coupling and Complementarity: compared with the traditional energy generations in which different energies are supplied independently, WEs have the capacity to transform various type of energy into the desired energy. According to energy pricing, peak shaving can be realized by transforming energy under meeting demand response. The strong coupling and complementarity of the energy production and the energy demand contribute to achieving the energy balance among WEs, reducing the cost of energy transmission, and improving the utility efficiency of renewable energy;
(3) Openness of energy trading: compared with the traditional energy monopolist, most of WEs are converted from terminal users, which make the energy trading more peer-to-peer, more renewable and less utilitarian, promoting the development of the non-fossil energy;
(4) Regionalization of Energy Consumption: the regional self-sufficient energy supply can be realized by increasing the amount of WEs, due to the broader renewable energy resources and other clean energy resources. Then the advantage of long-distance transmission of the traditional fossil energy is weakened greatly. As the energy production and the energy consumption of WEs are more convenient and flexible for the common users than that of the professional and large-scale traditional major energy suppliers, the development of the WEs can gradually change the structure of the traditional energy consumption and reduce the dependence on the fossil energy.

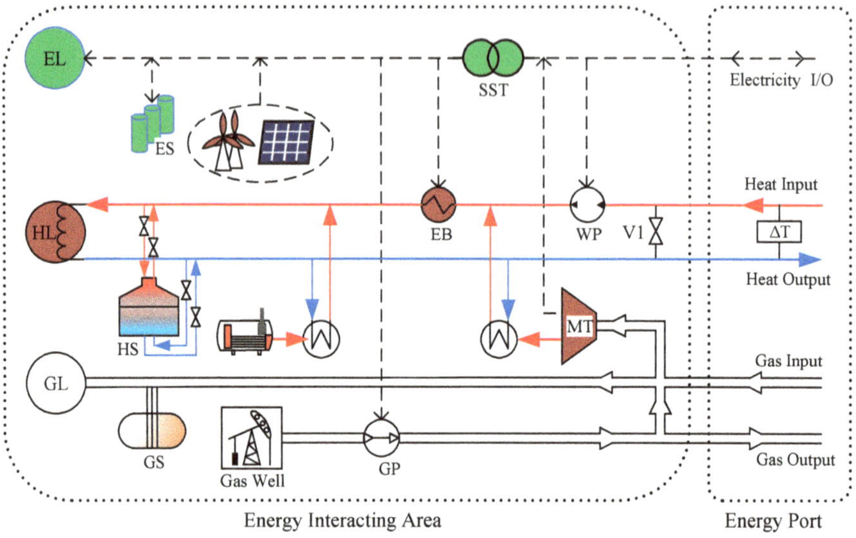

Fig. 3.1 Example of a hybrid We-Energy

3.2.2 *Structure of We-Energy*

In a WE with various energy carriers, electricity, heat and gas, energy is converted to various forms in energy interacting area for meeting the network requirements and self-interest. And a WE exchanges energy with other WEs via energy port. For example, the WE in Fig. 3.1 exchanges electricity, natural gas, and district heating hot water at the energy port (EP). A number of advanced instruments with communication are applied on EP (e.g., smart electricity meter, infrared thermometer and smart gas meter) to measure trading volume and upload data to EMS. And WE represents a random combination of energy production devices, energy storage devices and user loads. In Fig. 3.1, we consider that the loads including electric load, heating load and gas load (EL, HL and GL) are supplied by energy production devices (EPDs) which include wind farms, photovoltaic stations, boilers, gas well and micro-gas turbines (MTs). Meanwhile, surplus energy produced by EPDs not only can be stored by Energy Storage Devices (ESDs) which include gas storages (GSs), heating storages (HSs) and electronic storages (ESs), but also can be exchanged via EP.

In previous studies, "Energy Cell" and "Prosumer" are proposed to describe the feature for sub-system of Energy Internet. "Energy Cell" is defined as not only the electricity consumer but can also be the electricity supplier by locally operating and managing their own distributed generators, distributed energy storage devices, and dispatchable loads [14]. "Prosumer", as the name described, refers to a kind of energy unit that integrates producer and consumer. However, both the models are the energy-economy interaction models under the framework of electricity market in Energy Internet, which consists of energy trading volume and cost. More specifically, compared with "Prosumer", distributed coordination control is used in normal state, which makes We-Energies collaboration and realizes the fault diagnosis, fault recovery and state monitoring. Compared with "Energy Cell", the synchronicity of multi-energy trading is exhibited on We-Energy that We-Energy play roles in not only energy producer but also energy supplier at the same time. In the Energy Internet, the interaction between different kinds of energy and how the energy transfer among subsystems are primarily considered.

Compared with other energy subsystem, the outstanding character of We-Energy is that We-Energy exchanges energies with others in synchronism. According to these characters, the energy subsystem can realize full duplex energy transmission, which can receive a kind of energy and send out another energy simultaneously. Considering the converter devices as indicated in Fig. 3.1, a general expression covering all types of couplings can be created as follows:

$$\underbrace{\begin{bmatrix} E_{i,e} \\ E_{i,h} \\ E_{i,g} \end{bmatrix}}_{E} = \underbrace{\begin{bmatrix} \lambda_{ee} & \lambda_{he} & \lambda_{ge} \\ \lambda_{eh} & \lambda_{hh} & \lambda_{gh} \\ \lambda_{eg} & \lambda_{hg} & \lambda_{gg} \end{bmatrix}}_{\mathrm{X}} \left(\underbrace{\begin{bmatrix} G_{i,e} \\ G_{i,h} \\ G_{i,g} \end{bmatrix}}_{G} + \underbrace{\begin{bmatrix} E_{i,se} \\ E_{i,sh} \\ E_{i,sg} \end{bmatrix}}_{E_s} - \underbrace{\begin{bmatrix} L_{i,e} \\ L_{i,h} \\ L_{i,g} \end{bmatrix}}_{L} \right) \tag{3.1}$$

The external energy performance E can be integrated by the total amount of energy generations G, energy storage E_s and loads L. The matrix X is called energy converting matrix, which can be expressed as:

$$\mathrm{X} = \begin{bmatrix} \gamma_{e2e}\eta_{e2e} & \gamma_{h2e}\eta_{h2e} & \gamma_{g2e}\eta_{g2e} \\ \gamma_{e2h}\eta_{e2h} & \gamma_{h2h}\eta_{h2h} & \gamma_{g2h}\eta_{g2h} \\ \gamma_{e2g}\eta_{e2g} & \gamma_{h2g}\eta_{h2g} & \gamma_{g2g}\eta_{g2g} \end{bmatrix} \tag{3.2}$$

where γ is the energy distribution coefficient, η is conversion efficiency.

3.3 Quaternary Model of We-Energy in Different Operation States

In this chapter, a quaternary interactive model is proposed to realize We-Energy system of real-time sensing, optimal control and information service, which makes the system more reliable, efficient and real-time collaborative.

In the Energy Internet, different control modes, distributed and centralized control, are required for different operational states. Combined with the state of the traditional energy system analysis and the related research foundation, the running states of Energy Internet also can be divided into the steady and the transient state. When the Energy Internet run in a normal state, the system is under a steady state; when the

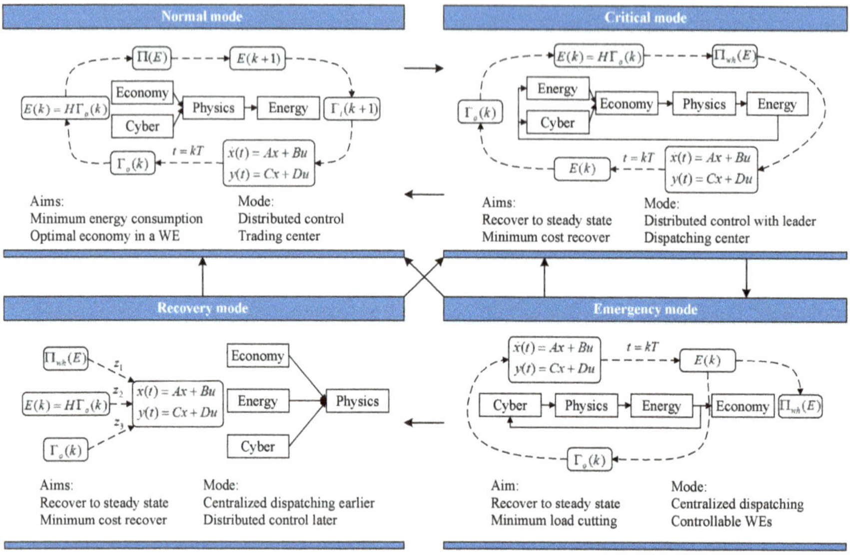

Fig. 3.2 Quaternary model of We-Energy

Energy Internet runs in the alert state, the emergency state and the recovery state, the system which is under the transient state needs to be restored to normal state.

System under different operating states, the status of WEs in the network and the role of WEs played are also different. The conversion among operating states can be described as a quaternary model, which is shown in Fig. 3.2.

(1) Under the normal state in which the status of all WEs are equal. A WE aims to maximize their own interests and compete with other WEs. And EI is in the distributed control mode with a trading center.
(2) Under the alert state in which qualities of energy fluctuate and the system will fall out of steady state. System needs a leader to dispatch surplus energy in order to restore to the normal state.
(3) Under the emergency state in which the energy qualities are beyond permissible range, system aims to realize the minimum cut cutting load and dispatches the controllable WEs in centralized control.
(4) Under the recovery state, the system aims to recover fast.

There is amount of equipment in a WE whose mechanisms are different, but all of them can be expressed as a continuous state space equation, denoted as Θ_i:

$$\begin{aligned} \dot{\boldsymbol{x}} &= \boldsymbol{A}(t)\boldsymbol{x}(t) + \boldsymbol{B}(t)\boldsymbol{u}(t) \\ \boldsymbol{y} &= \boldsymbol{C}(t)\boldsymbol{x}(t) + \boldsymbol{D}(t)\boldsymbol{u}(t) \end{aligned} \tag{3.3}$$

where $\boldsymbol{x}$ is the state of system, concluding the voltage U_i, current I_i, phase angle θ_i, pressure of heat-supply network and gas network P_h & P_g, mass flow rate of heat-supply network and gas network M_h and M_g, inlet temperature τ_S and outlet temperature τ_R, which can be expressed as $\boldsymbol{x} = [U_i, I_i, \theta_i, P_h, M_h, \tau_S, \tau_R, P_g, M_g]$, and $\boldsymbol{u}$ is the input variable which can be expressed as $\boldsymbol{u} = [U_G, \tau_H, M_G]$.

The sampling control ensures a large number of measurement information of the system which can denoted as $\Gamma_{i\_o}(k)$. According to the data, the output energy $E_i(k)$ can be calculate from:

$$E_i(k) = H\Gamma_{i\_o}(k) \tag{3.4}$$

where H is output matrix, and we denote (4) as Ξ_i.

According to the energy output of all equipment, the revenue of WE $\pi_i(k)$ can be expressed as

$$\pi(k) = p_E E(k) - c_i E(k) \tag{3.5}$$

where p_E is the energy price, c_i is the cost of WE$_i$.

Subjected to the system operation constraints Ψ_j, the revenue optimal control method based on (3.5), denoted as Π, can give the output energy of next moment $E_i(k+1)$. Through the related control algorithm, we can get physical device control quantity of next moment, denoted as $\Gamma_{i\_i}(k+1)$, which feedbacks to state space equation Θ_i. At this time, the system updates physical state of equipment and continues to the next control input. Therefore, WEs model can be expressed as

$$\Sigma = \{\mathrm{M}, I, \mathrm{K}, \mathrm{P}, \Psi, \mathrm{T}, \Lambda, \Lambda_0, \Phi\} \tag{3.6}$$

where M is the finite set of objective function, I is the physical finite set of continuous variables, K is finite set of discrete information which includes state of equipment energy input and output information, T is the finite set of physical device sampling cycle, control and optimization algorithm, Λ is finite set of physical state after the discrete, $\Lambda_0 \subseteq \Lambda$ is the initial set of discrete state, Φ is the relationship set of energy conversion, where $\varphi = \left\{\kappa, \sigma, \psi, \kappa'\right\}$, $\kappa, \kappa' \in \Lambda$, $\sigma \in \mathrm{P}$ are the conversion equations, $\psi \in \Psi$ is the constraints of corresponding conversion equations, P is the conversion equation, which can be expressed as

$$\mathrm{P} = \{\Theta, \Xi, \Pi\} \tag{3.7}$$

3.3.1 We-Energy Modeling in Normal State

In the normal state operation of the We-Energy, the system commits to decreasing the general energy consumption, and each We-Energy aims at the economic optimum (achieving best benefits of We-Energy). Then the model should use the economic (price) and information measurement to control the physical state of the system and ultimately to let the total energy output meet the requirements of operation. In this process, the We-Energies play game with the aim of maximizing their interests under the motivation mechanism designed by the energy trading center. And the system is based on distributed optimization control method, in which the We-Energies are coordinated to complete the overall goal and the status of them is basically equal.

In this state, for the whole Energy Internet, the We-Energy is similar to each other and can be represented by the basic model Σ. There is a global energy objective function with constraints, which depends on the We-Energy's input and output and trading strategies.

In the market transactions of the energy Internet, the We-Energy first submits the energy output to the transaction center, then the transaction center develops the energy clearing price according to the power balance equation. According to the inverse load demand curve, the energy clearing price can be developed as:

$$\begin{aligned} pc_{E,P} &= -\theta_{E,P}\sum P_{E,i} + \delta_{E,P} \\ pc_{E,Q} &= -\theta_{E,Q}\sum Q_{E,i} + \delta_{E,Q} \\ pc_{Q,P} &= -\theta_{Q,P}\sum P_{Q,i} + \delta_{Q,P} \\ pc_{G,Z} &= -\theta_{G,Z}\sum Z_{g,i} + \delta_{G,Z} \end{aligned} \tag{3.8}$$

where, pc is the energy clearing price, $\theta > 0$ is the energy regulation factor, $\delta > 0$ is the standard energy price.

Thus, the revenue objective function of We-Energy can be expressed as:

$$\min \pi_i = \begin{bmatrix} v_{E,P} & v_{E,Q} & v_{Q,P} & v_{G,Z} \end{bmatrix} \begin{bmatrix} P_{EL,i} \\ Q_{EL,i} \\ P_{QL,i} \\ Z_{GL,i} \end{bmatrix} + \begin{bmatrix} pc_{E,P} & pc_{E,Q} & pc_{Q,P} & pc_{G,Z} \end{bmatrix} \begin{bmatrix} P_{E,i} \\ Q_{E,i} \\ P_{Q,i} \\ Z_{G,i} \end{bmatrix} - C_{i,\Sigma} \tag{3.9}$$

where v is the We-Energy load demand utility and $C_{i,\Sigma}$ is the total cost of energy production for We-Energy i. And $C_{i,\Sigma}$ can be expressed as:

$$C_{i,\Sigma} = C_i(P_{E,DG}) + C_i(Q_{E,DG}) + C_i(P_{Q,Gen}) + C_i(Z_{G,Gen}) \tag{3.10}$$

where $C_i(P_{E,DG})$, $C_i(Q_{E,DG})$, $C_i(P_{Q,Gen})$ and $C_i(Z_{G,Gen})$ are the cost functions of the active power, reactive power, thermal power and bulk pressure, respectively, and can be defined as the quadratic function of $P_{E,DG}$, $Q_{E,DG}$, $P_{Q,Gen}$ and $Z_{G,Gen}$:

$$\begin{aligned} C_i(P_{E,DG}) &= a_{E1}P_{E,DG}^2 + b_{E1}P_{E,DG} \\ C_i(Q_{E,DG}) &= a_{E2}Q_{E,DG}^2 + b_{E2}Q_{E,DG} \\ C_i(P_{Q,Gen}) &= a_Q P_{Q,Gen}^2 + b_Q P_{Q,Gen} \\ C_i(Z_{G,Gen}) &= a_G Z_{G,Gen}^2 + b_G Z_{G,Gen} \end{aligned} \tag{3.11}$$

where, $a > 0$ and $b > 0$ are the fitting coefficient of the cost function of We-Energy energy production equipment.

In the objective function shown in Eq. (3.9), there are certain constraints on the We-Energy network transmission line and the included energy production equipment, including the active and reactive power of the distributed power supply, the thermal power of the coal-fired boiler, the gas well output pressure, We-Energy power transmission line voltage and frequency, the output port pressure of heat pipe network, the output port pressure of natural gas:

$$
\begin{aligned}
&0 \le P_{i,E,DG} \le P_{i,E,DG}^{\max}, \quad \forall i \in n_E \\
&0 \le Q_{i,E,DG} \le Q_{i,E,DG}^{\max}, \quad \forall i \in n_E \\
&0 \le P_{i,Q,Gen} \le P_{i,Q,Gen}^{\max}, \quad \forall i \in n_Q \\
&0 \le Z_{i,G,Gen} \le Z_{i,G,Gen}^{\max}, \quad \forall i \in n_G \\
&U_i^{\min} \le U_i \le U_i^{\max}, \quad \forall i \in n_E \\
&f_i^{\min} \le f_i \le f_i^{\max}, \quad \forall i \in n_E \\
&p_{i,Q}^{\min} \le p_{i,Q} \le p_{i,Q}^{\max}, \quad \forall i \in n_Q \\
&p_{i,G}^{\min} \le p_{i,G} \le p_{i,G}^{\max}, \quad \forall i \in n_G
\end{aligned} \tag{3.12}
$$

In this state, for the whole network, the objective functions of We-Energy are similar. The optimal benefit of We-Energy can be realized according to We-Energy's output energy trading strategies and the relevant constraints.

3.3.2 We-Energy Modeling in Alert State

When the system is running in the alert state, the energy quality fluctuates in the system which reaching or approaching the critical value of the system in the normal state operation. At this time, the scheduling center needs to dispatch the stable We-Energy with margin in the network thus the system can return to the steady state. The system aims to restore stabilization aiming the minimum cost (load shedding is not allowed, mainly to adjust the output energy). In this process, Energy Internet has a distributed control with leaders, and the control center can determine the network leader based on the operation of We-Energy. The status of We-Energy is not equal that the stable We-Energy with margin acts as leader. We-Energy determines the economic operation according to input energy and output information, following the leader to adjust energy consumption and changing energy output.

In this state, We-Energy can be divided into the leader We-Energy Σ_l and followers We-Energy Σ_f, which aims to minimize global cost function. In the condition of the system stabilization, We-Energies are out of the alert state.

3.3.3 We-Energy Modeling in Emergency State

If there exists serious interference, such as a short circuit of a power subsystem or the pump or compressor terminated abnormally, the We-Energy system in a normal state and an alert state will enter a state of emergency. In the emergency state, the energy quality deteriorates to exceed the allowable operating range of the system and the system concentrates on the centralized dispatching of controllable We-Energy with the minimum net load as the target, thus the system energy quality can be restored to the allowable operating range. We-Energies are controlled uniformly, and there is no distributed control. The controllable We-Energies carry out the instructions of control center and output energy according to the physical state and the load demand.

In the process of We-Energy returning from the emergency state to the normal state, the load demand is the main control of the We-Energy system and the energy demand response can be categorized into controllable and uncontrollable loads. In the power subsystem, controllable loads are generated by appliances that have the ability to shift their power consumption over time without compromising their provision of energy services, such as cooling and heating appliances. Uncontrollable loads are mainly composed of some rotating equipment, such as pumps and compressors, and the control method of them will cause huge economic losses and energy system solution. In the thermal sub-system, the control of the thermal load is mainly achieved through the switch of the pipeline branch and the thermal energy storage device. When the value of the pipeline pressure collected by the collection device is less than the alert state of the lower limit or greater than the alert state limit, the control center first gives the instructions to the heat storage valve to adjust its degree of closure to increase or decrease the pipeline pressure. If the pipeline pressure can not be restored to normal state or alert state, then We-Energy control center needs to control the branch pipe valve, which will make part of the pipeline to regulate the pipe pressure. In natural gas subsystems, the controllable loads are mainly composed of natural gas users, and uncontrollable loads are devices that generate mechanical energy by burning natural gas. The sudden outage of these devices will affect the energy sources such as the gas compressors, gas compressors and micro-turbines.

In the electric subsystem, when changes in the connection state occur for these devices, the temperature changes slightly but still remains within the target range due to their inherent thermal capacities and kinetic energy [15]. This flexibility can be used to shift the appliances' power demand to stabilize the grid frequency. Similarly, when changes of pressure in heating pipeline occur for pumps due to the continuous thermal energy in pipeline, some of branches can be closed but the heat stored in the pipeline can still meet user needs. And in the natural gas network, the compressibility of natural can shift a part of gas user by the pressure difference. The remainder of the demand is considered uncontrollable. These loads are combined into a single equivalent load with an equivalent damping coefficient (D). The consumption of the uncontrollable load depends on We-Energy system frequency deviations, fluid pressure deviations and gas pressure deviations, which are represented in D_{fe}, D_{pf} and D_{pg}.

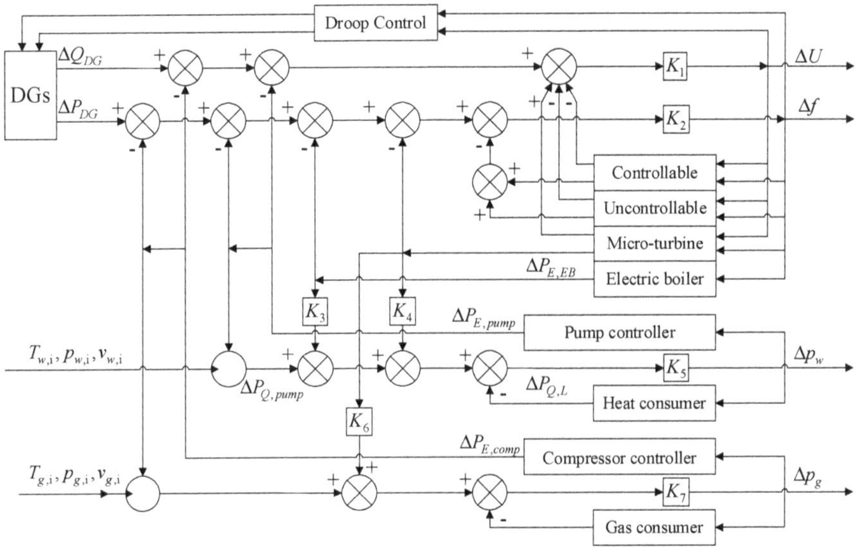

Fig. 3.3 Centralized control schematic of We-Energy system

A larger deviation of voltage, frequency, fluid pressure and gas pressure will cause the centralized control to initiate sooner, whereas small deviations are allowed for a longer period of time, thus delaying controller actuation. Figure 3.3 provides a centralized control schematic of We-Energy system.

In this state, the non-control We-Energy is not considered, only the controllable We-Energy Σ_c in the system to be considered, and the economic problem of this kind of We-Energy at this time is not considered, that is to say the conversion equation set $P = \{\Theta, \Xi\}$ does not include the collection Π; there is a system revenue function Π_{wh} in the global control.

3.3.4 We-Energy Modeling in Recovery State

When the system is in the state to be restored, the system will restore the system to the stable operation state with the fastest recovery and the minimum cost of recovery. At this time, the system takes the fastest restore and the minimum cost as the goal, and the control mode is first centralized control then the distributed control (We-Energy can be plug and play when the system restored). In the recovery process, We-Energy integrated the self output, energy, economic information, commonly play the role in the physical state.

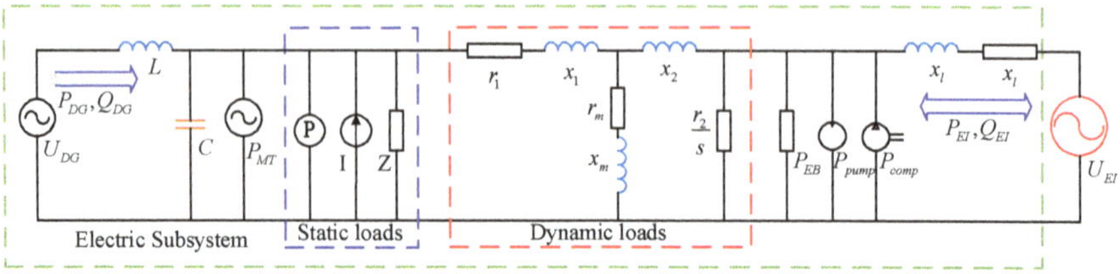

Fig. 3.4 Power subsystem model in We-Energy system

In this model, in the We-Energy base model, the transformation relation set Φ is changed greatly, and a new weight collection Z is made to coordinate the influence of information, energy and economy on the physical state, re-register the We-Energy model as Π_{qu}.

3.4 Dynamic Equation of We-Energy

3.4.1 Dynamic Equation in Power Subsystem

(1) Steady-state equation of power subsystem

According to the We-Energy structure shown in Fig. 3.4, the power subsystem model includes the distributed power supply, energy storage equipment, loads and energy conversion unit. Among them, there is a certain degree of planning in the charging and discharging process of the energy storage equipment, and it can be regarded as the power sub-network PQ nodes combined with distributed power. The distributed power supply module is equivalent to the output power control micro-power and the load unit is composed of many electrical equipment and users in the network. It is nonlinear and heterogeneous, and can be divided into static load and dynamic load according to the load characteristics. The load cell is divided into static load and dynamic load. The energy conversion units involved in this chapter mainly include electric boilers, water pumps, natural gas compressors. The energy input fluctuations of these conversion units affect the energy output of the coupled network, so the model cannot be regarded as a conventional load or power supply.

From the We-Energy energy port, the distributed power reserve module can be described as:

$$\tilde{S}_{DG} = -(P_{DG} + jQ_{DG}) \tag{3.13}$$

where $\tilde{S}_{DG}$ is the complex power output from the distributed power supply module, P_{DG} is the active power and Q_{DG} is the reactive power.

In order to make the distribution of coordinate the active and reactive power better, in the We-Energy system, the system uses the droop control method to control the inverter power output:

$$P_{DG} = P_{DG,0} - \frac{1}{m}(f - f_0)$$
$$Q_{DG} = Q_{DG,0} - \frac{1}{n}(U_{DG} - U_0) \quad (3.14)$$

where m is the frequency droop coefficient, n is the frequency dropping coefficient, f_0 and f are the inverter rated frequency and output frequency, respectively, U_0 and U are the rated voltage, and output voltage, respectively, $P_{DG,0}$ and $Q_{DG,0}$ are the inverter is the active and reactive power rated capacity.

In We-Energy, the micro-gas turbines generate heat by burning natural gas where high-grade heat with high pressure and temperature is converted to electrical energy. Low-grade heat is supplied to the heat load through the heat exchanger, and the relationship between the energy generated per unit time and gas volume is:

$$P_{E,MT} = \eta_{g2e} H_u m_g \quad (3.15)$$

where $P_{E,MT}$ is the micro-gas turbine output power, η_{g2e} is the micro-gas turbine power generation efficiency, H_u is the natural gas combustion low calorific value, m_g is the natural gas intake of unit time.

The static load of the power system in We-Energy can be divided into constant impedance (Z), constant current (I) and constant power (P) according to the characteristics. Therefore, the static load model can express the combination of these three kinds of loads according to a certain proportion, which is:

$$\tilde{S}_{L0} = P_{L0} + jQ_{L0} \quad (3.16)$$

$$\begin{cases} P_{L0} = P_0\left[P_Z\left(\frac{U}{U_0}\right)^2 + P_I\left(\frac{U}{U_0}\right) + P_P\right] \\ Q_{L0} = Q_0\left[Q_Z\left(\frac{U}{U_0}\right)^2 + Q_I\left(\frac{U}{U_0}\right) + Q_P\right] \end{cases} \quad (3.17)$$

where P_0 and Q_0 are respectively the active and reactive power absorbed by the load at the reference voltage, U and U_0 are respectively the actual voltage of the working bus as the reference voltage, (P_Z, P_I, P_P) and (Q_Z, Q_I, Q_P) are respectively the percentage of the load reflected:

$$\begin{cases} P_Z + P_I + P_P = 1 \\ Q_Z + Q_I + Q_P = 1 \end{cases} \quad (3.18)$$

The dynamic load of the power system in We-Energy is mainly composed of induction motor. According to the mechanical transient process of induction motor, the model can be described as:

$$\tilde{S}_L = \frac{R_L}{R_L^2 + X_L^2} U^2 + j \frac{X_L}{R_L^2 + X_L^2} U^2 \tag{3.19}$$

where $Z_L = R_L + jX_L$ is the equivalent impedance of the induction motor, the following equation can be obtained:

$$Z_L = r_1 + jx_1 + \frac{(r_m + jx_m)(r_2/s + jx_2)}{(r_m + r_2/s) + j(x_m + x_2)} \tag{3.20}$$

where x_1 and x_2 are respectively the motor stator and rotor leakage resistance, $r_m + jx_m$ is the mutual impedance between the stator and rotor, r_2/s is the equivalent resistance of rotor.

The electric boiler in the We-Energy can convert the electrical power into heat to provide heat for the heat pipe, the electrical power $P_{E,EB}$ can be expressed as follows:

$$P_{E,EB} = UI_{EB} \tag{3.21}$$

where $P_{E,EB}$ is the input power of the electric boiler, I_{EB} is the current of the electric boiler.

The pumps and air compressors in We-Energy convert the electrical energy into mechanical energy to increase the delivery capacity of the heat pipe network and the natural gas pipeline network for the medium. The power consumption can be expressed as:

$$\begin{aligned} P_{E,pump} &= UI_{pump} \\ P_{E,comp} &= UI_{comp} \end{aligned} \tag{3.22}$$

where $P_{E,pump}$ and $P_{E,comp}$ respectively are the input power for the pumps and compressors, I_{pump} and I_{comp} respectively are the current for the pump and compressor electric boilers.

According to the system structure, the power loss in the power subsystem can be expressed as $\tilde{S}_l = P_l + jQ_l$

$$\begin{aligned} P_l &= \frac{r_l}{r_l^2 + x_l^2}(U_{El} - U)^2 \\ Q_l &= \frac{x_l}{r_l^2 + x_l^2}(U_{El} - U)^2 \end{aligned} \tag{3.23}$$

where $r_l + jx_l$ is the line impedance from the We-Energy port to the load.

Since the input power of the power conversion device in the system affects the state variables of the coupled network, at the same time, the natural gas mass flow rate input to the micro-gas turbine in the natural gas network also affects the output of its electrical power and these coupling variables need to be separate in the power subsystem. Considering that, according to the power subsystem structure established in this chapter, the output power of We-Energy power subsystem is:

$$\begin{aligned} P_E &= P_{L0} + P_L + P_{E,EB} + P_{pump} + P_{comp} + P_l - P_{DG} - P_{E,MT} \\ Q_E &= Q_L + Q_{L0} + Q_l - Q_{DG} \end{aligned} \tag{3.24}$$

(2) Transient-state equation of power subsystem

The dynamic part can be described as:

$$\begin{cases} \frac{d\omega}{dt} = -\frac{1}{2H}\left[\left(A\omega^2 + B\omega + C\right)T_0 - \left(E_d' I_d + E_q' I_q\right)\right] \\ \frac{dE_q'}{dt} = -\frac{1}{T'}\left[E_q' - (X - X')I_d\right] + (\omega - 1)E_d' \\ \frac{dE_d'}{dt} = -\frac{1}{T'}\left[E_d' - (X - X')I_q\right] + (\omega - 1)E_q' \end{cases} \tag{3.25}$$

$$\begin{cases} I_d = \frac{1}{R_s^2 + X'^2}\left[R_s\left(U_d - E_d'\right) + X'\left(U_q - E_q'\right)\right] \\ I_q = \frac{1}{R_s^2 + X'^2}\left[R_s\left(U_q - E_q'\right) + X'\left(U_d - E_d'\right)\right] \end{cases} \tag{3.26}$$

where, $T' = \frac{X_r + X_m}{R_r}$, $X = X_s + X_m$, $X' = X_s + \frac{X_r X_m}{X_r + X_m}$, $A + B + C = 1$.

The static part can be described as:

$$\begin{cases} P_s^* = P_Z^*\left(\frac{U}{U_0}\right)^2 + P_I^*\left(\frac{U}{U_0}\right) + P_P^* \\ Q_s^* = Q_Z^*\left(\frac{U}{U_0}\right)^2 + Q_I^*\left(\frac{U}{U_0}\right) + Q_Q^* \end{cases} \tag{3.27}$$

where, $P_Z^* + P_I^* + P_P^* = 1 - \frac{P_{motor}}{P_0}$, $Q_Z^* + Q_I^* + Q_Q^* = 1 - \frac{Q_{motoer}}{Q_0}$.

The whole electric subsystem can be described as:

$$\begin{cases} \frac{dP_{EI,e}}{dt} = \frac{\sin\delta_{12}}{\sqrt{R_{12}^2 + X_{12}^2}}\left(U_{EI}\frac{dU_{DG}}{dt} + U_{DG}\frac{dU_{DG}}{dt} + \frac{dU_{DG}}{dt}\frac{dU_{EI}}{dt}\right) \\ \frac{dQ_{EI}}{dt} = \frac{\cos\delta_{12}}{\sqrt{R_{12}^2 + X_{12}^2}}\left(U_{EI}\frac{dU_{DG}}{dt} + U_{DG}\frac{dU_{EI}}{dt} + \frac{dU_{DG}}{dt}\frac{dU_{EI}}{dt}\right) + \frac{1}{\sqrt{R_{12}^2 + X_{12}^2}}\left[\left(\frac{dU_{EI}}{dt}\right)^2 + 2U_{EI}\frac{dU_{EI}}{dt}\right] \end{cases} \tag{3.28}$$

where δ_{12} is included angle between U_{DG} and U_{EI}, $X_{12} = X_{l-1} + X_s + X_r + X_{l-2}$, $R_{12} = R_{l-1} + R_s + R_{l-2}$.

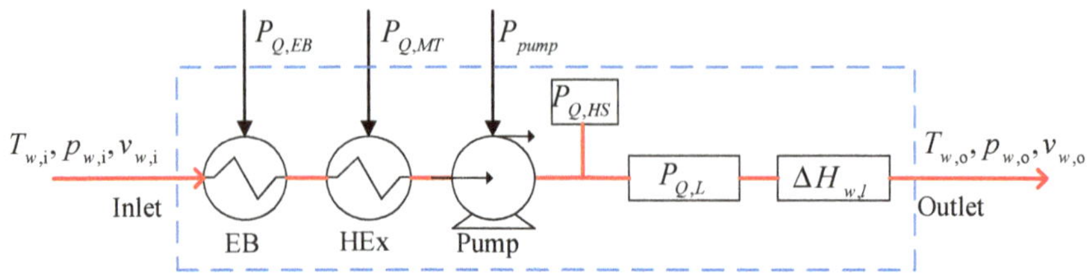

Fig. 3.5 Heating subsystem model in We-Energy system

3.4.2 Dynamic Equation in Heating Subsystem

(1) steady-state equation of heating subsystem

In the heating system, the water provided by the Energy Internet with a certain temperature and mass flow, pressurized and heated through the pump, electric boilers and heat exchangers and other equipment, to provide heat for the thermal load in the We- Energy, and finally return back to Energy Internet through the return pipe. At present, most of the economic benefits of central heating system are from fixed payment of heating costs of the heating users each year, rather than the users' real-time heating price function. Therefore, this chapter models We-Energy from the point of view of the power balance of the heating subsystem. By adjusting the state variables in the input power control system of the equipment, we control the We-Energy heating network system, and according to the model of real-time economic analysis of the heating system, the system structure is shown in Fig. 3.5.

In the heating subsystem shown in Fig. 3.5, the fluid-state variables $T_{w,i}$, $p_{w,i}$ and $v_{w,i}$ are measured when the energy is not coupled to the equipment. Because the water in the pipeline is fluid, in order to calculate the influence of the thermal power of the heating network on the temperature of the fluid, this chapter defines the thermal power from the thermodynamic point of view which is as follows:

Definition 1 In the idea state, the thermal power of the pipeline is the amount of heat in the pipeline when the fluid passes through a section in the unit time, unit of measurement W, and the thermal power characterizes the speed of the heat transportation. The formula is as follows:

$$P_Q = c\dot{m}T. \tag{3.29}$$

In the formula, c is the specific heat capacity $(J/kg \bullet K)$, $\dot{m} = \rho v S$ is the mass flow rate (kg/s) of the fluid, T is the fluid temperature (K), ρ is the fluid density, v is the velocity, S is the cross-sectional area of the pipe.

Electric boiler is the heating equipment in the heating pipe network, it can convert the electrical power into heat to provide heat for the heat pipe network, when the input power of the electric boiler is $P_{E,EB}$, the thermal power $P_{Q,EB}$ can be expressed as follows:

$$P_{Q,EB} = \eta_{EB} P_{E,EB} \tag{3.30}$$

where $P_{E,EB}$ is the electric boiler input power, η_{EB} is the electric boiler thermal efficiency.

In We-Energy, micro-gas turbine provide electricity to the power subsystem and provide the thermal energy to the thermal sub-system at the same time, the relationship between the thermal energy generated in the unit time and air intake of the micro-gas turbine is:

$$P_{Q,MT} = \eta_{g2h} H_u \dot{m}_g \tag{3.31}$$

where $P_{Q,MT}$ is the micro-gas turbine output heat power, η_{g2h} is the micro-gas turbine heat production efficiency, H_u is the low heating value for natural gas.

In the heat pipe network, the input power of the pump P_{pump} is proportional to the cubic of pump speed ω, and the pump speed change will directly affect the flow rate and pressure of the fluid in the heat pipe network:

$$P_{pump} = \frac{\dot{m}_w H_w}{1000\eta_{pump}} \tag{3.32}$$

where H_w is the pump lift, η_{pump} is the pump efficiency.

In practical engineering, the Bernoulli equation with mechanical energy input can be expressed as:

$$\frac{p_1}{\rho g} + \frac{v_1^2}{2g} + h_1 + H_w = \frac{p_2}{\rho g} + \frac{v_2^2}{2g} + h_2 \tag{3.33}$$

where p is the pipeline pressure, h is the pipe height.

Assuming that the inlet of the pump is equal to the outlet, i.e. $h_1 = h_2$, if the input power of the pump P_{pump} is due to the change of the flow rate of the inlet and outlet pipes, the fluid flow rate in the pipeline does not change abruptly. Therefore, the pump lift is instantaneously converted into fluid pressure, the Bernoulli equation can be obtained after the pressurized water pump pressure p_w:

$$p_w = p_{w,i} + \frac{1000\eta_{pump}\rho_w g P_{pump}}{\dot{m}_{w,1}} \tag{3.34}$$

where $\dot{m}_{w,1}$ is the fluid mass flow rate before the pump is pressurized.

For the entire heat pipe network, assuming that the pressure after the outlet of water $p_{w,o}$ will not change, according to (3.33), (3.34), the pressure of the fluid in the pipeline through the pump is:

$$v_w^2 = v_{w,1}^2 + \frac{2000\eta_{pump} g P_{pump}}{\rho_w S_{pipe} v_{w,1}} + \frac{2p_{w,i} - 2p_{w,o}}{\rho_w} \tag{3.35}$$

Based on the above analysis, when the pipeline with added power P_{pump} in the pump, the output heat power is:

$$P_{Q,pump} = c_w \rho_w S_{pipe}\left(v_w - v_{w,1}\right)T_w \tag{3.36}$$

In the actual project, due to the building energy and indoor and outdoor temperature, building structure and other factors, the users' heat load is difficult to get a precise model. In this chapter, the unit area index method is used for modeling thermal load. Due to the predictability of heating systems, its control can be equivalent to a control of the building heating area. The boiler user energy storage joint model can be expressed as:

$$\begin{gathered} P_{Q,L} = \chi_{Q,L} F_{Q,L} \\ F_{Q,L} \in \left(\frac{\chi_{Q,L} F_{Q,L}^{\min} - P_{Q,B}^{\max} - P_{Q,HS}^{\max}}{\chi_{Q,L}}, \frac{\chi_{Q,L} F_{Q,L}^{\max} + P_{Q,HS}^{\max}}{\chi_{Q,L}} \right) \end{gathered} \tag{3.37}$$

where $P_{Q,L}$ is the building heating load, $\chi_{Q,L}$ is the heat index of the building area, $F_{Q,L}$ is the controllable building area, $P_{Q,B}^{\max}$ and $P_{Q,HS}^{\max}$ are the maximum thermal power and thermal storage of coal-fired boiler maximum storage (release) power.

Due to the viscosity in the movement toward the fluid, friction is produced during the flow in the pipeline, which makes a part of the mechanical energy into heat, the flow rate is reduced due to the frictional resistance in the pipeline, but the temperature has increased, from the perspective of power balance, the thermal power loss of the fluid is very small and can be ignored.

According to the thermal sub-system structure established in this chapter, the output power of We-Energy heats sub-system is:

$$P_{Q,o} = P_{Q,i} + P_{Q,EB} + P_{Q,MT} + \Delta P_{Q,pump} - P_{Q,L} \tag{3.38}$$

(2) transient-state equation of heating subsystem

According to the principle of mass conservation:

$$\rho v A dt - \left(\rho v A + \frac{\partial}{\partial x}(\rho v A)dx \right)dt = \frac{\partial}{\partial t}(\rho A dx)dt \tag{3.39}$$

Assuming that the inner pipe radius *dt* is constant, the pipe mass conservation formula can be expressed as:

$$\frac{\partial \rho}{\partial t} + \frac{\partial(\rho v)}{\partial x} = 0 \tag{3.40}$$

According to Newton's second law, the momentum equation can be expressed as: (momentum conservation)

$$\frac{\partial(\rho v)}{\partial t}+\frac{\partial\left(\rho v^2\right)}{\partial x}=\rho g H_{pump}-\rho g h_f-\frac{\partial p}{\partial x} \quad (3.41)$$

According to energy conservation, energy conservation equation can be expressed as:

$$\frac{\partial}{\partial t}\left[\rho\left(u+\frac{v^2}{2}\right)\right]+\frac{\partial}{\partial x}\left[\rho v\left(h+\frac{v^2}{2}\right)\right]=\dot{q} \quad (3.42)$$

where u is the internal energy of fluid, h is enthalpy.

Since the fluid density ρ is a function (x, t) of, that is $\rho=\rho(x, t)$, therefore:

$$\frac{d\rho}{dt}=\frac{\partial \rho}{\partial t}+\frac{\partial \rho}{\partial x}\frac{\partial x}{\partial t}=\frac{\partial \rho}{\partial t}+v\frac{\partial \rho}{\partial x} \quad (3.43)$$

Put (3.40) into the continuous Eq. (3.39), the following equation can be obtained:

$$\frac{\partial \rho}{\partial t}+\frac{\partial(\rho v)}{\partial x}=\frac{\partial \rho}{\partial t}+v\frac{\partial \rho}{\partial x}+\rho\frac{\partial v}{\partial x}=\frac{d\rho}{dt}+\rho\frac{\partial v}{\partial x}=0 \quad (3.44)$$

From the elastic modulus of fluid available $K=\frac{dp}{d\rho/\rho}$, then $\frac{d\rho}{dt}=\frac{\rho}{K}\frac{dp}{dt}$, from the above formula:

$$\frac{dp}{dt}+K\frac{\partial v}{\partial x}=0 \quad (3.45)$$

Since the pipe pressure p is a function of (x, t), that is $p=p(x, t)$, Eq. (3.43) can be expressed as:

$$\frac{\partial p}{\partial t}+\frac{\partial p}{\partial x}\frac{\partial x}{\partial t}+K\frac{\partial v}{\partial x}=\frac{\partial p}{\partial t}+v\frac{\partial p}{\partial x}+K\frac{\partial v}{\partial x}=0 \quad (3.46)$$

where $\frac{\partial x}{\partial t}\frac{\partial p}{\partial x}$ is the trace higher than $\frac{\partial p}{\partial t}$, which is negligible. According to the water wave velocity of the pipeline $a=\sqrt{K/\rho}$, the pipeline continuous equation can be expressed as:

$$\frac{1}{\rho}\frac{\partial p}{\partial t}+a^2\frac{\partial v}{\partial x}=0 \quad (3.47)$$

The momentum equation in Eq. (3.39) can be expressed as

$$\rho_h\left(\frac{\partial v_h}{\partial t}+2\frac{\partial v_h}{\partial x}\frac{\partial x}{\partial t}\right)+v_h\left(\frac{\partial \rho_h}{\partial t}+\frac{\partial \rho_h}{\partial x}\frac{\partial x}{\partial t}\right)+\frac{\partial p_h}{\partial x}=\rho g H_{pump}-\frac{f\rho}{2d}v|v|. \quad (3.48)$$

In the formula, $\frac{\partial v_h}{\partial x}\frac{\partial x}{\partial t}$ is the trace higher than $\frac{\partial v_h}{\partial t}$, and $\frac{\partial \rho_h}{\partial x}\frac{\partial x}{\partial t}$ is the trace higher than $\frac{\partial \rho_h}{\partial t}$, which is negligible. Therefore, the momentum equation in this chapter is simplified as:

$$\rho_h \frac{\partial v_h}{\partial t} + \frac{\partial p_h}{\partial x} = \rho g H_{pump} - \frac{f\rho}{2d} v|v| \tag{3.49}$$

According to the internal energy of the fluid is very complicated, this chapter only considers the effect of temperature change on the internal energy of the fluid, that is $u_h = Q_h = cm\Delta T$, Eq. (3.42) can be expressed as:

$$\frac{\partial T_h}{\partial t} = \frac{\dot{q}_h}{cm\rho_h} - \frac{a^2}{cm}\frac{\partial v_h}{\partial x} - (T_h - T_1)\frac{\partial v_h}{\partial x} - cmv_h\frac{\partial T_h}{\partial x} \tag{3.50}$$

From (3.47), (3.49) and (3.50), heating system control equation is:

$$\begin{cases} \frac{\partial p_h}{\partial t} = -\rho_h a^2 \frac{\partial v_h}{\partial x} \\ \frac{\partial v_h}{\partial t} = -\frac{1}{\rho_h}\frac{\partial p_h}{\partial x} + gH_{pump} - gh_f \\ \frac{\partial T_h}{\partial t} = \frac{\dot{q}_h}{cm\rho_h} - \frac{a^2}{cm}\frac{\partial v_h}{\partial x} - (T_h - T_1)\frac{\partial v_h}{\partial x} - cmv_h\frac{\partial T_h}{\partial x} \end{cases} \tag{3.51}$$

According to the heat calculation formula $Q = cm\Delta T$, the transient thermal power H_i dynamic equation in the pipeline i can be obtained:

$$\frac{\partial \mathrm{H}_i}{\partial t} = \frac{\partial (cm\Delta t)}{\partial t} = cmT_1 + c\Delta t\frac{\partial m}{\partial t} - cm\frac{\partial T_i}{\partial t} \tag{3.52}$$

3.4.3 Dynamic Equation in Natural Gas Subsystem

(1) steady-state equation of gas subsystem

In the natural gas pipeline network, the natural gas well combined with the gas tank as a controllable gas source output of the natural gas with a certain $v_{g,1}, p_{g,1}$ and $\rho_{g,1}$. It is similar to the power flow in the power grid, when the natural gas pressure on the compressor is higher than the pressure of the energy Internet, We-Energy outputs natural gas to the energy Internet. On the other hand, the Energy Internet inputs natural gas to We-Energy. The structure of We-Energy's natural gas subsystem is shown in Fig. 3.6, $v_{g,1}, p_{g,1}$ and $\rho_{g,1}$ are the natural gas state variables of We-Energy port, $\dot{m}_{g,L}$ is the natural gas load.

Based on the conservation of total energy and the change of gas state variables in natural gas pipeline network, the flow of natural gas is studied. The input power of the natural gas compressor is coupled with the power network, and its working principle is similar to that of the water pump. However, since the natural gas is compressible, the density and flow rate of the natural gas after pressurization will change. Therefore, the impact of the compressor input power on the natural gas network is different from that of the heat pipe network. Assuming that the input power of the compressor is P_{comp}, due to the gas flow rate and the density cannot be

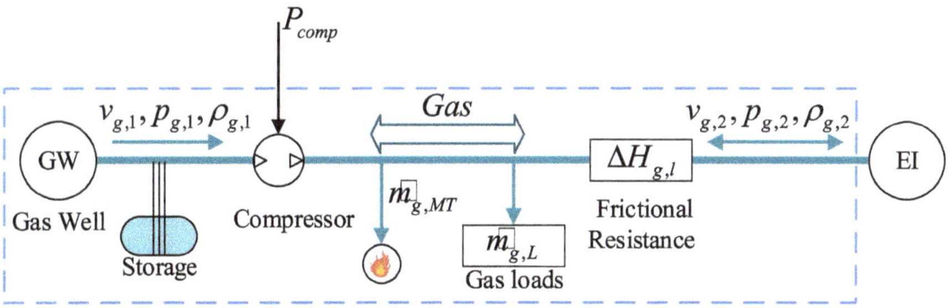

Fig. 3.6 Structure of We-Energy's natural gas subsystem

transient, the power variation is all converted into H_g. From (3.33), the instantaneous change in natural gas pressure $\Delta p_{g,1}$ can be obtained:

$$\Delta p_{g,1} = \rho_{g,1} g H_g \tag{3.53}$$

Lemma 1 *In the rational gas, the input power P_{comp} of the compressor in the pipeline is related to the change of the gas flow rate Δv_g and the change of the gas pressure Δp_g in the pipeline:*

$$p_{g,1}\Delta v_g + v_{g,1}\Delta p_g + \Delta p_g \Delta v_g = \gamma_1 P_{comp} \tag{3.54}$$

where, $\gamma_1 = \frac{1000 g \eta_{comp}}{1+c/RZ}$.

Prove according to the energy conservation equation, the part of the work done by the compressor on the gas is passed to the theoretical head of the gas in the form of mechanical energy, and the other part is passed to the gas in the form of heat energy without considering the external heat loss of the gas. The total power of the gas $P_{g,tot}$ is:

$$P_{g,tot} = P_M + P_Q \tag{3.55}$$

where P_M is the mechanical power of the gas.

When the compressor power is P_{comp}, according to Eq. (3.53), the total gas power change is:

$$\Delta P_{g,tot} = \Delta p_{g,1} \bullet S \bullet v_1 \tag{3.56}$$

The power balance equation of the gas in the pipeline can be obtained from the Eqs. (3.53), (3.55) when the input power of the compressor is P_{comp}:

$$\left(p_{g,1} + \Delta p_{g,1}\right) v_{g,1} S + c\dot{m}_{g,1} T_{g,1} = p_{g,2} v_{g,2} S + c\dot{m}_{g,2} T_{g,2} \tag{3.57}$$

According to the ideal gas equation:

$$\rho_g T_g = \frac{M}{R} p_g \tag{3.58}$$

where M is the molar mass of the gas, R is the gas constant. Put the (3.53), (3.58) and $p_{g,2} = p_{g,1} + \Delta p_g$ into (3.57) to get the relationship of P_{comp} and Δv_g, Δp_g. According to Eq. (3.54), when the natural gas pipe network access to the compressor of the input power P_{comp}, and the system is stable, the product of the changes of natural gas pressure and the speed in the pipeline can be expressed as:

$$\Delta p_g \Delta v_g = \gamma_1 P_{comp} + 2p_{g,1}v_{g,1} - p_{g,1}v_g - v_{g,1}p_g \tag{3.59}$$

In the natural gas pipeline network, both the micro-gas turbine and the natural gas outlet have the speed measuring device, and the output natural gas is controllable. Assuming that the micro-gas turbine and the natural gas load outlet pressures are both standard atmospheric pressure, the air flow rate of the micro-gas turbine and the natural gas load can be expressed as:

$$\begin{aligned} V_{g,MT} &= S_{g,MT} v_{g,MT} \\ V_{g,L} &= S_{g,L} v_{g,L} \end{aligned} \tag{3.60}$$

Assuming the gas state variables after being boosted by the compressor are $v_{g,s}$, $p_{g,s}$ and $\rho_{g,s}$, the natural gas pipe network is placed horizontally, and the Bernoulli equation of the natural gas flow can be listed according to the change of the natural gas pipeline shown in Fig. 3.6.

$$\begin{aligned} &\dot{m}_{g,2}\left(\frac{p_{g,2}}{\rho_{g,2}g} + \frac{v_{g,2}^2}{2g}\right) = \dot{m}_{g,s}\left(\frac{p_{g,s}}{\rho_{g,s}g} + \frac{v_{g,s}^2}{2g}\right) - \\ &\dot{m}_{g,MT}\left(\frac{p_{g,MT}}{\rho_{g,MT}g} + \frac{v_{g,MT}^2}{2g}\right) - \dot{m}_{g,L}\left(\frac{p_{g,L}}{\rho_{g,L}g} + \frac{v_{g,L}^2}{2g}\right) - \dot{m}_{g,L}\Delta H_{g,l} \end{aligned} \tag{3.61}$$

In the formula, $\dot{m}_{g,s} > 0$ indicating that the compressor transport the gas into the We-Energy port, and vice versa for the storage of tank; $\dot{m}_{g,2} > 0$ indicating that the Energy Internet transfer gas to the We-Energy, and vice versa for the We-Energy transfer gas to the energy Internet.

In the natural gas pipeline, the natural gas flow will produce friction to reduce the flow rate, thus increase the pressure of the pipeline. This chapter ignores the natural gas viscosity on the pipeline state variables. At the same time, in the actual project, the pressure potential of natural gas is far greater than its kinetic energy, so it can be assumed that the flow rate of each node in the natural gas at stable state is equal. Based on the above assumptions, Eq. (3.61) can be simplified as:

$$p_{g,2}\dot{V}_{g,2} = p_{g,s}\dot{V}_{g,s} - p_{g,MT}\dot{V}_{g,MT} - p_{g,L}\dot{V}_{g,L} \tag{3.62}$$

where $\dot{V}_g = v_g S_g$ is the volume flow rate of the natural gas.

Here, the electric power in the analog electricity is the product of the voltage and current, in the Energy Internet, we make the following definition for the product of the natural gas pressure and volume flow rate in the natural gas network:

Definition 2 At standard atmospheric pressure, the **vopress** of the gas is the amount of gas volume passing through a section of the pipe in a unit time. The unit of measurement is the $bar \bullet m^3/s$, the pressure of the pipeline is to characterize the capacity of the pipeline. The formula is:

$$Z_g = \frac{p_g \dot{V}_g}{1 \times 10^5}. \tag{3.63}$$

According to the natural gas pipeline structure constructed in this chapter, the We-Energy natural gas subsystem model can be expressed as:

$$Z_{g,2} = Z_{g,1} - Z_{g,MT} - Z_{g,L} + \Delta Z_{g,comp} \tag{3.64}$$

(2) Transient-state equation of gas subsystem

According to the natural gas pipeline continuity, momentum and energy balance equations for the expression of pipeline pressure, mass flow rate and temperature, transient-state equation of gas subsystem can be expressed as follows:

$$\frac{\partial p}{\partial t} = -\frac{ART}{A_g}\frac{\partial Q}{\partial x} \tag{6.65}$$

$$\frac{\partial \dot{m}_g}{\partial t} = -\frac{\partial}{\partial x}\left(\frac{Q^2 ZRT}{p_g A_g}\right) - A_g \frac{\partial p}{\partial x} - \left(\frac{f_g ZRT}{2DA_g}\right)\frac{Q|Q|}{p} - \frac{A_g pg}{ZRT}\sin\theta_g \tag{3.66}$$

$$\frac{\partial T_g}{\partial t} = v\frac{\partial T_g}{\partial x} + \frac{V_w^2}{C_p}\left[1 + \frac{T}{Z}\left(\frac{\partial Z}{\partial T_g}\right)_{p_g}\right]\frac{\partial v_g}{\partial x} - \frac{V_w^2}{C_p p_g}\left[1 - \frac{p}{Z}\left(\frac{\partial Z}{\partial p_g}\right)_{T_g}\right]\frac{\Omega + \tau_w \pi v D}{A_g} \tag{3.67}$$

where, $\tau_w = \frac{f_g \rho v|v|}{8}$, $V_w^2 = \frac{ZRT}{1 - \frac{p}{Z}\left(\frac{\partial Z}{\partial p_g}\right)_{T_g} - \frac{p}{\rho C_p T_g}\left[1 + \frac{T}{Z}\left(\frac{\partial Z}{\partial T_g}\right)_{p_g}\right]^2}$.

In the natural gas pipeline, the density of gas ρ_g is not only a function of time but a function of distance:

$$\begin{cases} \frac{\partial \rho_g}{\partial t} = \left(\frac{\partial \rho_g}{\partial p_g}\right)_{T_g}\frac{\partial p_g}{\partial t} + \left(\frac{\partial \rho_g}{\partial T_g}\right)_{p_g}\frac{\partial T_g}{\partial t} = \frac{1}{ZRT_g}\frac{\partial p_g}{\partial t} - \frac{p_g}{ZRT_g^2}\frac{\partial T_g}{\partial t} \\ \frac{\partial \rho_g}{\partial x} = \left(\frac{\partial \rho_g}{\partial p_g}\right)_{T_g}\frac{\partial p_g}{\partial x} + \left(\frac{\partial \rho_g}{\partial T_g}\right)_{p_g}\frac{\partial T_g}{\partial x} = \frac{1}{ZRT_g}\frac{\partial p_g}{\partial x} - \frac{p_g}{ZRT_g^2}\frac{\partial T_g}{\partial x} \end{cases} \tag{3.68}$$

Meanwhile, the internal energy of gas h_g is also a function of time and distance:

$$\begin{cases} \frac{\partial h_g}{\partial t} = \left(\frac{\partial h_g}{\partial p_g}\right)_{T_g} \frac{\partial p_g}{\partial t} + \left(\frac{\partial h_g}{\partial T_g}\right)_{p_g} \frac{\partial T_g}{\partial t} = \frac{1}{\rho_g} \frac{\partial p_g}{\partial t} + \frac{3R}{16} \frac{\partial T_g}{\partial t} \\ \frac{\partial h_g}{\partial x} = \left(\frac{\partial h_g}{\partial p_g}\right)_{T_g} \frac{\partial p_g}{\partial x} + \left(\frac{\partial h_g}{\partial T_g}\right)_{p_g} \frac{\partial T_g}{\partial x} = \frac{1}{\rho_g} \frac{\partial p_g}{\partial x} + \frac{3R}{16} \frac{\partial T_g}{\partial x} \end{cases} \tag{3.69}$$

According to the equation of state of gas $\rho = \frac{p}{ZRT}$, $\left(\frac{\partial \rho_g}{\partial p_g}\right)_{T_g} = \frac{1}{ZRT_g}$ and $\left(\frac{\partial \rho_g}{\partial T_g}\right)_{p_g} = -\frac{p_g}{ZRT_g^2}$ can be obtained.

According to the formula for internal energy $u_g = \frac{iR}{2M_{CH_4}} T_g$, $h_g = \frac{iR}{2M_{CH_4}} T_g + \frac{1}{\rho_g} p_g$ can be obtained in which $M_{CH_4} = 16\,\text{g/mol}$ is molar mass of Methane, $i = 6$ is gas constant of rigid polyatomic molecule, $R = 8.314472\,\text{J/mol} \bullet K$ is molar gas constant. So $\left(\frac{\partial h_g}{\partial p_g}\right)_{T_g} = \frac{1}{\rho_g}$ and $\left(\frac{\partial h_g}{\partial T_g}\right)_{p_g} = \frac{iR}{2M_{CH_4}}$ can be derived.

Put (3.68) and (3.69) into $m_g = \rho_g A_g v_g$, the following equation about T_g, p_g and m_g can be obtained:

$$\frac{\partial m_g}{\partial t} = A_g \rho_g \frac{\partial v_g}{\partial t} + A_g v_g \frac{\partial \rho_g}{\partial t} \tag{3.70}$$

3.4.4 Integral Model of We-Energy System

According to the power balance equation, thermal power balance equation and vopress balance equation, the whole model of We-Energy can be obtained:

$$\begin{aligned} P_E &= \theta_{11} U^2 + \theta_{12} U + \theta_{13} f - \theta_{14} v_{g,MT} + \theta_{15} \\ Q_E &= \theta_{21} U^2 + \theta_{22} U + \theta_{23} U_{DG} + \theta_{24} \\ P_Q &= \theta_{31} v_w T_w - \theta_{32} T_w + \theta_{33} v_{g,MT} - \theta_{34} F_{Q,L} + \theta_{35} + \eta_{EB} P_{EB} \\ Z_g &= \frac{\theta_{41} v_g + \theta_{42} p_g + \theta_{43} v_{g,MT} + \theta_{44} v_{g,L} + \theta_{45} P_{comp} + \theta_{46}}{1 \times 10^5} \end{aligned} \tag{3.71}$$

where, θ are parameters of system model. The parameters of active power can be expressed as:$\theta_{11} = \frac{R_L}{R_L^2+X_L^2} + \frac{P_0 P_Z}{U_0^2} + \frac{r_l}{r_l^2+x_l^2}$, $\theta_{12} = \frac{P_0 P_I}{U_0} - \frac{2r_l U_{El}}{r_l^2+x_l^2}$, $\theta_{13} = \frac{1}{m}$, $\theta_{14} = \eta_{g2e} H_u \rho_{g,MT} S_{g,MT}$, $\theta_{15} = P_0 P_P + P_{EB} + P_{pump} + P_{comp} + \frac{2r_l U_{El}^2}{r_l^2+x_l^2} - \frac{f_0}{m} - P_{DG,0}$. The parameters of reactive power can be expressed as: $\theta_{21} = \frac{X_L}{R_L^2+X_L^2} + \frac{Q_0 Q_Z}{U_0^2} + \frac{x_l}{r_l^2+x_l^2}$, $\theta_{22} = \frac{Q_0 Q_I}{U_0} - \frac{2x_l U_{El}}{r_l^2+x_l^2}$, $\theta_{23} = \frac{1}{n}$, $\theta_{24} = Q_0 Q_P + \frac{2x_l U_{El}^2}{r_l^2+x_l^2} - \frac{U_0}{m} - Q_{DG,0}$. The parameters of thermal power can be expressed as: $\theta_{31} = c_w \rho_w S_{w,pipe}$, $\theta_{32} = c_w \rho_w S_{w,pipe} v_{w,i}$, $\theta_{33} = \eta_{g2h} H_u \rho_{g,MT} S_{g,MT}$, $\theta_{34} = \chi_{Q,L}$, $\theta_{35} = \rho_w v_{w,\mathrm{i}} S_{w,1} T_{w,\mathrm{i}}$. The parameters of vopress can be expressed as: $\theta_{41} = -S_{g,pipe} p_{g,1}$, $\theta_{42} = -S_{g,pipe} v_{g,1}$, $\theta_{43} = p_{g,0} S_{g,MT}$, $\theta_{44} = p_{g,0} S_{g,L}$, $\theta_{45} = \gamma_1 S_{g,pipe}$, $\theta_{46} = p_{g,1} v_{g,1} S_{g,pipe} + 2S_{g,pipe} p_{g,1} v_{g,1}$.

3.5 Simulation and Analysis

We-Energy system is as an integrated energy interactive system, which is different with other energy supply systems (such as power system, heating system and gas system). In We-Energy, response time of different energy on physical devices is different. According to the structure of We-Energy shown in Fig. 3.1, in the power subsystem, the fluctuation of the electromagnetic process at any point is transmitted at the speed of light, which will influence and spread to the whole system. Therefore, rapid information collection and real-time monitoring are required by We-Energy control center in the power subsystem. In direct heating subsystem, thermal energy production, transmission and consumption belong to a continuous process. Compared to the response speed of state variables in the power subsystem (voltage, frequency), the response speed of state variables in heating supply network (temperature, pressure and velocity) are slow whose change in a relatively short time interval is not obvious. It is not necessary to use the same frequency, which is used in power subsystem the energy system, to collect information and monitor the state of heating subsystem. In the natural gas subsystem, natural gas pipelines not only have the ability to convey natural gas, but also have the capacity for storing natural gas. Therefore, the production and transportation process of natural gas are similar with direct heating system. The consumption of natural gas is a random and intermittent process which is similar with the power subsystem. Meanwhile, the transmission speed of natural gas in the pipeline is more quickly than the fluid in the heating pipeline. The state variables of natural gas pipeline (pressure, temperature, flow velocity) is faster than the response speed in heating network response. In conclusion, as a complex multiple time scale dynamic system, energy transmission speed (electricity, gas, heating) is different in We-Energy. Information collection and optimization control are required by different time scales. The proposed multi-time scale method (see Fig. 3.7) is composed of the following steps:

Step (1) Solve the heating subsystem to obtain the solution, use it as initial guess of heating subsystem and calculate the gas and electricity powered unit consumption (such as MT, pump and electric boiler).
Step (2) Solve the natural gas subsystem to obtain the solution based on the gas consumption of the gas-powered unit, calculate the gas compressor consumption and present an initial guess of the gas pipeline pressure.
Step (3) Solve the power subsystem to obtain the solution based on the power consumption and power generation by the gas-powered unit.
Step (4) Divide state variables into three groups based on their time scale and obtain initial states of the whole We-Energy system.
Step (5) Solve electricity pipe network state equation, with $t_e = t_e + \Delta t_e$ until $t_e \geq t_g$.
Step (6) Solve natural gas network state equation with $t_g = t_g + \Delta t_g$ until $t_g \geq t_h$.
Step (7) Solve heating network state equation until $t_h = t_h + \Delta t_h$.
Step (8) Exchange data among three solvers, and return to Step 5, until $t_h \geq t_{end}$.

In the natural gas subsystem, the pipeline length is 60 km, inside diameter is 0.5 m, friction coefficient is 0.045, gas compression factor $Z = 1$ in Medium and

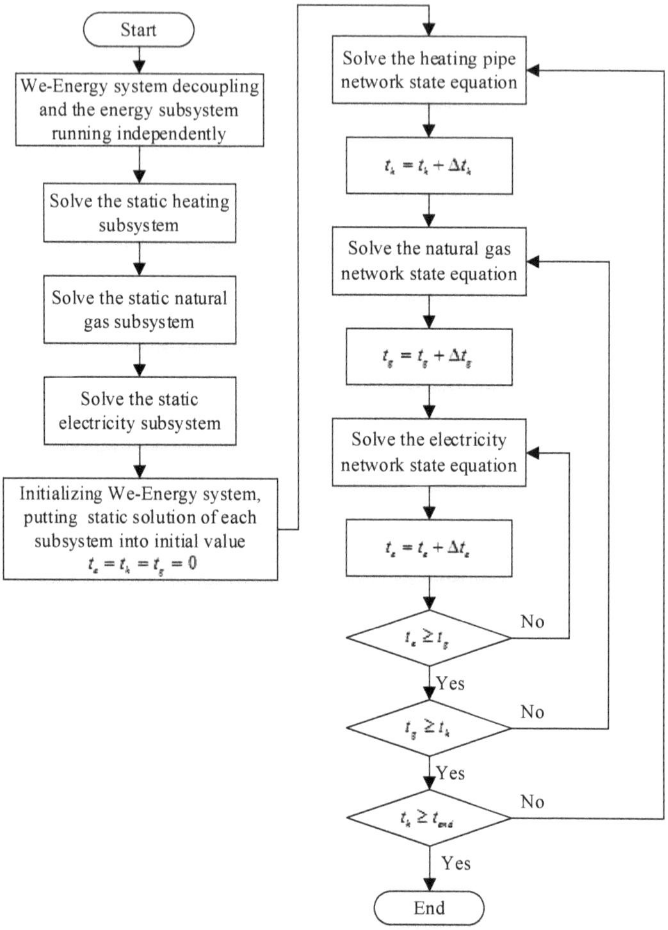

Fig. 3.7 Proposed multi-time scale method

low pressure gas pipe network, gas constant $R = 8.3141$ J/(mol • K), soil temperature $T_{amb} = 278.5$ K. In the heating subsystem, direct heating pipeline network is assumed as pressure bury steel pipe, the pipeline length is 800 m, water-hammer velocity is 1200 m/s, inside diameter is 0.5 m, pipeline fluid density is 1000 kg/m^3, friction coefficient is 0.25, water flow rate is 30 kg/s and pressure is 1.6 MPa in the water inlet, and the maximum output water temperature is 90 °C, the user heating indicators of We-Energy is 45 W/m^2, maximum supply area is 8.6×10^4 m^2. The parameters of devices in We-Energy system are shown in Table 3.1. Due to restriction of distributed generation technology and energy storage technology and the limit of the distributed generation penetration, the capacity of DGs and energy storage is not enough to support the load of We-Energy system alone. Therefore, thermal power, gas turbine and gas boiler capacity are required to occupy a larger of proportion in total capacity.

Table 3.1 Parameter of equipment in we-energy system

We-energy system	Capacity (KW)	Lower limit (KW)	Upper limit (KW)
Thermal power	40×2	20	80
Photovoltaic power	4×3	0	12
Wind power	30	0	30
Electricity storage	5×2	-10	10
Micro-turbine	80	20	80
Gas boiler	40×2	20	80
Electric boiler	5×4	0	20
Heating storage	5×2	-10	10
Pump	0.5×4	0.4	0.6
Compressor	0.3×2	0.25	0.35

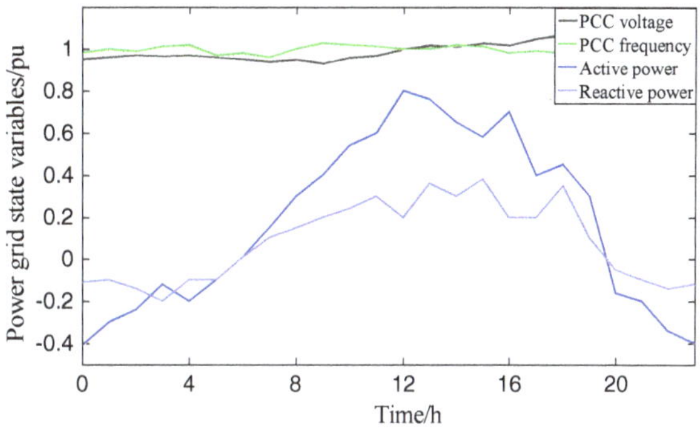

Fig. 3.8 Operating state of power subsystem in WE

3.5.1 Case 1: Operation of We-Energy in Normal State

In this chapter, the actual winter demand for electric / natural gas / heating loads are provided in a region of the north of China. In the north of China, the heating loads are provided by regional boilers which supply more heat energy at night. However, the electrical loads and gas loads are required mostly in day. The forecast curve of electric / natural gas / heating loads are shown in Table 3.2.

In this chapter, a We-Enengy is studied in normal state, so the market clearing price can be decided with multi-We-Energies. Here, a set of fixing price is designde that the electricity price is 0.5 Yuan/KW, the heating price is 1 Yuan/KW, and the gas price is 1.5 Yuan$/bar \bullet m^3$. According to Eq. (3.9), the benefit of We-Energy is shown as Table 3.3

According to the benefit of optimal operation, the operating state of power subsystem, heating network and natural gas network are shown in Figs. 3.8, 3.9 and 3.10.

Table 3.2 Forecast curve of electrical/thermal load

Time	Electricity (KW)		Heating load (KW)	Gas load $bar \bullet m^3/s$
	Active	Reactive		
0	65	24	105	33
1	60	26	97	35
2	57	23	110	38
3	55	25	112	37
4	53	33	118	39
5	50	25	127	50
6	62	33	122	55
7	80	55	120	60
8	90	60	110	65
9	115	65	100	53
10	125	68	97	45
11	120	70	93	42
12	110	66	82	40
13	105	68	70	43
14	100	70	85	45
15	103	75	90	47
16	110	77	97	46
17	115	80	105	50
18	117	76	110	57
19	125	60	120	62
20	130	55	135	65
21	120	43	127	57
22	100	35	122	52
23	80	30	114	46

Table 3.3 Benefit of optimal operation

Time	0–1	1–2	2–3	3–4	4–5	5–6
Benefit	23	25.5	27	30.5	38	32.5
Time	6–7	7–8	8–9	9–10	10–11	11–12
Benefit	20	–10.5	–27	–29	–26	–24.5
Time	12–13	13–14	14–15	15–16	16–17	17–18
Benefit	–19.5	–25.5	–36	–40.5	–36.5	–37
Time	18–19	19–20	20–21	21–22	22–23	23–24
Benefit	–28	–23.5	–17.5	–7	10.5	23.5

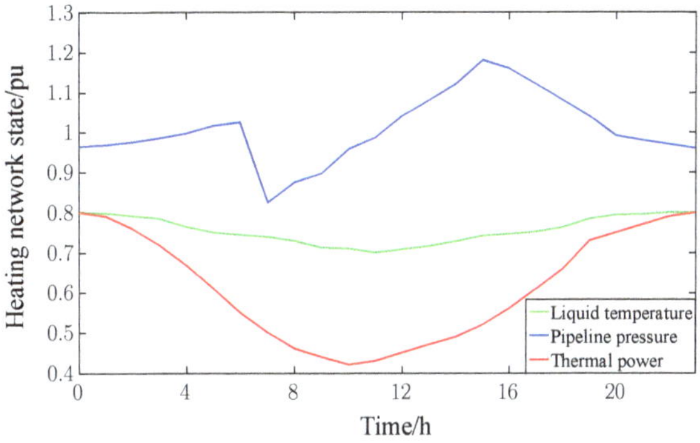

Fig. 3.9 Operating state of heating network in WE

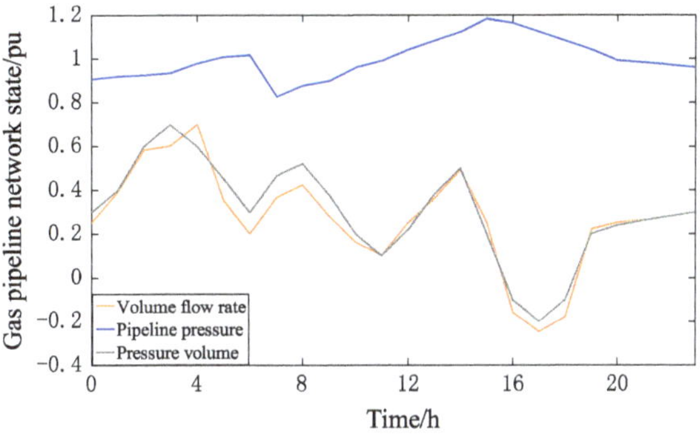

Fig. 3.10 Operating state of natural gas network in WE

3.5.2 Case 2: State Switch of We-Energy Under Abnormal Condition

In this chapter, we focus on the impact of state switch of We-Energy on the output of We-Energy under abnormal condition. In order to ensure the accuracy of the model, the process is studied from transient to steady state. Frequency variation of the electric subsystem is shown in Fig. 3.11. As depicted in Fig. 3.11, the frequency variation has obvious effects on electric, heat and natural gas output, which are shown in Fig. 3.12.

3.6 Summary

This chapter proposed a mechanism model of We-Energy based on its structure, which embodies the distinguishing features of bi-directional power transformation and energy coupling. A quaternary model of WE is established under steady and transient state, which can be divided into normal state, alert state, emergency state and recovery state. And the interaction process of quaternary model is described

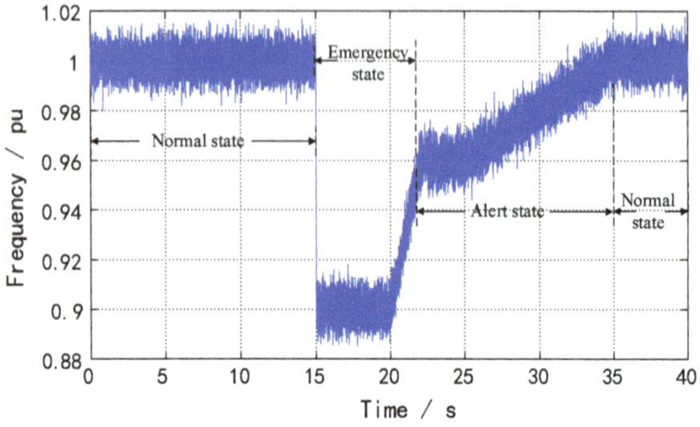

Fig. 3.11 Frequency state switch of energy port in We-Energy

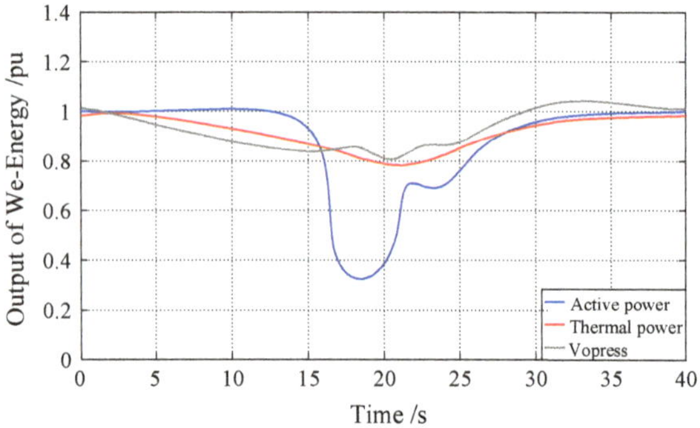

Fig. 3.12 Impact of state switch on output of energy in We-Energy

as Cyber-Physics-Economy-Energy. Simulation results validate that the proposed model is of high identification accuracy and has better generalization performance, and can effectively fit the state variation of each node of the whole system under different operation modes.

References

1. T. Krause, G. Andersson, K. Frohlich et al., Multiple-energy carriers: modeling of production, delivery, and consumption. Proc. IEEE **99**(1), 15–27 (2011)
2. P. Mancarella, MES (multi-energy systems): an overview of concepts and evaluation models. Energy **65**(2), 1–17 (2014)
3. S. Kim, H. Zhang, Q. Sun et al., Consensus-based distributed control for accurate reactive, harmonic and imbalance power sharing in microgrids. IEEE Trans. Smart Grid **99**, 1 (2016)
4. M. Chaudry, N. Jenkins, G. Strbac, Multi-time period combined gas and electricity network optimization. Electr. Power Syst. Res. **78**(7), 1265–1279 (2008)
5. A. Zlotnik, L. Roald, S. Backhaus et al., Coordinated scheduling for interdependent electric power and natural gas infrastructures. IEEE Trans. Power Syst. **32**(1), 600–610 (2017)
6. Q. Zeng, J. Fang, J. Li et al., Steady-state analysis of the integrated natural gas and electric power system with bi-directional energy conversion. Appl. Energy **184**, 1483–1492 (2016)

7. X. Chen, C. Kang, M. O'Malley et al., Increasing the flexibility of combined heat and power for wind power integration in China: Modeling and implications. IEEE Trans. Power Syst. **30**(4), 1848–1857 (2015)
8. Z. Li, W. Wu, M. Shahidehpour et al., Combined heat and power dispatch considering pipeline energy storage of district heating network. IEEE Trans. Sustain. Energy **7**(1), 12–22 (2016)
9. Y. Dai, L. Chen, Y. Min et al., Dispatch model of combined heat and power plant considering heat transfer process. IEEE Trans. Sustain. Energy **99**, 1 (2017)
10. A. Shabanpour-Haghighi, A.R. Seifi, An integrated steady-state operation assessment of electrical, natural gas, and district heating networks. IEEE Trans. Power Syst. **31**(5), 3636–3647 (2016)
11. M. Moeini-Aghtaie, A. Abbaspour, M. Fotuhi-Firuzabad et al., A decomposed solution to multiple-energy carriers optimal power flow. IEEE Trans. Power Syst. **29**(2), 707–716 (2014)
12. P. Mancarella, G. Andersson, J.A. Peças-Lopes, et al., Modelling of integrated multi-energy systems: drivers, requirements, and opportunities. in *Power Systems Computation Conference (PSCC)*, 2016. (IEEE, 2016), pp. 1–22
13. W. Su, A.Q. Huang, A game theoretic framework for a next-generation retail electricity market with high penetration of distributed residential electricity suppliers. Appl. Energy **119**(119), 341–350 (2014)
14. A. Molina-García, I. Muñoz-Benavente, A.D. Hansen et al., Demand-side contribution to primary frequency control with wind farm auxiliary control. IEEE Trans. Power Syst. **29**(5), 2391–2399 (2014)
15. X. Xu, H. Jia, H.D. Chiang et al., Dynamic modeling and interaction of hybrid natural gas and electricity supply system in microgrid. IEEE Trans. Power Syst. **30**(3), 1212–1221 (2015)

Chapter 4
Coordinated Power Management Control Strategy for Interconnected AC and DC Microgrids

Abstract This chapter investigates the issue of active power sharing among a cluster of microgrids formed by a set of ac and dc microgrids network-interconnected through a set of interlinking converters. First, we investigate the power sharing problem for two interconnected AC/DC microgrids. An appropriate control strategy is developed to control the interlinking converter (IC) to realize proportional power sharing between ac and dc microgrids, which includes two parts: the primary outer-loop dual-droop control method along with secondary control; the inner-loop data-driven model-free adaptive voltage control. Using the proposed scheme, the interlinking converter have the ability to regulate and restore the dc terminal voltage and ac frequency and the design of the controller is only based on input/output (I/O) measurement data but not the model any more. Second, we investigate the same problem for more than two microgrids connected by ICs. An event-based distributed consensus control approach is proposed to address this issue. We first construct the agent system for each IC and design its consensus control protocol, which uses a cooperative approach to indirectly adjust the active power load of individual microgrid. Then, an event-based control scheme is utilized to design the consensus protocol to reduce the communication between interlinking converters. The proposed distributed control method allows a sparse communication structure and higher reliability and flexibility operation. Simulation results are presented to demonstrate the proposed control method.

4.1 Introduction

Microgrid is a small-scale power system, and can provide a promising solution to integrate renewable and distributed energy resources as well as distributed energy storage systems. It has gained significant attention recently. Due to the presently dominant role of ac systems and advantages of dc microgrids, a more likely scenario is the presence of both ac and dc microgrids in the future Energy Internet [1, 2]. Therefore, ac, dc and ac/dc microgrids have been widely studied and a variety of surveys have been reported particularly on the subject of structure [3], modeling [4],

Q. Sun, *Energy Internet and We-Energy*, Renewable Energy Sources & Energy Storage, https://doi.org/10.1007/978-981-13-0523-8_4

stability analysis and enhancement [5–7], power quality improvement [8–10], power sharing [1–5, 11], synchronization [12, 13].

Despite the progress mentioned above, some drawbacks of the previous methods can also be found. (i) The majority of the existing inner loop control techniques are greatly dependent on mathematical model. These techniques cannot give satisfactory results when suffering poor model. Uncertainty dynamics and disturbances [14, 15] widely exist in inverter-based microgrids, and it is difficult to obtain the accurate model. Although robust control [16], predictive control [17], variable-structure control [18], and neural network [19] —based control have been proposed for power converters, some challenges still exist. Partial mathematical model and uncertainty dynamics should be known for design of robust controller and variable-structure controller. While predictive control has good performance and strong robustness, the model or structure of the plant also should be known. (ii) Proportional power sharing and voltage (frequency) regulation cannot be achieved at the same time. Interlinking converters in [20, 21] can be viewed as voltage sources, but proportional power sharing between two microgrids cannot be achieved accurately. On the other hand, in [22–24], interlinking converters can be viewed as current sources since they are current controlled converters, which implies these interlinking converters cannot participate in voltage and frequency regulation. (iii) Although various secondary control schemes have been developed to restore the frequency and voltage to their nominal values, the restoration of ac frequency and dc voltage has not been considered in the previous literatures for the interlinked ac and dc microgrids, such as [21–25]. Therefore, a new appropriate control scheme should be further developed for interlinking converts to address these issues mentioned above. Obtaining the system model information that is accurate enough is very difficult in such complicated interlinked microgrid systems. It is more important and meaningful to take advantage of the large amount of the process data produced by the system to boost the operating efficiency and cut the costs. Data-driven model-free adaptive control (DDMFAC) does not require any model information of the controlled plant and the required control performance can be achieved by using the input/output (I/O) data. It is of great significance to take advantage of the process data in such complex system particularly for the future smart grid and energy internet. Therefore, this chapter firstly presents a data-driven control (DDC) structure for interlinking converters in the two interlinked ac and dc hybrid microgrids. Also, a dual-droop controller is also proposed to realize coordinate power sharing between microgrids and allow IC to participate in voltage and frequency regulation.

Following, we discuss the power sharing issue among clusters of microgrids. As known, a key topic of interest within the microgrid community is accurate power sharing among a bank of inverters operated in parallel. A large amount of research has been conducted on this topic, where the goal is to achieve power sharing proportional to DGs' capacities while keeping the desired frequency and voltage values. For instance, several centralized and distributed secondary control strategies have been reported in [26–37], which are mainly based on multi-agent theory. Developments mentioned above are, however, directed at power sharing among DGs mainly within one ac or dc microgrid based on hierarchical control frame-work. Enforcing ac and dc microgrids intertied by an interlinking power converter is a promising topology in future power networks. Recently, several authors have also researched the power sharing problem between ac and dc microgrids, where the microgrids are interconnected by interlinking converters. With respect to this topic, P. C. Loh et al. [38–41] have done some pioneering works. Models and co-ordination control schemes were proposed and verified for smooth power exchange between dc and ac grid and for stable system operation under various generation and load conditions for the first time [38]. Then, a normalized bidirectional droop control scheme was developed for controlling the ICs with the defined common per-unit range of the droop characteristics of ac and dc microgrids [19], resulting in proper power sharing between the two ac and dc microgrids. Following, similar methods were developed in [40, 41] where an energy storage system was integrated into the hybrid microgrids and better performance was obtained. Another droop control scheme has been followed in [42] for bidirectional power flow regulation between the interconnected microgrids. Additionally, a dual-droop scheme and data-driven controller has also been developed in [1]. Droop-based methods are mainly developed and primarily one ac and one dc microgrids are discussed in the above mentioned works. In this paper, we will mainly focus on the more general network-interconnected microgrid cluster which can be a promising topology in future Energy Internet [43]. And droop-free control method will be proposed. Power sharing among multiple interconnected microgrids is also important and necessary since the power ratings and loads of microgrids are usually different in practice. The peak electricity consumption time can be different in different areas. The interconnected microgrids allow back-up reserve with each microgrid to be avoided as well, resulting in greater reliability and flexibility. Recently, a few results have been reported in literature [44–47] about power sharing among microgrids within a cluster. A distributed tertiary control scheme based on multi-agent consensus has been proposed to handle the power sharing among a cluster of dc microgrids in [44], where each microgrid has its own controller used to update the set points of each microgrid. However, only dc microgrids were considered, and IC is not used between two dc microgrids. In [45], power sharing among only ac microgrids has been discussed, where multiple ac microgrids were connected to the utility grid by parallel back-to-back converters. The system stability was improved with a decentralized control. In [46], a three-port AC/DC/DS hybrid microgrid was investigated by using a droop-based method, where the subsystems were connected by a two-stage IC together. A centralized control method was proposed to control the ICs to exchange power between microgrids where the dc microgrids were connected to ac microgrid through the parallel ICs [47]. In this paper, we mainly focus on the power sharing among networked interconnected microgrids which is a more general structure. And a distributed control method will be developed.

Regarding solving the power sharing issues by using distributed approach, communication between agents is utilized and computations are locally performed. Large scale IT infrastructure will be developed for future smart grids and energy internet [48], in which tremendous data exchange would rapidly exhaust the network resources, leading to unreliable operations and bad and even failed performance of the distributed approach. Also, to reduce the control costs, the distributed scheme may be embedded with micro-processors which are usually run with limited energy resources and computing capability. Previous work on power sharing mainly fails to take the communication issue into account. Therefore, it would be desirable to reduce the communication burden. The event-based control can be an alternative providing cited advantage on communication reduction [49]. Recently, a few results on the application of event-based control have been reported in the literature involving economic dispatch in smart grids [50] and load frequency control of multi-area power systems [51].

With mainly the aforementioned inspirations, this chapter further focuses on the power sharing control problem of networked interconnected ac and dc microgrids, where an event-based distributed consensus control strategy is proposed to address this issue. The main contribution and salient features of this chapter can be summarized as follows:

(1) A novel data-driven model-free adaptive voltage control (DDMFAVC) scheme is introduced for interlinking converters in interlinked ac and dc hybrid microgrids. The model, structure, uncertainty dynamics, and unmodeled dynamics are not required in this scheme.
(2) Dynamic dual-droop control scheme is proposed to achieve proportional power sharing between ac and dc microgrids. This droop scheme, along with the voltage controller, enables proportional power sharing and voltage/frequency regulation realized simultaneously like DGs in microgrids.
(3) Active power sharing among multiple ac and dc microgrids is investigated, where the interconnection form of the microgrids are networked and more general compared with previous reported works.
(4) An event-based distributed consensus control strategy with time delays is developed for ICs to realize proportional active power sharing among these microgrids, which can reduce communication among agents dramatically and is more suitable for the interconnected microgrids.

Reactive power support for ac microgrids, as an ancillary service, is developed by using the available IC rating without affecting proportional active power sharing. The system has plug-and-play ability and can effectively operate subject to islanding and communication link lost.

4.2 System Structure

A possible layout of two interlinked ac and dc microgrids and networked microgrids is shown in Fig. 4.1 and Fig. 4.2, respectively, in which an interlinking converter (IC) (shown in Fig. 4.1) is utilized to link the ac and dc microgrids together. Each microgrid has its own sources, storages and loads. A ring-shape communication topology is also illustrated in Fig. 4.2. The interlinking converter, between the two microgrids, is to provide bidirectional power flow depending on present generating and loading conditions of each microgrid. The formed interconnected microgrids can be tied to

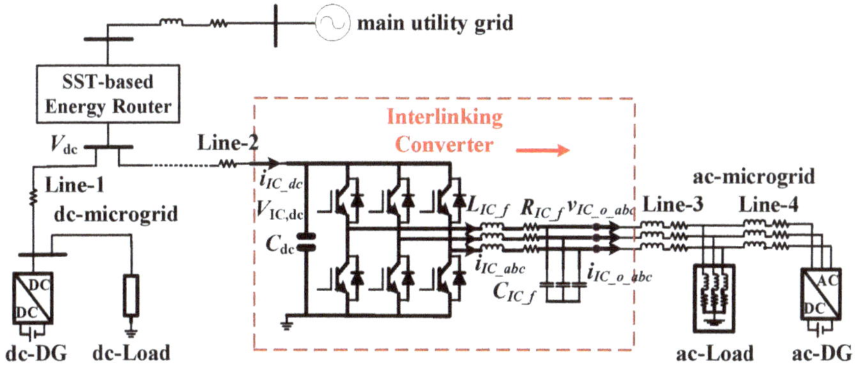

Fig. 4.1 An example of interlinked ac/dc microgrids

the utility grid through a solid state transformer (SST) based energy router [28] (not shown in Fig. 4.2). Note that DC/DC (or AC/AC) converters can also be utilized to interconnect two dc (or ac) microgrids, which is also not shown in Fig. 4.2. In the grid-connected mode, the energy router can operate as a constant power source seen from the main utility grid side, injecting (or absorbing) constant active power to (or from) the utility grid so as to not to disturb the main utility grid unnecessarily. It means that for the main utility grid, the interconnected microgrids will become "controllable", which is greatly beneficial to the stable operation of the main utility grid. The interlinking converter, in both grid-connected and interconnected modes, will provide bidirectional power flow to participate in p roper power sharing among microgrids. In this chapter, we only consider the interconnected mode. In this mode, all the microgrids interconnected by ICs should cooperate with each other to enhance the power supply reliability and flexibility and improve the utilization of the distributed energy resources. The most important task of each IC is to determine and control the amount of active power that should be transferred. And sources should decide on the right amount of energy to produce and to meet the load demand according to the load within each microgrid itself. The idea of the proposed distributed control strategy is to control the ICs to regulate the active power flows between microgrids to indirectly regulate the load power of individual microgrid, providing higher-level active load power sharing among the microgrids. For the two interconnected microgrids, three operation modes of the interlinking converter are considered as follows.

Mode-1: If the determined active power is negative, it means that the $V_{IC,dc} - P$ droop is selected and that the interlinking converter will absorb the power from the ac microgrid and then inject into the dc microgrid. The interlinking converter, seen from the dc-link side, just acts as a "dc DG" unit in this mode.

Mode-2: If the determined active power is positive, it means that the $\omega_{IC} - P$ droop is selected and that the interlinking converter will inject the power to the corresponding ac microgrid. The interlinking converter takes the same role of an "ac DG" as that in the ac microgrid in this mode.

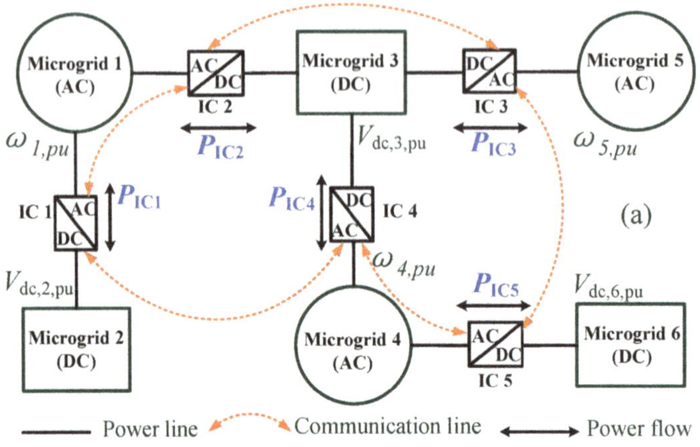

Fig. 4.2 A possible layout of a cluster of interconnected ac and dc microgrids with a ring-shape communication topology spanned across the network

Mode-3: There will be no power transferred by the interlinking converter when both the ac and dc microgrids are under-loaded or over-loaded, or some faults occur, or deviation of the per-unit values of dc voltage and ac frequency is less than threshold.

4.3 Modeling of the System

4.3.1 Dynamic Linearization Data Model of IC

Usually, a voltage source inverter (VSI) can be adopted for the interlinking converter. And a sample configuration is shown in Fig. 4.1. In natural reference frame, considering the dc-link voltage dynamics and ignoring conducting resistances of the switching devices in the IC, the complete average switching dynamics of the interlinking converter can be given by

$$\begin{aligned}
\dot{V}_{IC,dc} &= (1/C_{dc})i_{IC_dc} - \left(1/C_{dc}V_{IC,dc}\right)\mathbf{u}^{T}_{abc_ave}\mathbf{i}_{IC_abc} \\
\dot{\mathbf{i}}_{IC_abc} &= \left(-R_{IC_f}/L_{IC_f}\right)\mathbf{i}_{IC_abc} + \left(1/L_{IC_f}\right)\left(V_{IC,dc}\mathbf{u}_{abc_ave} - \mathbf{v}_{IC_o_abc}\right) \\
\dot{\mathbf{v}}_{IC_o_abc} &= \left(1/C_{IC_f}\right)\left(\mathbf{i}_{IC_abc} - \mathbf{i}_{IC_o_abc}\right)
\end{aligned} \tag{4.1}$$

where $\mathbf{i}_{IC\_abc}$, $\mathbf{v}_{IC\_o\_abc}$, $\mathbf{u}_{abc\_ave}$, and $\mathbf{i}_{IC\_o\_abc}$ are ac currents of filter inductance, ac voltages of filter capacitor, average switching signals and ac output currents of the interlinking converter, respectively. $V_{IC,dc}$ and i_{IC_dc} are dc-link voltage and input current, respectively. L_{IC_f}, R_{IC_f}, C_{IC_f} and C_{dc} are the filter inductance, filter resistance, filter capacitance, and dc-link capacitance, respectively. We choose $V_{IC,dc}$ and $\mathbf{v}_{IC\_o\_abc}$ as the system outputs when the interlinking converter operates in mode-1 and mode-2, respectively, and $\mathbf{u}_{abc\_ave}$ as the system control inputs. Then the dynamics, for digital implementation, can be expressed in a discrete-time domain with the conversion $H(s) = (1 - e^{-sT})/s$ as

$$\begin{aligned}
\mathbf{v}_{IC-o-abc}(k+1) &= \mathbf{f}_1(\mathbf{v}_{IC-o-abc}(k), \ldots, \mathbf{v}_{IC-o-abc}(k-d_{y1}), \\
&\quad \mathbf{u}_{abc-ave}(k),), \ldots, \mathbf{u}_{abc-ave}(k-d_u)) \\
V_{IC,dc}(k+1) &= f_2(V_{IC,dc}(k) \ldots, V_{IC,dc}(k-d_{y2}), \\
&\quad \mathbf{u}_{abc-ave}(k), \ldots, \mathbf{u}_{abc-ave}(k-d_u))
\end{aligned} \tag{4.2}$$

where $\mathbf{v}_{\mathrm{IC\_o\_abc}} = [\mathrm{v}_{\mathrm{IC\_o\_a}}, \mathrm{v}_{\mathrm{IC\_o\_b}}, \mathrm{v}_{\mathrm{IC\_o\_c}}]^{\mathrm{T}}$, $\mathbf{u}_{\mathrm{abc\_ave}} = [\mathrm{u}_{\mathrm{a\_ave}}, \mathrm{u}_{\mathrm{b\_ave}}, \mathrm{u}_{\mathrm{c\_ave}}]^{\mathrm{T}}$. Let $\mathrm{V}_1 = \mathbf{v}_{\mathrm{IC\_o\_abc}}$, $\mathrm{V}_2 = \mathrm{V}_{\mathrm{IC,dc}}$, $\mathbf{u} = \mathbf{u}_{\mathrm{abc\_ave}}$, and $\mathbf{f}_2 = f_2$, then the dynamics in (4.2) can be expressed as

$$\mathbf{V}_i(k+1) = \mathbf{f}_i\big(\mathbf{V}_i(k), \cdots, \mathbf{V}_i\big(k-d_y\big), \mathbf{u}(k), \cdots, \mathbf{u}(k-d_u)\big), i = 1, 2 \tag{4.3}$$

where *dy* and *du* are the unknown orders, and $\mathbf{f}_i(\cdot)$ are unknown nonlinear function vectors.

As mentioned in the previous section, uncertainty dynamics, unmodeled dynamics, and disturbances widely exist in the interlinked microgrids, and it is difficult to obtain the unknown nonlinear function vector $\mathbf{f}_i(\cdot)$. Therefore, data-driven-based partial form dynamic linearization (PFDL) can be the best to be adopted in this paper to obtain the equivalent dynamic linearization data model of system (4.3).

The implementation of the data-driven MFAC method is usually based on two assumptions: (1) The partial derivatives of with respect to control inputs are continuous; (2) System (4.2) is generalized Lipschitz. These assumptions imposed on the controlled system are reasonable and acceptable from a practical viewpoint. Assumption 1 is a typical condition of control system design for general nonlinear systems. Assumption 2 limits the rates of changes of the system outputs driven by the changes of the control inputs. From the 'energy' point of view, the output energy change rates inside a system cannot go to infinity if the changes of the control input energy are in a finite altitude. According to these assumptions and Theorem 1 in [26], for the nonlinear system (4.3), there must be parameters $\mathbf{\Phi}_i(k), \forall i = 1, 2$ called (pseudo partitioned Jacobian matrix, PPJM), and system (4.3) can be transformed into the following PFDL description when $\|\Delta\mathbf{U}(k)\| \neq 0$:

$$\Delta\mathbf{V}_i(k+1) = \Phi_i(k)\Delta\mathbf{U}(k), \forall i = 1, 2 \tag{4.4}$$

where each variable is given in Appendix A.

4.3.2 Modeling of the AC and DC Microgrids

In this paper, both ac and dc microgrids are considered for the interconnected system. Generally, in a microgrid, three control loops, namely, the voltage control loop, current control loop, and droop control loop, are adopted in the local DG controller. It should be noted that the dynamics of the *LCL* filter, the *RL* output connector, and the voltage and current control loop are much faster than that of the droop control. Therefore, only the dynamics of the droop control are considered in remodeling the primary controller. Hence, the widely accepted $\omega - P$ and $V_{dc} - P$ droop control equations representing each distributed energy resource in the ac and dc microgrids respectively, are given by

$$\begin{aligned} \omega_{k,r,pu} &= \omega^*_{k,r,pu} - k_{ac,k,r}\left(P_{ac,k,r} - P^*_{ac,k,r}\right) \\ V_{dc,k,r,pu} &= V^*_{dc,k,r,pu} - k_{dc,k,r}\left(P_{dc,k,r} - P^*_{dc,k,r}\right) \end{aligned} \tag{4.5}$$

where $\omega^*_{k,r,pu}$ and $V^*_{dc,k,r,pu}$ are the desired frequency and dc voltage, respectively; $P_{ac,k,r}$ and $P_{dc,k,r}$ are the output active powers, $P^*_{ac,k,r}$ and $P^*_{dc,k,r}$ are the desired active powers of DGs in ac and dc microgrids, respectively; and $k_{ac,k,r}$ and $k_{dc,k,r}$ are the droop coefficients. Notice that the subscript "*pu*" represents the per-unit values that are defined by applying the expressions in [26], and "*r*" and "*k*" represent the *r* th DG unit in microgrid "*k*". The $E - Q$ droop equation is not included in Eq. (4.5) since the active power sharing is mainly considered in this paper. Based on the linear Eq. (4.5), the combined droop Eq. (4.6) of the whole DGs in one ac or dc microgrid can be obtained. This means that all the DGs in one ac or dc microgrid are viewed as an equivalent larger DG.

$$\begin{aligned} \omega_{k,pu} &= \omega^*_{k,pu} - k_{ac,k}\left(P_{ac,k} - P^*_{ac,k}\right) \\ V_{dc,k,pu} &= V^*_{dc,k,pu} - k_{dc,k}\left(P_{dc,k} - P^*_{dc,k}\right) \end{aligned} \tag{4.6}$$

where $\omega^*_{k,pu} = \omega^*_{k,r,pu}$, $V^*_{dc,k,pu} = V^*_{dc,k,r,pu}$, $P_{ac,k} = \sum_{r=1}^{n} P_{ac,k,r}$, $P_{dc,k} = \sum_{r=1}^{n} P_{dc,k,r}$, $P^*_{ac,k} = \sum_{r=1}^{n} P^*_{ac,k,r}$, $P^*_{dc,k} = \sum_{r=1}^{n} P^*_{dc,k,r}$, $k_{ac,k} = 1/\sum_{r=1}^{n}\left(1/k_{ac,k,r}\right)$, and $k_{dc,k} = 1/\sum_{r=1}^{n}\left(1/k_{dc,k,r}\right)$. $\omega^*_{k,pu}$ and $V^*_{dc,k,pu}$ are the desired ac frequency and dc voltage, respectively; $P_{ac,k}$ and $P_{dc,k}$ are the output active powers, $P^*_{ac,k}$ and $P^*_{dc,k}$ are the desired active powers of DGs in ac and dc microgrids, respectively; and $k_{ac,k}$ and $k_{dc,k}$ are the droop coefficients. Note that regarding the dc microgrids, the effects of the line impedance mismatches are omitted for simplifying the analysis. This will not affect the controller design.

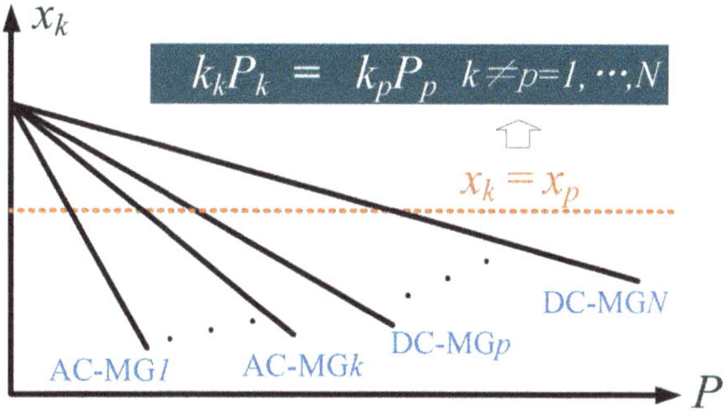

Fig. 4.3 The illustration of the normalized consolidated droop lines of the interconnected ac and dc microgrids

For simplicity and readability, let x_k represents $\omega_{k,pu}$ and $V_{dc,k,pu}$, P_k represents $P_{ac,k}$ and $P_{dc,k}$, and k_k reprsents $k_{ac,k}$ and $k_{dc,k}$, then the combined droop Eq. (4.6) from microgrid \$k\$ can be rewritten as a unified form

$$x_k = x_k^* - k_k\left(P_k - P_k^*\right). \tag{4.7}$$

The active power outputs P_k in Eq. (4.7) can also be given by

$$P_k = P_{L_k} + \sum\nolimits_{i \in \mathcal{C}_k} P_{IC,i}$$

where P_{L_k} is the active power load within the kth dc microgrid ($P_{dc-MGk-L}$) or ac microgrid ($P_{ac-MGk-L}$), $\mathcal{C}_k$ is the set that indexes the interlinking converters connected to the kth microgrid, and $P_{IC,i}$ is the active power transferred by IC "i".

The droop lines (4.7) representing the normalized consolidated droop responses of the ac and dc microgrids are given in Fig. 4.3. Applying the principle of droop control of DGs to the interconnected microgrids to achieve proportional power sharing means maintaining $x_k = x_p (or\ k_kP_k = k_pP_p)(k \neq p)$ for all the microgrids. Despite the well-recognized droop control strategies in standalone ac or dc microgrids, proper power sharing among multiple microgrids tied together through the interlinking converters can be difficult by using the conventional droop methods. Therefore, this paper aims at designing a distributed controller to control the interlinking converter to properly exchange active power $P_{IC,i}$ between microgrids, achieving proportional active power sharing among the interconnected microgrids by keeping the per-unit values of frequencies and dc voltages equal. Thus it will benefit the utilization of the distributed energy resources, and enhance the power supply reliability and flexibility.

4.4 Power Management Control Strategy of Interconnected AC and DC Microgrids

4.4.1 *Data-Driven- and Dual-Droop-Based Control Strategy for Hybrid AC/DC Microgrids*

4.4.1.1 Data-Driven Model-Free Adaptive Voltage Controller

Based on the PFDL system (4.4), the DDMFAVC controller for the interlinking converter can be designed in the following. Before giving out the controller, an observer is proposed to estimate the parameters $\Phi_i(k)$ (PPJM), and the observer and the adaptive update law for $\Phi_i(k)$ are given by (1) Steady-state equation of power subsystem

$$
\begin{aligned}
\hat{\mathbf{V}}_i(k+1) &= \hat{\mathbf{V}}_i(k) + \hat{\Phi}_i(k)\Delta\mathbf{U}(k) + \mathbf{K}_i\tilde{\mathbf{V}}_i(k) \\
\hat{\Phi}_i^T(k+1) &= \hat{\Phi}_i^T(k) + \Gamma_i(k) \\
&\left(\tilde{\mathbf{V}}_i(k+1) - \mathbf{F}_i\tilde{\mathbf{V}}_i(k)\right)\Delta\mathbf{U}(k)\mathbf{I}_{1\times 3L}
\end{aligned}
\tag{4.8}
$$

where all the variables are given in Appendix A.

Upon the parameters $\Phi_i(k)$ estimated, the DDMFAVC controller can be designed as

$$
\begin{aligned}
\mathbf{u}(k) &= \mathbf{u}(k-1) + \hat{\Phi}_i^T(k)\left[\alpha_i + \hat{\Phi}_i(k)\hat{\Phi}_i^T(k)\right]^{-1} \\
&\times\left[\mathbf{V}_i^*(k+1) - \hat{\mathbf{V}}_i(k) - \mathbf{K}_i\tilde{\mathbf{V}}_i(k)\right], for\|\Delta\mathbf{U}(k)\| \le \delta \\
\mathbf{u}(k) &= \mathbf{u}(k-1) + \delta\cdot sign(\Delta\mathbf{u}(k)), \text{ for}\|\Delta\mathbf{U}(k)\| > \delta
\end{aligned}
\tag{4.9}
$$

where $\alpha_i = diag(\alpha_1, \alpha_2, \alpha_3)$, $\alpha_2 = \alpha_4$, and $\mathbf{V}_i^*(k)$ are the reference trajectories.

The stability of the proposed DDMFAVC closed-loop control system (4.9) can be guaranteed by using the Lyapunov-based stability theory. Detailed proof can be found in [44].

It is worthwhile to remark here that the designed voltage controllers (4.8) and (4.9), unlike the robust controller [36], predictive controller et al. [37] can be obtained and implemented easily only by using input-output data through the data-driven control theory. Mathematical models are not required in the design of the proposed controller. The controller is a lower cost controller since it does not require any external testing signals and any training process. It is simple and easy to be implemented and flexible with small computational burden. It is also suitable to complex and large-scale practical systems particularly for the interlinked microgrids since the structure of the plant is often difficult to determine and the parameters are hard to identify and neces-

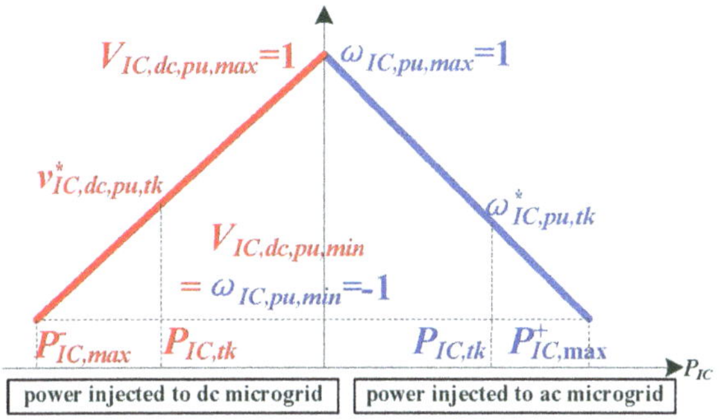

Fig. 4.4 The proposed dual-droop control characteristics of the interlinking converter

sary process information that the data-driven MFAC needs can be directly extracted from huge amounts of process data. On the other hand, the volt-age controller can be implemented flexibly under different operation modes that are determined by the dual-droop controller discussed in detail in the following subsection. Additionally, with respect to the voltage controller design, complex coordinate transformation can be avoided.

4.4.1.2 Dual-Droop Controller

Proportional power sharing is necessary. In this paper, all the DGs in each microgrid are seen as a larger equivalent controllable distributed generator by summing all their respective source characteristics. The power ratings and loads of microgrids are usually different in practice. Consequently, this allows back-up reserve with each microgrid to be reduced considerably and overstress of each microgrid to be avoided as well, resulting in greater reliability.

Despite the well-recognized droop control strategies in standalone ac or dc microgrids, proper power sharing among multiple microgrids tied together through the interlinking converter cannot be achieved by the conventional droop methods. Considering the statements, to achieve proportional power sharing between the interlinked microgrids and participate in voltage and/or frequency regulation simultaneously just like DGs in microgrids, a dynamical dual-droop control scheme with power management and distribution is proposed in this paper. The proposed dual-droop control characteristics of the interlinking converter for active power sharing are drawn in Fig. 4.4. Their mathematical representations are given as

$$P_{IC,t_k} = \begin{cases} \left(\omega^*_{IC,pu,t_k} - \omega_{IC,pu,\max}\right)/\tilde{\sigma}_{IC,t_k} \\ \left(V^*_{IC,dc,pu,t_k} - V_{IC,dc,pu\max}\right)/\tilde{k}_{IC,t_k} \end{cases} \tag{4.10}$$

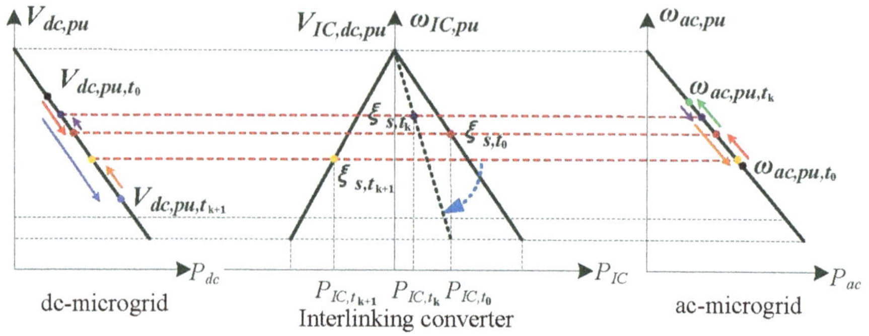

Fig. 4.5 Illustration of proportional power sharing process realized within the intertied ac and dc microgrids

where $\tilde{\sigma}_{IC,t_k}$ and $\tilde{k}_{IC,t_k}$ are the active droop coefficients, and "*pu*" represents the per-unit values that are defined by applying the expressions in [23].

For the appropriate power flow decisions using only variables measured locally, different thresholds can then be set for the frequencies and voltages to distinguish when the microgrids are under-loaded (UL), normal-loaded (NL), or over-loaded (OL) in terms of active powers. Using the proposed dual-droop control, the interlinking converter will have three operation modes defined in above, and at any instant it just operates at one mode.

Conventionally, within the ac microgrid for example, active and reactive powers at the source terminals are measured for determining reference values for its frequency and voltage magnitude. However, the active power command of the interlinking converter in this paper is determined by the proposed power management and distribution module. Underlying principles of this new droop control scheme can better be understood by referring to the example drawn in Fig. 4.5. In that figure, the droop lines drawn in the left and right sides are for representing the normalized consolidated droop responses of the ac and dc microgrids given as

$$\begin{aligned}\omega_{ac,pu} &= \omega_{o,pu} - k_{ac}P_{ac}\\ V_{dc,pu} &= V_{o,pu} - k_{dc}P_{dc}\end{aligned} \tag{4.11}$$

where k_{ac} and k_{dc} are droop coefficients.

When the interlinking converter starts to operate at first time, the interlinking converter will "know" the operating conditions of the two microgrids by measuring the local terminal ac frequencies and dc voltages. For instance, they initially operate at ω_{ac,pu,t_0} and V_{dc,pu,t_0}, respectively, corresponding to the black dots shown in Fig. 4.5. According to droop control principle, achieving proportional power sharing between the ac and dc microgrids means maintaining $\omega_{ac,pu} = V_{dc,pu} = \xi_s$, corresponding to the red dashed horizontal line drawn in Fig. 4.5. Therefore, Eq. (4.12) can be obtained.

$$\omega_{ac,pu,t_0} + k_{ac}P_{IC,t_0} = V_{dc,pu,t_0} - k_{dc}P_{IC,t_0} \tag{4.12}$$

where P_{IC,t_0} is the determined active power to be transferred through the interlinking converter for proportional power sharing at $t = t_0$. P_{IC,t_0} can be rewritten as

$$P_{IC,t_0} = \frac{V_{dc,pu,t_0} - \omega_{ac,pu,t_0}}{k_{ac} + k_{dc}} \tag{4.13}$$

Upon reaching the steady state, proportional active power sharing between different types of microgrids in the interlinking converter enabled system can be realized due to the same vertical axis values ($\omega_{ac,pu} = V_{dc,pu} = \xi_s$) of the consolidated droop lines.

After that, when the loads in the ac and dc microgrids changes, the active power command will be updated using the following equation

$$\begin{aligned} P_{IC,t_{k+1}} &= P_{IC,t_k} - \underbrace{\frac{k_{ac}}{k_{ac}+k_{dc}}\Delta P_{Lac}}_{\varphi_{dc}} + \underbrace{\frac{k_{dc}}{k_{ac}+k_{dc}}\Delta P_{Ldc}}_{\varphi_{ac}} \\ &= \frac{1}{k_{ac}+k_{dc}}\left(V_{dc,pu,t_{k+1}} - \omega_{ac,pu,t_{k+1}}\right), k \geq 0 \end{aligned} \tag{4.14}$$

where P_{IC,t_k} represents the active power to be transferred by the interlinking converter when loads changed at the time $t = t_k$, ΔP_{Lac} and ΔP_{Ldc} represent the increased active powers of the ac and dc microgrids at the time $t = t_{k+1}$, respectively, φ_{ac} and φ_{dc} represent the active power that should be shared by the ac and dc microgrids, respectively, and $\omega_{ac,pu,t_{k+1}}$, $V_{dc,pu,t_{k+1}}$ are the measured terminal ac frequency and dc voltage at the present time $t = t_{k+1}$.

Considering the defined operation modes and thresholds, a more general expression of (4.14) can be given as

$$P_{IC,t_k} = \begin{cases} 0, \; when \, V_{dc,pu,t_k}, \omega_{ac,pu,t_k} \\ \quad \in (\xi_{UL}, \xi_{\max}] \cup [\xi_{\min}, \xi_{OL}] \\ \frac{1}{k_{dc}+k_{ac}}\left(V_{dc,pu,t_k} - \omega_{ac,pu,t_k}\right), \; others \end{cases} . \tag{4.15}$$

Seen from Eq. (4.10) and (4.15), the interlinking converter monitors the operating of the ac and dc microgrids, and updates the active power command in real time only using the measured ac frequency and dc microgrid voltage. Upon the determined active power $P_{IC,t_{k+1}}$ transferred by the interlinking converter, proportional active power sharing between the ac and dc microgrids can be realized.

However, it should be noted that using Eq. (4.15) will cause almost continuous operation of the interlinking converter for any load variations that will result in more power loss in the converter. Moreover, when the deviation is small enough, it is not necessary for the interlinking converter to transfer the active power. The main reasons can be summarized as: (1) the determined active power is too small and much of that will be loosed in the converter under this condition; (2) reliable operation of the system cannot be affected even if the determined active power is not transferred by the interlinking converter. Therefore, to avoid this, threshold of the deviation is introduced into Eq. (4.15), and then Eq. (4.15) can be rewritten as follows:

$$\begin{aligned} P_{IC,t_k} = & \tfrac{1}{k_{ac}+k_{dc}}\left(V_{dc,pu,t_k} - \omega_{ac,pu,t_k}\right) \\ & u\left(\left|V_{dc,pu,t_k} - \omega_{ac,pu,t_k}\right| - \eta\right) \end{aligned} \tag{4.16}$$

where $u(\cdot)$, is the unit step function, η is the threshold of the deviation and it can be expressed as $\eta = f\left(V_{dc,pu,t_k}, \omega_{ac,pu,t_k}\right)$. Equation (4.16) can be better and more easily illustrated by using Fig. 4.8 which is a three-dimensional space Cartesian coordinate reference frame constructed by $V_{dc,pu}$, $\omega_{ac,pu}$ and η. As shown in Fig. 4.8, the surface S_{ABCD} represents the absolute value of the deviation $\eta_e = \left|V_{dc,pu,t_k} - \omega_{ac,pu,t_k}\right|$, and S_{BEJI} and S_{FCHG} represent the over-load and light-load conditions of both the ac and dc microgrids, respectively. S_{EFGHIJ} can be the considered area in some other operation conditions, where active power could not be transferred due to the small deviation. Taking the above into consideration, the surface $S_{B_1E_1F_1C_1H_1I_1}$ is designed to be the threshold $\eta = f\left(V_{dc,pu,t_k}, \omega_{ac,pu,t_k}\right)$ of the deviation in this paper. Thus, when both the ac and dc microgrids are operating in light-load (S_{FCHG}) or over-load (S_{BEJI}) condition, none active power would be transferred from one microgrid to the other due to $\eta_e < \eta$. On the other hand, when the microgrids are operating in the area of S_{EFGHIJ}, the interlinking converter would not transfer any active power due to the small deviation and $\eta_e < \eta$. Therefore, continuous operation of the interlinking converter can be avoided. It is worthy to remark here that the threshold η ($S_{B_1E_1F_1C_1H_1I_1}$) can be flexibly designed according to the requirements of practical applications by using Fig. 4.7.

As discussed in the literature [29, 30, 32], current-source, rather than voltage-source, characteristics are exhibited by the interlinking converter. In order to participate in voltage and frequency regulation, data-driven model-free adaptive voltage controller is proposed in this paper for the interlinking converter. Therefore, dynamical tuning of the proposed dual-droop lines is indispensable, making the determined reference value $\omega^*_{IC,pu,t_{k+1}}$ or $V^*_{IC,dc,pu,t_{k+1}}$ equal to $\xi_{s,t_{k+1}}$. This can be realized by updating the coefficients of the dual-droop lines using the following equation (Fig. 4.6)

$$\begin{aligned} \tilde{\sigma}_{IC,t_{k+1}} &= \Lambda_\sigma \tilde{\sigma}_{IC,t_k}, P_{IC,t_{k+1}} > 0 \\ \tilde{k}_{IC,t_{k+1}} &= \Lambda_k \tilde{k}_{IC,t_k}, P_{IC,t_{k+1}} < 0 \end{aligned} \tag{4.17}$$

where Λ_σ and Λ_k are given in Appendix B.

The updating process of the active power command and coefficients is also illustrated in Fig. 4.5.

It can be seen that using the proposed dynamical dual-droop control strategy (4.10), along with the active power command management and distribution method (4.16) and the coefficients updating scheme (4.17), proportional power sharing between the ac and dc microgrids can be realized by the interlinking converter as

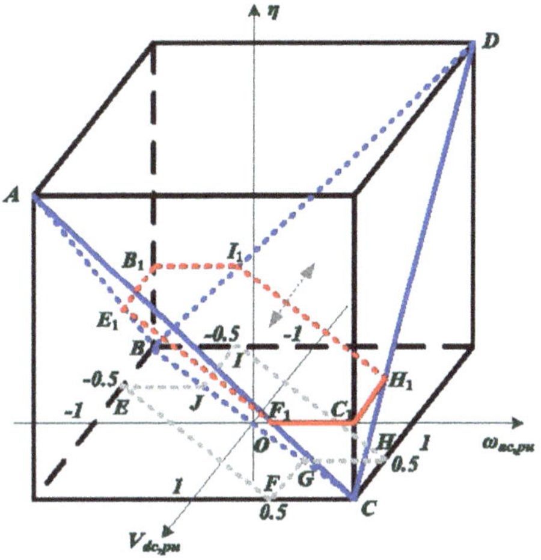

Fig. 4.6 The schematic diagram of design of the threshold

well as participating in voltage and frequency regulation. The interlinking converter will update the active power command $P_{IC,t_{k+1}}$ and coefficients $\tilde{\sigma}_{IC,t_{k+1}}$ and $\tilde{k}_{IC,t_{k+1}}$ if it "finds" the frequency $\omega_{ac,pu,t_{k+1}}$ of the ac microgrid and/or dc voltage $V_{dc,pu,t_{k+1}}$ of the dc microgrid deviate from the present consensus value ξ_{s,t_k} mainly due to load changing and source changing, reaching at a new consensus state $\xi_{s,t_{k+1}}$. And if the frequency $\omega_{ac,pu,t_{k+1}}$ and dc voltage $V_{dc,pu,t_{k+1}}$ don't change, the active power command $P_{IC,t_{k+1}}$ and coefficients $\tilde{\sigma}_{IC,t_{k+1}}$, $\tilde{k}_{IC,t_{k+1}}$ will be maintained at P_{IC,t_k}, $\tilde{\sigma}_{IC,t_k}$ and $\tilde{k}_{IC,t_k}$, respectively.

Therefore, we could find that the dual-droop control proposed in our paper mainly includes two stages. The first stage is to guarantee the proportional power sharing which can be realized by calculating P_{IC} using Eq. (4.16) according to the present load condition of each microgrid. And the second stage is to realize accurate dc voltage and ac frequency regulation which can be achieved by using Eq. (4.10) and (4.17). We admit that droop characteristics similar to [24, 32] may be obtained if we eliminate P_{IC} from the equation, but it really has some differences. These could be the reason that we use the determined active power P_{IC} as a medium to calculate the references $\omega^*_{IC,pu}$ and $V^*_{IC,dc,pu}$ of the IC. This concept is shown in the flow chart depicted in Fig. 4.4.

4.4.1.3 Overall Control Diagram

The overall control block diagram showing the realization of the proposed dual-droop control for the interlinking converter can be found in Fig. 4.7, in which the secondary controller is not included. Firstly, the interlinking converter control is realized by measuring the local ac frequency and dc terminal voltage. These measured variables upon normalized by using the method proposed in [22] are then used to determine the active power command $P_{IC,t_{k+1}}$ (Eq. 4.16) for the interlinking converter.

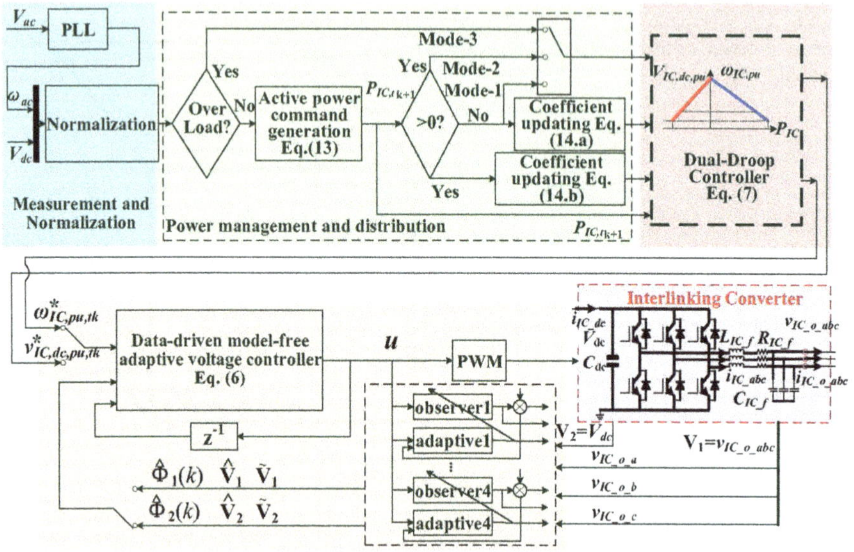

Fig. 4.7 Overall control diagram for the interlinking converter

According to the determined active power command $P_{IC,t_{k+1}}$, the operation mode of the interlinking converter can be selected and then the corresponding coefficient updating signal $\tilde{\sigma}_{IC,t_{k+1}}$ or $\tilde{k}_{IC,t_{k+1}}$ (Eq. (4.17)) can also be generated. Secondly, using the proposed dual-droop controller (4.10), along with the active power commands $P_{IC,t_{k+1}}$ and coefficient updating signal $\tilde{\sigma}_{IC,t_{k+1}}$ or $\tilde{k}_{IC,t_{k+1}}$, the references $\omega^*_{IC,pu}$ or $V^*_{IC,dc,pu}$ will be determined. Finally, these references will be given into the inner data-driven model-free adaptive voltage controller.

According to the control diagram, a PLL is used to detect ac frequency for the dual-droop controller. For the inner loop controller, a data driven model-free adaptive controller is designed, which, unlike the conventional PI controller, does not require the ac voltage and current obtained from the coordinate transformation by using the frequency from the PLL. The dynamic I/O data that the DDMFAC requires could be extracted directly from huge amounts of the recorded process data or observer. Maybe, this has nothing to do with PLL directly. The system and controller would not be evidently affected by the PLL that just provides the frequency for the dual-droop controller.

4.4.2 Event-Triggered Distributed Power Sharing Control for Interconnected Microgrids

4.4.2.1 Distributed Active Power Sharing Control Strategy

In this subsection, we address the question of "how" to achieve proportional active power sharing among the interconnected microgrids, which means how to achieve the following objective

$$k_1 P_1 = \cdots = k_k P_k = \cdots = k_p P_p = \cdots = k_N P_N,$$
$$P_k = P_{L_k} + \sum\nolimits_{i \in C_k} P_{IC,i}, k = 1, \cdots, N, i = 1, \cdots, M. \tag{4.18}$$

As shown in (4.18), to achieve this proportional power sharing among microgrids, we can control the converter IC i to properly regulate the bidirectional active power flow $P_{IC,i}$ dynamically between the interconnected microgrid k and p according to different load level P_{L_k} in each microgrid at different time slots. Therefore, the key issue is that how each IC determines the appropriate amount of active power that should be transferred according to the generation and consumption scenarios.

Unlike previous works reported in [29–32], a distributed control scheme is developed to regulate the active power flow $P_{IC,i}$ by utilizing the multi-agent consensus theory. Therefore, we first construct the first-order linear integral multi-agent dynamics by using the active power sharing deviations among interconnected microgrids.

Let $e_i(t) = k_k P_k - k_p P_p$, where "$k$" and "$p$" represent microgrid "$k$" and "$p$" interconnected through IC "i", then the multi-agent system can be constructed as

$$\dot{e}_i(t) = p_{IC,i}(t), i = 1, \ldots, M, \tag{4.19}$$

where $p_{IC,i}(t)$ is the auxiliary control inputs for the active power, $e_i(t)$ is the active power sharing deviation between microgrid k and p and can be considered as the state variable from the perspective of control theory. Thus, the objective is to design the control input $p_{IC,i}(t)$ to achieve the consensus of $e_i(t)$, i.e., $e_i(t) = e_j(t) = 0, i \neq j$, as $t \to \infty$, achieving the objective of proportional active power sharing among microgrids (4.18). This problem is now transformed into a consensus problem of the first-order linear integral multi-agent system. To achieve the consensus of the system (4.19), the active powers of microgrids are utilized to construct the auxiliary control inputs

$$p_{IC,i}(t) = -c e_{IC,i}(t) = -c \sum\nolimits_{j \in \mathcal{N}_i} a_{ij} \big(e_i(t) - e_j(t)\big), \tag{4.20}$$

where c is the coupling again and $e_{IC,i}(t)$ is the sum of the errors of the per-unit values of dc voltages and frequencies from the local IC and its neighboring ICs. Further, $e_{IC,i}(t)$ also represents the sum of the deviations of the active power sharing among microgrids interconnected by the local and neighboring ICs. According to the literature [38], all the agents with the protocol in (4.20) are asymptotically stable. This means $ei(t) = e j(t) = 0$ as $t \to \infty$, where $i\ l = j$, resulting in proportional active power sharing $k1P1 = k2P2 = \cdots = kN\ PN$ among the microgrids.

Thus, the active power $P_{IC,i}$ transferred by IC "i" can be obtained by using a PI controller $G_{IC,i}(s)$

$$P_{IC,i} = K_{IC,i}^{P} p_{IC,i}(t) + K_{IC,i}^{I} \int_{0}^{t} p_{IC,i}(\tau) d\tau,$$

where $K_{IC,i}^{P}$ and $K_{IC,i}^{I}$ are the proportional and integral gains.

4.4.2.2 Event-Based Distributed Consensus Control Strategy

It can be found that continuous communication between neighboring agents is used for the protocol (4.20), which may result in heavy communication burden. It is undesirable and unnecessary to update the control actions for all the agents at the same time. Recently, the event-based control method has been employed in control systems to reduce the communication and computation, which can achieve satisfactory performance while significantly save the usage of bandwidth and signal processing resources. Motivated by this observation, the event-based control method is further introduced to design the consensus protocol (4.20) to reduce communication in the interconnected systems. To reduce the clutter in the notation, define $\hat{e}_i(t) = e_i(t_k^i)$ for $t_k^i \le t < t_{k+1}^i$, where t_k^i is the kth event instant for agent i, and $e_i(t_k^i)$ is the difference of the active power sharing between two microgrids connected by IC "i" at the event instant t_k^i. With this defined notation and following the proposed distributed controller (4.20), an event-based consensus protocol is proposed for $t \in [t_k^i, t_{k+1}^i)$

$$p_{IC,i}(t) = -c \sum\nolimits_{j \in \mathcal{N}_i} a_{ij} \left(e_i(t_{ki(t)}^i) - e_j(t_{kj(t)}^j) \right), \tag{4.21}$$

where $ki(t) = \arg\max_{k} \{t_k^i | t_k^i \le t\}$ represents the subscript of the latest event time instant to the current time t. To apply this event-based consensus protocol, an event detector is configured for each agent to determine when the sampled local information should be used to update the control actions of itself and its neighbors. The proposed event condition has the following form

$$\left\| e_{e_i}(t) \right\|_2^2 \le \sigma_i \left\| \hat{z}_i(t) \right\|_2^2, t \in [t_k^i, t_{k+1}^i), \tag{4.22}$$

where σ_i is a positive scalar to be determined, $e_{e_i}(t)$ is the measurement error of active power sharing difference at the latest event time instant t_k^i and the current time t, and $\hat{z}_i(t)$ is the sum of the deviation of active power sharing difference between agent IC "i" and its neighboring ICs at the latest event time instant. They are respectively defined as

$$e_{e_i}(t) = e_i\left(t^i_{ki(t)}\right) - e_i(t)$$
$$\hat{z}_i(t) = \sum\nolimits_{j\in N_i} a_{ij}\left(e_i\left(t^i_{ki(t)}\right) - e_j\left(t^j_{kj(t)}\right)\right).$$

In fact, the definition of $e_{e_i}(t)$ along with the definition of $e_i(t)$ reflects the active power sharing deviation between interconnected microgrids indirectly.

Combining the definition of $e_{e_i}(t)$, the dynamics of the closed-loop system for agent i can be written as the following compact form for any $t \in [t^i_k, t^i_{k+1})$

$$\dot{e}(t) = -c\mathcal{L}\hat{e}(t) = -c\mathcal{L}(e(t) + e_e(t)), \tag{4.23}$$

where $e(t) = [e_1(t), \cdots, e_M(t)]^T$, $e_e(t) = [e_{e_1}(t), \cdots, e_{e_M}(t)]^T$, $\hat{e}(t) = [\hat{e}_1(t), \cdots, \hat{e}_M(t)]^T$, and $\mathcal{L}$ is the Laplacian matrix of the communication topology. Given a connected graph $\mathcal{G}$, we consider the following Lyapunov functional candidate:

$$V(e(t)) = \frac{1}{2}e^T(t)e(t). \tag{4.24}$$

Then the time evolution of the function $V(e(t))$ along the trajectory generated by (4.23) for any $t \in [t^i_k, t^i_{k+1})$ is given by

$$\begin{aligned}\dot{V}(t) &= -ce^T(t)\mathcal{L}\hat{e}(t)\\ &= -c(\hat{e}^T(t) - e^T_e(t))\mathcal{L}\hat{e}(t)\\ &= -c\hat{e}^T(t)\mathcal{L}\hat{e}(t) + ce^T_e(t)\mathcal{L}\hat{e}(t).\end{aligned}$$

Using the inequality

$$e^T_e(t)\mathcal{L}\hat{e}(t) \le \frac{1}{2}e^T_e(t)\mathcal{L}e_e(t) + \frac{\hat{e}^T(t)\mathcal{L}\hat{e}(t)}{2}, \tag{4.25}$$

$\dot{V}(t)$ can be bounded as

$$\begin{aligned}\dot{V}(t) &\le -c\hat{e}^T(t)\mathcal{L}\hat{e}(t) + \frac{c}{2}(\hat{e}^T(t)\mathcal{L}\hat{e}(t) + e^T_e(t)\mathcal{L}e_e(t))\\ &= -\frac{c}{2}\hat{e}^T(t)\mathcal{L}\hat{e}(t) + \frac{c}{2}e^T_e(t)\mathcal{L}e_e(t).\end{aligned} \tag{4.26}$$

Combining the event condition (4.22), we can get

$$\dot{V}(t) \le -\frac{c}{2}(1 - \lambda^2_n\sigma_{\max})\hat{e}^T(t)\mathcal{L}\hat{e}(t), \tag{4.27}$$

where $\sigma_{\max} = max\{\sigma_i, i = 1, \cdots, M\}$ and λ_n is the maximum eigenvalue of Laplacian matrix $\mathcal{L}$. Thereby $\dot{V}(t) \le 0$ for any $t \in [t^i_k, t^i_{k+1})$ if $c > 0$ and $0 < \sigma_{\max} < \frac{1}{\lambda^2_n}$. From LaSalle's invariance principle, $\dot{V}(t) \le 0$ for $\forall t \ge 0$ implies consensus for all agents, i.e.,$t \to \infty$, $e_i(t) = k_k P_k - k_p P_p = \frac{1}{M}\sum_{i=1}^{M} e_i(0) = 0$. That is to say, $k_1 P_1 = k_2 P_2 = \cdots = k_N P_N$. Therefore, according to the principle of droop control, active power will be properly shared among microgrids no matter where the loads are located and no matter how the microgrids are interconnected by the interlinking converters. Thus, we propose the distributed event condition for each agent (IC) "i", $i = 1, \ldots, M$:

$$\left\| e_{e_i}(t) \right\|_2^2 > \sigma_i \left\| \sum_{j \in \mathcal{N}_i} a_{ij} \left(e_i\left(t_{ki(t)}^i\right) - e_j\left(t_{kj(t)}^j\right) \right) \right\|_2^2 \tag{4.28}$$

where $t \in \left[t_k^i, t_{k+1}^i\right)$, and $0 < \sigma_i \le \sigma_{\max} < \frac{1}{\lambda_n^2}$. Therefore, the active power sharing results of the interconnected microgrids under the event-based distributed controller can be described in Theorem 1.

Theorem 1 *Consider the first-order agent control system for active power sharing among the multiple interconnected AC and DC microgrids in (4.19). Assume that the communication graph G is undirected and connected. Then, the proportional active power sharing problem is solved asymptotically by the event-based distributed consensus controller (4.21), where the event time instants are determined by distributed event condition (4.28) with* $c > 0$ *and* $0 < \sigma_{\max} < \frac{1}{\lambda_n^2}$.

It is worthy to remark here that in this distributed control scheme, the IC is assigned to a particular agent which is different from most distributed schemes found in the literature. We use the ICs in the interconnected hybrid ac and dc microgrids to autonomously regulate the active power flow between microgrids but not to set the set points of microgrids like that in [29]. This method is more general and suitable for the interconnected system no matter how the microgrids are interconnected by ICs, which is also different from the most of reported works where multiple dc microgrids are connected to ac microgrid via ICs in parallel. The reason is that the dynamics of agents and the consensus protocol are designed by using the difference of active power sharing between microgrids interconnected by IC directly.

In the proposed approach, both the controller (4.21) and the event detector (4.22) are fully distributed. Each agent does not need to have a global model of the interconnected microgrid systems, and does not have any knowledge of the dynamics of its neighbors, but can communicate freely among them to reach an agreement. Only the information from its local and neighboring ICs is required for IC "i" to decide the updating and communicating instants. The implementation of the event-based control provides an advantage of communication reduction among ICs since the updating of control actions of the ICs and the communication between ICs are occurred only at some specific event instants. Specifically, an event of IC i occurs if (4.28) is satisfied. Then, IC i will take the new state sampling of $e_i(t) = k_k P_k(t) - k_p P_p(t)$ to update its controller (4.21) by replacing $e_i(t_{ki(t)}^i) = k_k P_k(t_{ki(t)}^i) - k_p P_p(t_{ki(t)}^i)$. At the same time, IC i will also send the new sampled state to its neighbors through communication. On the other hand, IC i will receive the sampled active power sharing deviation state from its neighboring ICs if any of them triggers an event. Thus the controller of IC

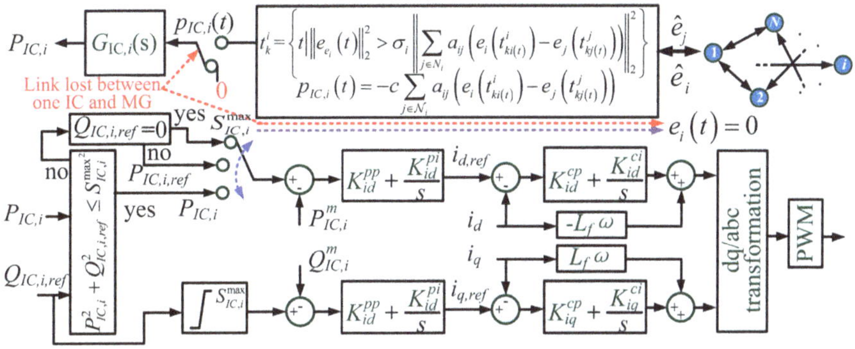

Fig. 4.8 The overall control block diagram of the proposed distributed control scheme

i will be updated to a new level. Following this idea, the communication will be therefore reduced significantly.

In addition, this distributed fashion has plug-and-play feature, better flexibility and reliability, which is important for microgrids operation. It can be immune to topological variations caused by communication links or physical failures. Thus, it has a greater tolerance to failures. Plug-and-play is also a potential benefit that if a new microgrid system is added into (or disconnected from) the system, only those agents affected by this new microgrid would have to be updated. Additionally, the system can operate flexibly. If ac microgrids need reactive power support, ICs can inject reactive power into ac microgrids by selecting proper operation mode.

The overall control block diagram showing the realization of the proposed control scheme can be found in Fig. 4.8 in detail. At each sampling instant, through the Cyber layer, each agent (interlinking converter (IC)) i broadcasts its state information to the neighbors and also receives state information from its neighbor ICs ($e\hat{}i(t)$, $e\hat{}\,j\,(t)$) for event detection. If the condition in (4.22) is satisfied, no further action is required for this IC; otherwise, this agent i will update its control action $p_{IC,i}(t)$ and notify its neighbors to update their control actions $p_{IC,j}(t), j \in \mathcal{N}_i$ by using its current state information. The control actions $p_{IC,i}(t)$ are then fed to the PI controllers $G_{IC,i}(s)$ to generate the corresponding active power command $p_{IC,i}(t)$ to adaptively regulate the bidirectional power flow between microgrids. It can be seen that the controller at each IC is totally distributed and that each controller only use the information of its local and neighboring units, which can be more flexible and reliable.

4.4.2.3 Reactive Power Support for AC Microgrids

Although the active power transfer is the prime task of the ICs, they in the interconnected microgrids are not operating at full rating all the time [52]. Therefore, several ancillary services can be developed by using the available IC rating in a smart way. As an example, reactive power support for ac microgrids is considered in this subsection. In practical application, the IC rating should be considered and is defined by a maximum apparent power rating $S_{IC,i}^{max}$ for IC "i". Thus the maximum reactive power is limited by $Q_{IC,i}^{max} = \sqrt{S_{IC,i}^{max2} - P_{IC,i}^{2}} \leq S_{IC,i}^{max}$. The command of reactive power support, $Q_{IC,i,ref}$, may come from the ac microgrids. In this paper, we consider that if $P_{IC,i}^{2} + Q_{IC,i,ref}^{2} \leq S_{IC,i}^{max2}$, $P_{IC,i}$ and $Q_{IC,i,ref}$ will be transferred by IC "i", respectively, where $P_{IC,i}$ comes from the proposed distributed controller. In this sense, proportional active power sharing among all the interconnected microgrids will not be affected and reactive power support for ac microgrids just can be seen as the load increase of dc microgrids. Otherwise, the communication link between IC "i" and its neighboring ICs will be disconnected from the communication network, and the reactive power, $Q_{IC,i,ref} \leq S_{IC,i}^{max}$, will be transferred through IC "i" and a small amount of active power $P_{IC,i,ref}$ as well. By using this method, the other ICs can "know" that IC "i" has some other things to do and cannot work in cooperation with them for the time being. This method, similar to the plug-and-play feature, is flexible and can benefit the stable operation of the interconnected microgrids. This is shown in Fig. 4.8.

4.5 Simulation and Results

To validate the performance of the proposed control scheme for the interlinked ac/dc microgrids, the interlinked systems, depicted in Figs. 4.1 and 4.9 have been simulated with the proposed data-driven based and event-triggered based methods, respectively, in MATLAB/Simulink environment. The ac and dc microgrids have its own DGs and loads, and are emulated with a dc-ac inverter and a boost converter, respectively. A six-switch dc-ac converter with *LC* filters serving as the interlinking converter is adopted to interface the ac and dc microgrids. To verify the feasibility of the proposed controller, different operating conditions have been considered. The ring-shape communication topology is also shown in Fig. 4.9. This topology is the sparsest network where the failure of a single link does not compromise the graphical connectivity. Therefore, it is chosen for data exchange in the cyber layer. Some results are presented and discussed in detail in the following.

4.5.1 Simulation for Two Interconnected AC/DC Microgrids

4.5.1.1 Case 1

The over load condition of both the ac and dc microgrids is considered in this case. The initial conditions are set to 8 kW for the ac microgrid and 6 kW for the dc microgrid, respectively. That means the ac microgrid is initially operating in over load condition while the dc microgrid is operating in normal load condition. In the steady state, the

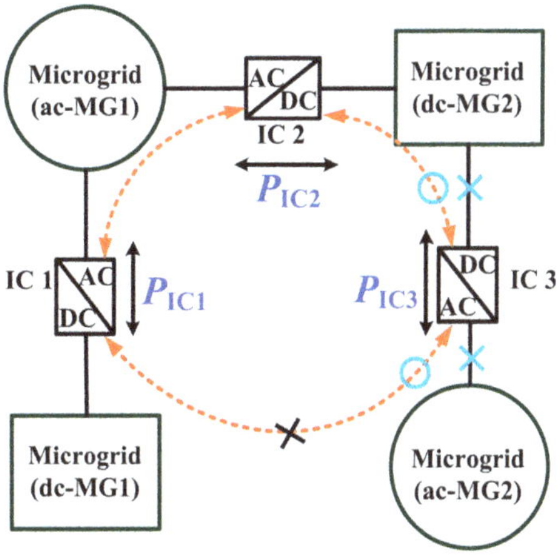

Fig. 4.9 The simulation test system with a ring-shape communication topology

interlinking converter transfers 1 kW from the dc microgrid to the ac microgrid, which is discussed in case 2 in detail. At t = 3 s, the ac and dc microgrids are changed to 9.5 kW and 9 kW, respectively, which makes both the ac and dc microgrids over loaded. This can also be demonstrated by the measured normalized values (−0.9 p. u and −0.8 p. u). According to (4.16), the active power to be transferred by the interlinking converter is updated to 0 kW, which means the interlinking converter transfers no power and each microgrid is responsible for the power sharing in this load condition. Figure 4.10 shows the power and the normalized values of the ac side frequency and the dc side voltage. It can be seen that the interlinking converter can reasonably manage the power sharing and has a good performance. It is necessary to mention that a loading shedding system may be activated in this condition in practice to guarantee the system stability.

4.5.1.2 Case 2

The secondary control scheme is added in this case. Figure 4.11 shows the simulation results, from which it can be seen that the per-unit values of the ac frequency and dc voltage are restored to zero corresponding to their nominal values with help of the secondary controllers. Specifically, at t = 2.1 s, synchronization is achieved between the ac and dc microgrids, and the per-unit values of dc voltage and ac frequency meet the condition of $\left|V_{dc,pu,t_k} - \omega_{ac,pu,t_k}\right| < \varepsilon = 0.1$. Therefore, the IC sends signals to the ac and dc microgrids to change the reference values of the secondary controllers. Then, the secondary controllers of the DGs in microgrids are started to restore the dc voltage and ac frequency to their nominal values. At t = 3 s, the ac microgrid is changed to 6.5 kW and the dc microgrid remains 6 kW. In this load condition, no active power is transferred by the interlinking converter since the deviation is less than the threshold ($\eta_e < \eta$), which means each microgrid is responsible for the power sharing. Therefore, the secondary controller of the interlinking converter is stopped

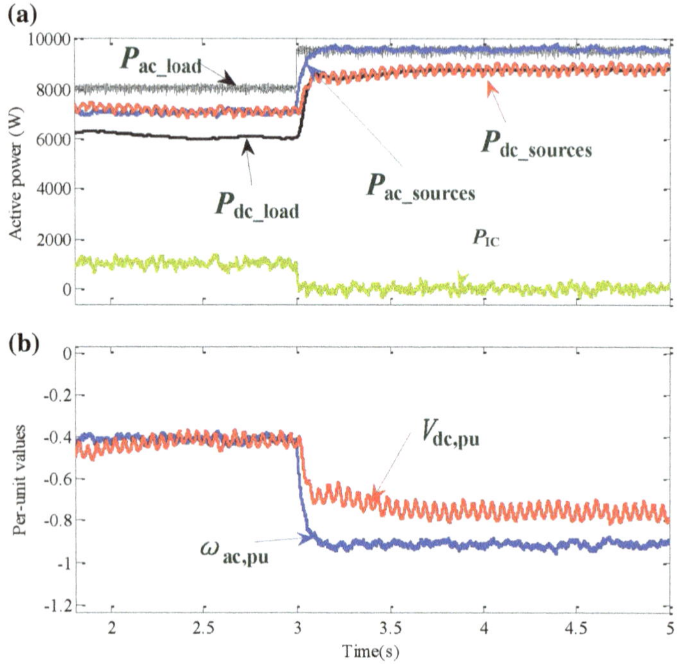

Fig. 4.10 Power responses and per-unit values. **a** Active power (W). **b** per-unit values of the dc side voltage and ac side frequency

to work and the secondary controllers in each ac and dc microgrid remain working. The dc voltages and the ac frequency in dc and ac microgrids can be kept at their nominal values, respectively.

4.5.2 Simulation for Networked AC/DC Microgrids

4.5.2.1 Case 1: Performance Evaluation with Load Change

In this case, different load conditions and the change of the load in different microgrids are considered to verify the performance of the proposed method. The simulation results are shown in Fig. 4.12 and the simulation process is as follows. Initially, the microgrids (dc-MG1, dc-MG2, ac-MG1 and ac-MG2) are experiencing a load demand of 10.5 kW, 10 kW, 30.5 kW and 25 kW, respectively, before the time of t = 6 s. Obviously, the active power is not properly shared among these four microgrids if any strategy is not adopted since $k_{dc1}P_{dc1} \neq k_{dc2}P_{dc2} \neq k_{ac1}P_{ac1} \neq k_{ac2}P_{ac2}$. In this paper, the proposed event-based distributed consensus control method is applied to the ICs in the interconnected microgrids. Thus, with the help of the consensus pro-

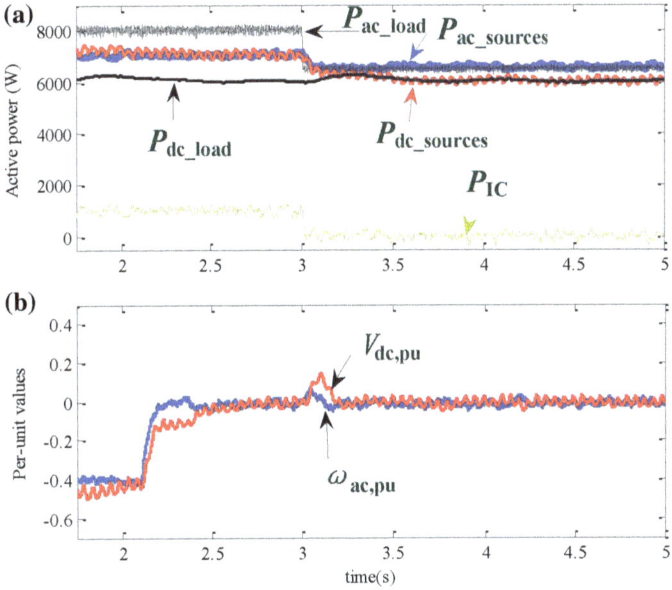

Fig. 4.11 Power responses and per-unit values. **a** Active power (W). **b** per-unit values of the dc side voltage and ac side frequency

tocol, IC1 transfers approximately 8.5 kW from dc-MG1 to ac-MG1, IC2 transfers about 3 kW from dc-MG2 to ac-MG1 and IC3 transfers about 6 k W from dc-MG2 to ac-MG2 (shown in Fig. 4.12a). Upon reaching steady-state, all the microgrid source generations are noted to be the same at about 19 kW each ($P_{dc1} = P_{dc2} = P_{ac1} = P_{ac2} = 19$ kW) shown in Fig. 4.12b. And at the same time, Eq. (4.18) is satisfied seen from Fig. 4.12c, i.e., $k_{dc1}P_{dc1} = k_{dc2}P_{dc2} = k_{ac1}P_{ac1} = k_{ac2}P_{ac2} = 0.95$, resulting in proportional power sharing of the total load among the interconnected ac and dc microgrids. At the time of t=6 s, the loads in dc-MG2 and ac-MG1 are changed to 12.5 kW–30 kW, respectively, but the loads in dc-MG1 and ac-MG2 are not changed. Power sharing mismatches caused by the load changes are sensed by the event-based consensus protocol (4.21). Up-on sensing these mismatches, each IC will immediately regulate the amount of the active power to be transferred adaptively by using the proposed strategy. As shown in Fig. 4.12a, due to the increasing of the load in dc-MG2, IC2 and IC3 reduced the active power transferred from dc-MG2 to ac-MG1 and ac-MG2, respectively. Meanwhile, IC1 increased the amount of active power transferred to ac-MG1. This means that dc-MG1 has ability to share more load active power for other micro-grids than dc-MG2. Under this operating condition, IC1, IC2 and IC3 transferred the active power of about 9 kW, 1.5 kW and 5.5 kW, respectively (shown in Fig. 4.12a). It can be seen that all the loads are approximately proportionally shared among the microgrids ($k_{dc1}P_{dc1} = k_{dc2}P_{dc2} = k_{ac1}P_{ac1} = k_{ac2}P_{ac2} = 0.975$) (see Fig. 4.12c, d). The source generations of all the microgrids are the same at about 19.5 kW (see Fig. 4.12a). The active power sharing mismatches have been eliminated (shown in Fig. 4.12d). Following, from $t = 11.5$ s, the load in dc-MG1 increased 3 kW arriving at 13.5 kW. The ICs need to regulate the active power they should transfer again according to the proposed strategy to achieve proportional active power sharing among the interconnected microgrids. Due to the load relationship among the microgrids ($P_{dc-MG2-L} < P_{dc-MG1-L} < P_{ac-MG2-L} < P_{ac-MG1-L}$), both IC1 and IC3 will reduce their transferred active power from dc-MG1 and dc-MG2 to ac-MG1 and ac-MG2, respectively, while IC2 will increase the active power transferred from dc-MG2 to ac-MG1. As shown in Fig. 4.12a, IC1, IC2 and IC3 transferred about 7 kW, 3 kW and 5 kW, respectively. Thus, all the active power loads in the system are proportionally shared among the microgrids since $k_{dc1}P_{dc1} = k_{dc2}P_{dc2} = k_{ac1}P_{ac1} = k_{ac2}P_{ac2} = 1$ (shown in Fig. 4.12c) and the total source generation of each microgrid is the same at about 20 kW (shown in Fig. 4.12b).The active power sharing mismatches have also been eliminated. Similar to the previous scenarios, we reduced the loads in both ac-MG1 and ac-MG2 at $t = 15.5$ s and $t = 19.5$ s, respectively. Finally, loads in ac-MG1 and ac-MG2 were reduced to 10 and 8 kW after $t = 19.5$ s, respectively. Because the loads of both ac-MG1 and ac-MG2 are less than that of both dc-MG1 and dc-MG2, ac-MG1 and ac-MG2 should share some active power for dc-MG1 and dc-MG2. The ICs reregulated the amount of their transferred active power, and in the final stage, they (IC1, IC2 and IC3) transferred about −2.5 kW, 1.5 kW, and −3 kW, respectively, resulting in proportional active power sharing among the interconnected microgrids. The total generation of each microgrid is noted to be the same at about 11 kW (shown in Fig. 4.12b). Note that the negative active power transferred by ICs means that the active power is transferred from ac microgrid to dc microgrid. Actually, we can find that ac-MG1 shared about 1 kW for dc-MG1 and ac-MG2 shared about 1.5 kW for dc-MG2 and 1.5 kW for dc-MG1 at the final stage. During the time of $t = 15.5$–19.5 s, IC1 transferred a very small amount of active power. In the practical application, ICs can stop running under this condition. Figure 4.12d shows the responses of power sharing mismatches (e_i). It can be seen that the mismatches were eliminated no matter what the load conditions were. Figure 4.13 is the zoom-in simulation results of Fig. 4.12b–d, which illustrates dynamic performance of the active power generation of each microgrid and the power sharing deviations when the load changed at $t = 11.5$ s. The simulation results demonstrate that the active power can be properly shared among the interconnected microgrids by using the proposed strategy. This could provide more reliable and flexible power supply.

It should be noted that as shown in Fig. 4.12e, while we have achieved proportional active power sharing among the interconnected microgrids, the per-unit values of the dc voltages and the ac frequencies didn't fully arrive at the same values at the steady state under different load conditions by using the proposed control strategy. This is mainly caused by the line impedances in the $V_{dc} - P$ droop controlled dc microgrids just like that of ac microgrids. Additionally, the total active power generation of each microgrid was used to design the controller in this paper, achieving proper active power sharing among the microgrids. This has a potential benefit that it is not affected by the secondary controllers in each microgrids if implemented.

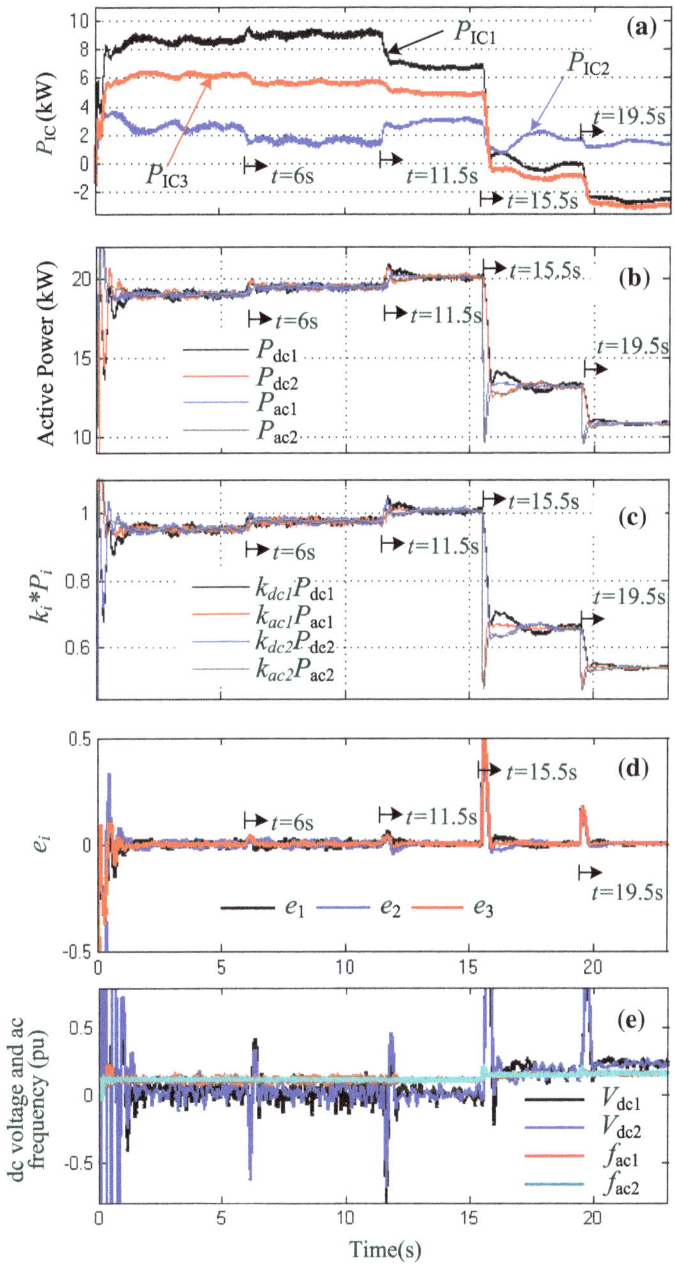

Fig. 4.12 Simulation results. **a** Active power transferred by ICs. **b** Active power generation of each microgrid. **c** Responses of k_iP_i. **d** Power sharing mismatches among microgrids. **e** Per-unit values of dc voltages and ac frequencies

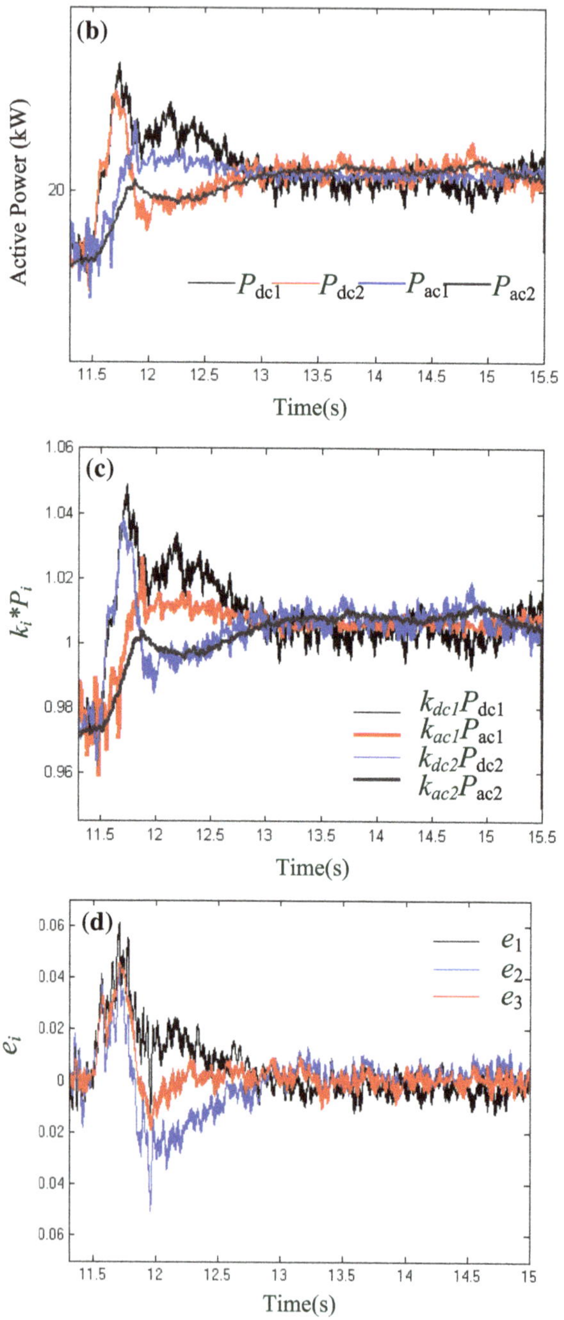

Fig. 4.13 Zoom-in simulation results of Fig. 4.4b–d when the load changed at $t = 11.5$ s

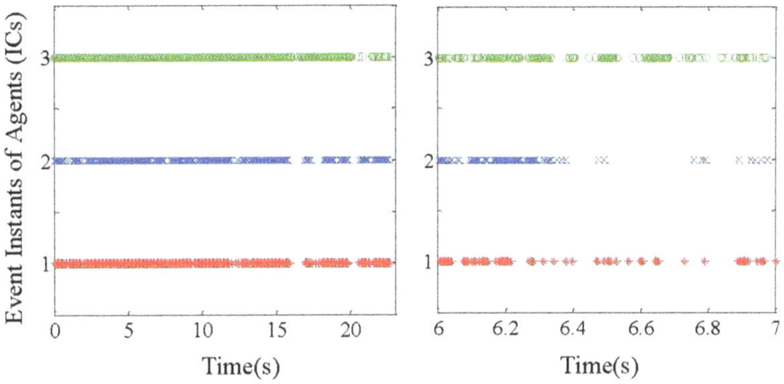

Fig. 4.14 The event time instants of ICs

To demonstrate that the event-based control can reduce the communication amount between ICs, we also present the event time instants for ICs given in Fig. 4.14. It can be seen that the number of IC control updates is greatly reduced to reach consensus compared with continuous communication scheme that requires communication at every step.

4.5.2.2 Case 2: Plug-and-Play

Plug-and-play is a key feature of the interconnected microgrid system with the proposed distributed controller. And in reality, islanding of one of microgrids may occur due to some different reasons, resulting in leaving-off of ICs connected to it. Therefore, stable and effective operation of the system should be guaranteed under this scenario. In this simulation case, we consider the scenario that islanding occurs in ac-MG2 which results in leaving-off of the interlinking converter, IC3, i.e., IC3 will stop running. It should be noted that leaving-off of IC3 means all communication edges connected to IC3 are removed. Therefore, the communication edges between IC1 and IC3, and IC2 and IC3 are automatically given up, but the link between IC1 and IC2 still maintains a connected graph with balanced Laplacian matrix. In this sense, proportional power sharing is only achieved among the microgrids except ac-MG2. Figure 4.15 shows the plug-and-play capability of the proposed method and its performance in case of leaving-off of IC3 due to the islanding of ac-MG2. As seen, after IC3 is disconnected from the interconnected microgrids at $t = 3$ s because of the islanding of ac-MG2, dc-MG1, dc-MG2 and ac-MG1 can still proportionally share the active power of the loads located in the microgrids except ac-MG2. This is illustrated by Fig. 4.15c, d. Figure 4.15a shows that IC1 and IC2 reassign the transferred active power $P_{IC,1}$ and $P_{IC,2}$ between microgrids to realize this proportional power sharing objective. In the islanding mode, the active power of loads in ac-MG2 is shared by DGs in ac-MG2 itself without the support from IC3. This simulation results show that the proposed control method has good plug-and-play ability and can effectively operate due to islanding of one of microgrids.

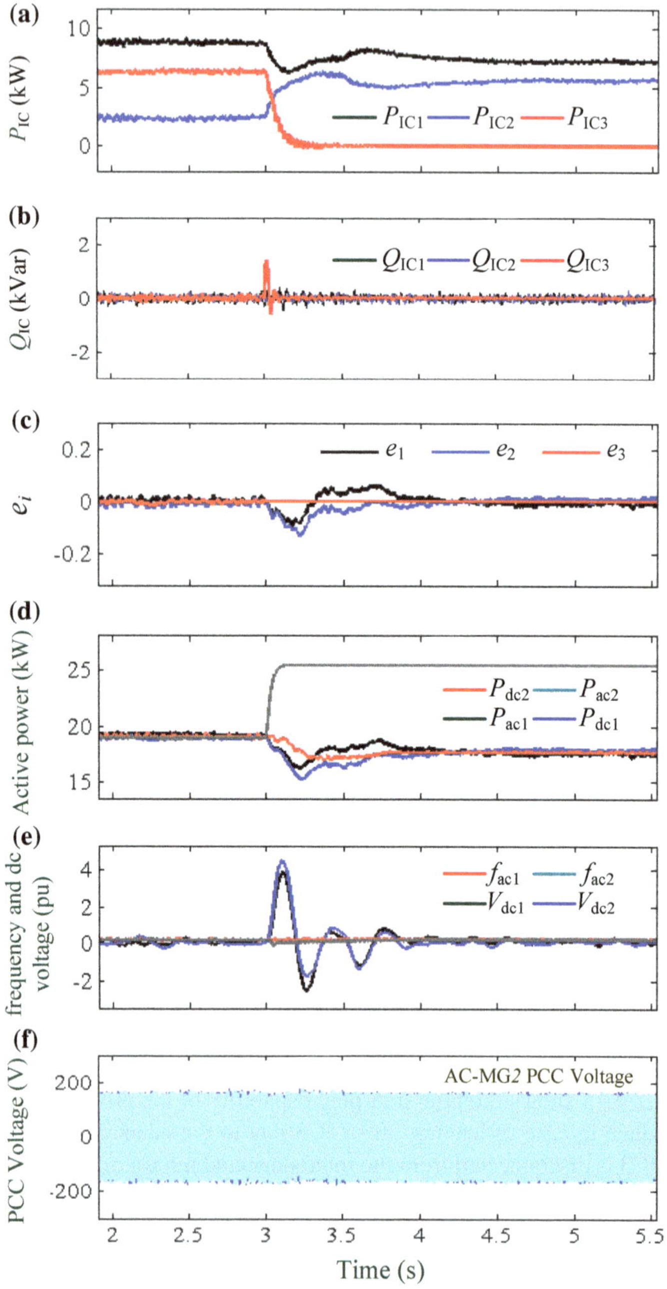

Fig. 4.15 Simulation results with consideration of plug and play and islanding. **a** Active power transferred by ICs. **b** Reactive power transferred by ICs. **c** Power sharing mismatches among microgrids. **d** Active power generation of each microgrid. **e** Per-unit values of dc voltages and ac frequencies. **f** PCC voltages of ac-MG2

4.5.2.3 Case 3: Different Ratings of Microgrids

In the previous five cases, we assume that the four interconnected microgrids have the same power capacity. In this case, different power ratings of microgrids are carried out since this is a more general scenario in a real world. The ratios of power ratings of microgrids are considered as dc-MG1: ac-MG1: dc-MG2: ac-MG2 = 4:1:3:2. Besides the proportional active power sharing among the interconnected microgrids, reactive power support for the ac microgrids (ac-MG1 and ac-MG2) is also considered in this simulation case. Figure 4.16 shows the results. As shown in Fig. 4.16a, d, we can see that IC1, IC2 and IC3 transferred the active power of about 18.6 kW, 2.75 kW and 7.7 kW from dc microgrids to ac microgrids, respectively, and that at the same time, IC1 and IC3 transferred reactive power of 3 kVar and 3.5 kVar to ac-MG1 and ac-MG2, respectively. Also, it can be found that active power of loads can be proportionally shared among the four microgrids (see Fig. 4.16b, e), i.e., $P_{dc,1}$:$P_{ac,1}$:$P_{dc,2}$:$P_{ac,2}$ = 4:1:3:2 ($P_{dc,1}$ = 34.6 kW, $P_{ac,1}$ = 8.64 kW, $P_{dc,2}$ = 25.96 kW, $P_{ac,2}$ = 17.3 kW) and $k_{dc,1}P_{dc,1} = k_{ac,1}P_{ac,1} = k_{dc,2}P_{dc,2} = k_{ac,2}P_{ac,2}$ (about 0.865). From the perspective of dc microgrids, reactive power support for ac microgrids just can be seen as the load increase of dc microgrids, which will not affect the proportional active power sharing among microgrids without consideration of IC ratings. The reactive power support can bring better voltage profiles of ac microgrids (see Fig. 4.16f). The results show that the proposed method has a good flexibility and effectiveness that can benefit the operation of the interconnected microgrids.

4.6 Conclusion

This paper investigates on coordinated power sharing issues of the interconnected ac and dc microgrids. To realize proportional power sharing between two ac and dc microgrids, a novel primary controller including a dual-droop controller and a data-driven model-free adaptive voltage controller has been proposed. The design of the controller is only based on input/output (I/O) measurement data but not the model any more. Following, an event-based distributed consensus control method also has been developed to realize proper power sharing among multiple interconnected microgrids. This distributed method allows each IC to adaptively regulate the active power flow between microgrids. Thus the active power load can be properly shared by DGs in microgrids no matter where the loads are located. And the proposed strategy is suitable to the interconnected microgrids no matter how they are connected by ICs. The event-based strategy can effectively reduce the communications between ICs. Also, the power supply flexibility and reliability can be enhanced. Simulation results have been given to verify the proposed power sharing strategy. In future research work, we will focus on the system stability analysis and design, interaction between microgrids and power quality improvement of ac microgrids.

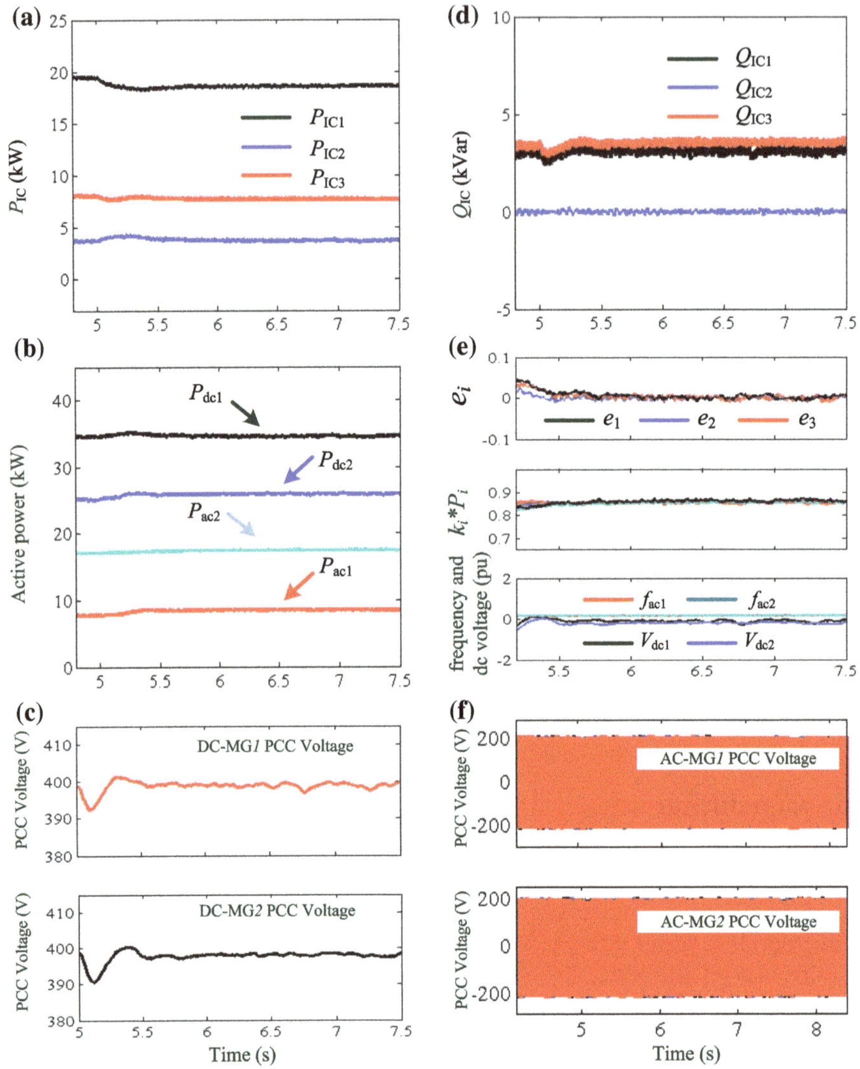

Fig. 4.16 Simulation results with consideration of different ratings of microgrids. **a** Active power transferred by ICs. **b** Active power generation of each microgrid. **c** PCC voltages of dc micro-grids. **d** Reactive power transferred by ICs. **e** Power sharing mismatches among microgrids, responses of $k_i P_i$, and per-unit values of dc voltages and ac frequencies. **f** PCC voltages of ac- MG1 and ac-MG2

References

1. H. Zhang, J. Zhou, Q. Sun, J.M. Guerrero, D. Ma, Data-driven control for interlinked AC/DC microgrids via model-free adaptive control and dual-droop control. IEEE Trans. Smart Grid **8**(2), 557–571 (2017)
2. C. Jin, P. Wang, J. Xiao, Y. Tang, F.H. Choo, Implementation of hierarchical control in dc microgrids. IEEE Trans. Ind. Electron. **61**(8), 4032–4042 (2014)
3. J.M. Guerrero, M. Chandorkar, T.-L. Lee, P.C. Loh, Advanced control architectures for intelligent microgrids-part I: decentralized and hierarchical control. IEEE Trans. Ind. Electron. **60**(4), 1254–1262 (2013)
4. N. Pogaku, M. Prodanovic, T.C. Green, Modeling, analysis and testing of autonomous operation of an inverter-based microgrid. IEEE Trans. Power Electron. **22**(2), 613–625 (2007)
5. R. Majumder, B. Chaudhuri, A. Ghosh, R. Majumder, G. Ledwich, F. Zare, Improvement of stability and load sharing in an autonomous microgrid using supplementary droop control loop. IEEE Trans. Power Syst. **25**, 796–808 (2010)
6. N. Bottrell, M. Prodanovic, T.C. Green, Dynamic stability of a microgrid with an active load. IEEE Trans. Power Electron. **28**(11), 5107–5119 (2013)
7. S.M. Ashabani, Y.A.R.I. Mohamed, A flexible control strategy for grid-connected and islanded microgrids with enhanced stability using nonlinear microgrid stabilizer. IEEE Trans. Smart Grid **3**(3), 1291–1301 (2012)
8. T.L. Lee, P.T. Cheng, Design of a new cooperative harmonic filtering strategy for distributed generation interface converters in an islanding network. IEEE Trans. Power Electron. **22**, 1919–1927 (2007)
9. Q.-C. Zhong, Y. Zeng, Control of inverters via a virtual capacitor to achieve capacitive output impedance. IEEE Trans. Power Electron. **29**(10), 5568–5578 (2014)
10. J. He, Y.W. Li, F. Blaabjerg, Flexible microgrid power quality enhancement using adaptive hybrid voltage and current controller. IEEE Trans. Ind. Electron. **61**(6), 2784–2794 (2014)
11. Q.-C. Zhong, Robust droop controller for accurate proportional load sharing among inverters operated in parallel. IEEE Trans. Ind. Electron. **60**(4), 1281–1290 (2013)
12. J.W. Simpson-Porco, F. Dorfler, F. Bullo, Synchronization and power sharing for droop controlled inverters in islanded microgrids. Automatica **49**(9), 2603–2611 (2013)
13. J. Schiffer, R. Ortega, A. Astolfi, J. Raisch, T. Sezi, Conditions for stability of droop controlled inverter-based microgrids. Automatica **50**(10), 2457–2469 (2014)
14. H. Zhang, C. Qin, B. Jiang, Y. Luo, Online adaptive policy learning algorithm for $H\infty$ state feedback control of unknown affine nonlinear discrete-time systems. IEEE Trans. Cybern. **44**(12), 2706–2718 (2014)
15. H. Zhang, Y. Quan, Modeling, identification, and control of a class of nonlinear systems. IEEE Trans. Fuzzy Syst. **9**(2), 349–354 (2001)
16. A.H. Etemadi, E.J. Davison, R. Iravani, A decentralized robust control strategy for multi-DER microgrids-part I: fundamental concepts. IEEE Trans. Power Del. **27**(4), 1843–1853 (2012)
17. T. Geyer, D.E. Quevedo, Multistep finite control set model predictive control for power electronics. IEEE Trans. Power Electron. **29**(12), 6836–6846 (2014)
18. M. Davari, Y.A.-R.I. Mohamed, Variable-structure-based nonlinear control for the master VSC in dc-energy-pool multiterminal grids. IEEE Trans. Power Electron. **29**(11), 6196–6213 (2014)
19. H. Zhang, C. Qin, Y. Luo, Neural-network-based constrained optimal control scheme for discrete-time switched nonlinear system using dual heuristic programming. IEEE Trans. Autom. Sci. Eng. **11**(3), 839–849 (2014)
20. U. Nutkani, P.C. Loh, F. Blaabjerg, Distributed operation of interlinked AC microgrids with dynamic active and reactive power tuning. IEEE Trans. Ind. Appl. **49**(5), 2188–2195 (2013)
21. P.C. Loh, D. Li, Y.K. Chai, F. Blaabjerg, Hybrid AC-DC microgrids with energy storages and progressive energy flow tuning. IEEE Trans. Power Electron. **28**(4), 1533–1543 (2013)
22. P.C. Loh, D. Li, Y.K. Chai, F. Blaabjerg, Autonomous operation of hybrid microgrid with AC and DC subgrids. IEEE Trans. Power Electron. **28**(5), 2214–2223 (2013)

23. P.C. Loh, D. Li, Y.K. Chai, F. Blaabjerg, Autonomous control of interlinking converter with energy storage in hybrid AC-DC microgrid. IEEE Trans. Ind. Appl. **49**(3), 1374–1382 (2013)
24. N. Eghtedarpour, E. Farjah, Power control and management in a hybrid AC/DC microgrid. IEEE Trans. Smart Grid **5**(3), 1494–1505 (2014)
25. X. Liu, P. Wang, P.C. Loh, A hybrid ac/dc microgrid and its coordination control. IEEE Trans. Smart Grid **2**(2), 278–286 (2011)
26. H. Zhang, S. Kim, Q. Sun, J. Zhou, Distributed adaptive virtual impedance control for accurate reactive power sharing based on consensus control in microgrids. IEEE Trans. Smart Grid (in publication)
27. J.W. Simpson-Porco, Q. Shafiee, F. Dorfler, J.C. Vasquez, J.M. Guerrero, F. Bullo, Secondary frequency and voltage control of islanded microgrids via distributed averaging. IEEE Trans. Ind. Electron. **62**(11), 7025–7038 (2015)
28. A. Bidram, A. Davoudi, F.L. Lewis, J.M. Guerrero, Distributed cooperative secondary control of microgrids using feedback linearization. IEEE Trans. Power Syst. **28**(3), 3462–3470 (2013)
29. A. Bidram, A. Davoudi, F.L. Lewis, S. Sam Ge, Distributed adaptive voltage control of inverter-based microgrids. IEEE Trans. Energy Convers. **29**(4), 862–872 (2014)
30. H. Behjati, A. Davoudi, F. Lewis, Modular dc-dc converters on graphs: cooperative control. IEEE Trans. Power Electron. **29**(12), 6725–6741 (2014)
31. V. Nasirian, A. Davoudi, F.L. Lewis, J.M. Guerrero, Distributed adaptive droop control for dc distribution systems. IEEE Trans. Energy Convers. **29**(4), 944–956 (2014)
32. V. Nasirian, S. Moayedi, A. Davoudi, F.L. Lewis, Distributed cooperative control of dc microgrids. IEEE Trans. Power Electron. **30**(4), 2288–2303 (2015)
33. Q. Shafiee, J.M. Guerrero, J.C. Vasquez, Distributed secondary control for islanded microgrids ł a novel approach. IEEE Trans. Power Electron. **29**(2), 1018–1031 (2014)
34. Q. Shafiee, C. Stefanovic, T. Dragievi, P. Popovski, J.C. Vasquez, J.M. Guerrero, Robust networked control scheme for distributed secondary control of islanded microgrids. IEEE Trans. Power Electron. **61**(10), 5363–5374 (2014)
35. F. Guo, C. Wen, J. Mao, Y.D. Song, Distributed secondary voltage and frequency restoration control of droop-controlled inverter-based microgrids. IEEE Trans. Ind. Electron. **62**(7), 4355–4364 (2015)
36. X. Liu, P. Wang, P.C. Loh, A hybrid ac/dc microgrid and its coordination control. IEEE Trans. Smart Grid **2**(2), 278–286 (2011)
37. P.C. Loh, D. Li, Y.K. Chai, F. Blaabjerg, Autonomous operation of hybrid microgrid with AC and DC subgrids. IEEE Trans. Power Electron. **28**(5), 2214–2223 (2013)
38. P.C. Loh, D. Li, Y.K. Chai, F. Blaabjerg, Autonomous control of interlinking converter with energy storage in hybrid AC-DC microgrid. IEEE Trans. Ind. Appl. **49**(3), 1374–1382 (2013)
39. P.C. Loh, D. Li, Y.K. Chai, F. Blaabjerg, Hybrid AC-DC microgrids with energy storages and progressive energy flow tuning. IEEE Trans. Power Electron. **28**(4), 1533–1543 (2013)
40. N. Eghtedarpour, E. Farjah, Power control and management in a hybrid AC/DC micro-grid. IEEE Trans. Smart Grid **5**(3), 1494–1505 (2014)
41. A.Q. Huang, M.L. Crow, G.T. Heydt, J.P. Zheng, S.J. Dale, The future renewable electric energy delivery and management system the Energy Internet. IEEE Proc. IEEE **99**(1), 133–148 (2011)
42. S. Moayedi, A. Davoudi, Distributed tertiary control of dc microgrid clusters. IEEE Trans. Power Electron. **31**(2), 1717–1733 (2016)
43. R. Majumder, G. Bag, Parallel operation of converter interfaced multiple microgrids. Int J. Electr. Power Energy Syst. **55**, 486–496 (2014)
44. P. Wang, C. Jin, D. Zhu, Y. Tang, P.C. Loh, F.H. Choo, Distributed control for autonomous operation of a three-port AC/DC/DS hybrid microgrid. IEEE Trans. Ind Electron. **62**(2), 1279–1290 (2015)
45. A.A.A. Radwan, Y.A.R.I. Mohamed, Networked control and power management of AC/DC hybrid microgrids. IEEE Syst. J. **99**, 1–12 (2014), https://doi.org/10.1109/jsys-T.2014.2337353
46. N. Liu, J. Chen, L. Zhu, J. Zhang, Y. He, A key management scheme for secure communications of advanced metering infrastructure in smart grid. IEEE Trans. Ind. Electron. **60**(10), 4746–4756 (2013)

47. X. Meng, T. Chen, Event based agreement protocols for multi-agent networks. Automatica **49**(7), 2125–2132 (2013)
48. C. Li, X. Yu, W. Yu, T. Huang, Z.-W. Liu, Distributed event-triggered scheme for economic dispatch in smart grids. IEEE Trans. Ind. Inform. **12**(5), 1775–1785 (2016)
49. S. Wen, X. Yu, Z. Zeng, J. Wang, Event-triggering load frequency control for multi-area power systems with communication delays. IEEE Trans. Ind. Electron. **63**(2), 1308–1317 (2016)
50. Z. Hou, S. Jin, Data driven model-free adaptive control for a class of MIMO nonlinear discrete-time systems. IEEE Trans. Neural Netw. **22**(12), 2173–2188 (2011)
51. D. Xu, B. Jiang, P. Shi, A novel model-free adaptive control design for multivariable industrial processes. IEEE Trans. Ind. Electron. **61**(11), 6391–6398 (2014)
52. F. Nejabatkhah, Y.W. Li, Overview of power management strategies of hybrid AC/DC micro-grid. IEEE Trans. Power Electron. **30**(12), 7072–7089 (2015)
53. X. Lu, J.M. Guerrero, K. Sun, J.C. Vasquez, R. Teodorescu, L. Huang, Hierarchical control of parallel ac-dc converter interfaces for hybrid microgrids. IEEE Trans. Smart Grid **5**(2), 683–692 (2014)
54. Q. Sun, R. Han, H. Zhang, J. Zhou, J.M. Guerrero, A multiagent-based consensus algorithm for distributed coordinated control of distributed generators in the energy internet. IEEE Trans. Smart Grid **6**(6), 3006–3019 (2015)
55. A.Q. Huang, M.L. Crow, G.T. Heydt, J.P. Zheng, S.J. Dale, The future renewable electric energy delivery and management system the Energy Internet. IEEE Proc. IEEE **99**(1), 133–148 (2011)
56. J. Zhou, S. Kim, H. Zhang, Q. Sun, R. Han, Consensus-based distributed control for accurate reactive, harmonic and imbalance power sharing in microgrids. IEEE Trans. Smart Grid (in publication)

Chapter 5
Distributed Coordinated Control for Energy Internet

Abstract With the consideration of the large-scale amount and the characteristics of the distributed renewable energy generation, how to achieve the proportional power-sharing among DGs is an important issue to guarantee the stability and safety of the Energy Internet. In this section, a multi-agent system-based distributed coordinated control scheme is studied, and the main content contains: (1) architecture of multiagent system-based distributed coordinated control for energy internet; (2) implementation of distributed coordinated control for energy internet; (3) analysis of circulating current and design of the primary energy agent based on nonlinear model of distributed generator; (4) design of distributed coordinated control strategy based on multi-agent consensus algorithm.

5.1 Introduction

To realize the envisioned Energy Internet conception, the critical design and control objectives should be considered: (1) maintaining flexible and proportional power-sharing among DGs; (2) maintaining system continuous synchronization with the Main-Grid considering load variations; (3) minimizing circulating currents between DGs, (4) achieving seamless energy transitions between Energy Internet and Main-Grid if necessary.

To realize proportional power-sharing among DGs in the Energy Internet, various control schemes have been proposed. The most important one is the droop control [1], an attractive distributed control scheme that has been widely studied recently since its operation does not require for high-bandwidth communication systems. In addition, a power sharing unit (PSU) is proposed to achieve power management strategy in hybrid microgrid architecture in [2]. However, the conventional droop controller has several drawbacks such as load-dependent frequencies and voltage amplitudes, large circulating currents among DGs [3], the tradeoff between power-sharing accuracy and voltage synchronization [4, 5]. To overcome these shortcomings, a modified droop control strategy [6] has been proposed to improve the power-sharing accuracy considering the line-impedance effect through designing a proper virtual impedance. To

Q. Sun, *Energy Internet and We-Energy*, Renewable Energy Sources & Energy Storage, https://doi.org/10.1007/978-981-13-0523-8_5

restore the voltage and frequency of DGs to nominal values, the so-called secondary control has been recently investigated in the literature [7, 8], which demonstrate that proper communication systems are necessary to realize the control goals. The secondary control can be classified into centralized control and distributed control in general. The centralized control [9, 10] requires a central unit to receive all the information and to broadcast the decisions. Due to the centralized nature, it presents a single point failure which can reduce the reliability and stability of the whole system. Alternatively, the multi-agent system as a kind of distributed control structure has drawn much attention due to its flexibility and computational efficiency [11, 12]. In [13], the participation of a multi-agent-system-based microgrid into the Energy Market is proposed. In order to charge electric vehicles (EVs) at low electricity prices, an agent-based control system that coordinates the battery charging of electric vehicles in distribution networks is presented in [14]. In [15], the multi-agent system for EV charging control is proposed based on the Nash Certainty Equivalence to solve the grid impact. Meanwhile, the multi-agent consensus algorithm has been applied into control system based on the multi-agent system structure [16, 17]. However, these applications only solved a single problem of voltage restoration or frequency restoration. Furthermore, Energy Internet always suffers from large circulating currents caused by the slight differences among phases and amplitudes of the output voltages [18], which cannot be eliminated only by maintaining the synchronization of the output voltage angles or amplitudes.

In order to overcome the aforementioned challenges, a novel distributed coordinated controller combined with the multi-agent-based consensus algorithm is proposed to control DGs in the Energy Internet. The multi-agent system-based distributed cooperative control algorithm is an important way to realize the intelligent coordinated control of the energy internet, by which the safety and stability of the energy Internet can be guaranteed. Two main control objectives are achieved by the proposed control scheme: (1) keep angles and amplitudes of all DGs' output voltages being synchronized with the Main-Grid information (restore to their nominal values), while keeping accurate proportional power-sharing; (2) eliminate (minimize) circulating currents among DGs in the Energy Internet. Specifically, $P - \delta$ and $Q - V$ droop controls are adopted to suppress the circulating currents, while achieving proportional power-sharing. This can be achieved by regulating both the angles (δ) and amplitudes (V) at the same time. Due to the nonlinear feature of DGs in the Energy Internet, the proposed control approach is designed based on the input-output feedback linearization control principle [19] and related stability analysis approach [20].

The main features and benefits of the proposed controller are given as follows: (1) the structure of the proposed controller related to the conception of the multi-agent system is proposed including tasks decomposition, types of agents and information flow; (2) the novel distributed coordinated controller combined with the multi-agent consensus algorithm is proposed to control DGs in the Energy Internet; (3) the power can be shared in proportional and the angles and amplitudes of output voltages can be synchronized with the Main-Grid; (4) the circulating currents among DGs can be suppressed effectively; (5) the control method requires only a sparse communication structure which means each DG only needs its local information and its neighbor's information to achieve control objectives, then the controller can be more reliable and less expensive.

It is worthwhile to remark here that the proposed controller can bring extra benefits: (1) Energy Internet can be operated as a spinning reserve system when the leader information arises from Main-Grid, then it can achieve seamless integration into Main-Grid; roughly speaking, the spinning reserve means that the bidirectional power flow between Energy Internet and Main-Grid could respond voluntarily to load disturbances within a given period of time [21, 22]; (2) facilitated by this advanced control and communication scheme, Energy Internet can provide opportunities to other smart end users (flexible loads such as Intelligence Data Centers) in our daily lives to satisfy the needs of their power demands while minimizing their energy cost [23]; (3) the system can supply other ancillary services such as market participation as well required by Main-Grid [24, 25]; (4) with the increasing number of renewable sources and the development of Energy Internet, the fuel crisis and environmental problems can be solved gradually.

5.2 Architecture of Multiagent System-Based Distributed Coordinated Control for Energy Internet

In order to achieve voltage synchronization, proportional power-sharing and spinning reserve requirements for Energy Internet, the distributed coordinated controller combined with multi-agent consensus algorithm is proposed.

5.2.1 Requirements of Distributed Coordinated Control

A typical architecture of Energy Internet integrated with the proposed controller is shown in Fig. 5.1. The energy sources in an Energy Internet consist of distributed renewable energy resources (DRERs), distributed energy storage devices (DESDs), Main-Grid (MG). In this paper, the designation of the controller structure used in an Energy Internet is based on the multi-agent system structure and each DRER, DESD, MG is controlled by each different agent taking advantages of the autonomous, intelligent, cooperative proactive and adaptive features of the multi-agent system.

The proposed Energy Internet structure and distributed coordinated controller should meet the following requirements:

(1) Achieve power-sharing in proportional among different DGs dynamically and restore the amplitudes and angles of output voltages to nominal value to keep tracking with the information from the leader (Main-Grid or one DG).

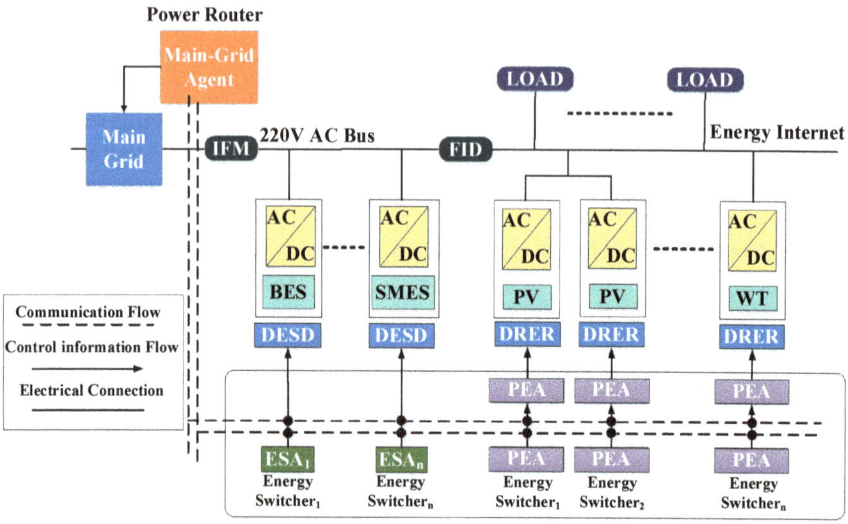

Fig. 5.1 Architecture of the energy internet

(2) Ensure the high reliability of the proposed algorithm and maintain seamless transition when Energy Internet switches between the grid-connected mode and islanded mode, which is called spinning reserve condition.
(3) Minimize the circulating current between different DGs to enhance efficiency of energy transmission.
(4) Achieve sparse communication structure to enhance the reliability and efficiency of the control system.

5.2.2 *Definitions of Agents*

According to the functional characteristics of the equipment in the Energy Internet, the following six types of agents are defined as the primary energy agent, the load agent, the energy storage agent, the main grid agent, the power router agent and the manager agent, as shown in Fig. 5.2. The functions of each agent are as follows:

(1) Primary energy agent (PEA): a static agent that corresponds to the generation equipment at the primary energy side, including renewable energy and clean energy, which can monitor the real power generation status of the primary energy. It can also response control instructions from the power router agent (PRA) or the manager agent (MA) to calculate the power generation capacity of each power supply unit. In accordance with the coordination control instructions from PRA or MA, it can cooperate with other PEAs to adjust the generation

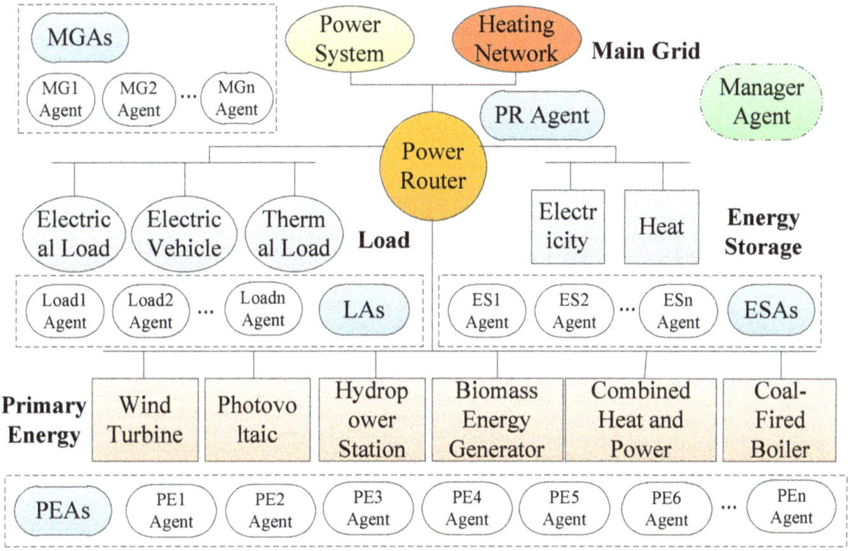

Fig. 5.2 Dynamic multi-agent system based energy coordinated control framework

status of power supply equipment to meet the energy need of the distribution network or the users of Energy Internet.

(2) Load agent (LA): a static agent that corresponds to the electricity load at the user side of the Energy Internet, which can monitor the real state of the load, and report the energy demand of the users to MA. It can also shut the redundant load according to the control instructions from power router agent.
(3) Energy storage agent (ESA): a static agent that corresponds to the energy storage device in Energy Internet, which can monitor the real status of the energy storage device. It can stabilize short-time load fluctuation according to the control instructions of PRA.
(4) Main grid agent (MGA): a static agent that corresponds to the traditional power system, which can exchange information of energy need with PRA.
(5) Power router agent (PRA): a static agent that corresponds to the power router in Energy Internet, based on the energy demand signal from MGA or MA, dispatching the primary energy, aiming at maximizing the energy efficiency of the renewable energy. Based on the multi-agent consensus algorithm, it can calculate the status of generation equipment at the primary energy side to keep power balance, keeping the frequency of electricity same as that of the traditional power system and the voltage at a proper range to guarantee high-quality electric energy output, in turns to meet the energy need of users in Energy Internet, to realize the energy sharing and information exchange with the traditional power system to support the safe and stable operation of the traditional power system.
(6) Manager agent (MA): a virtual dynamic agent that generates a dynamic load demand response team (DLDRT) when it detects an energy demand signal

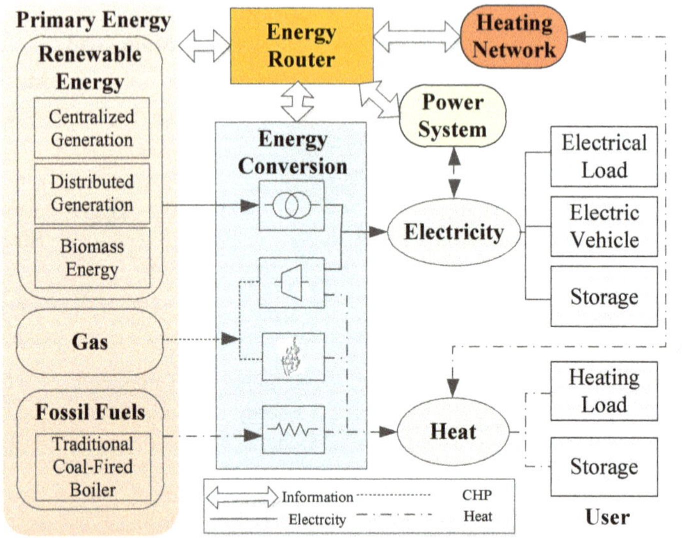

Fig. 5.3 Presentation of primary energy side in the energy internet

from the users of Energy Internet, and communicates with the neighbor PEA. In DLDRT, PEAs cooperate with each other, based on multi-agent consensus algorithm to calculate the status of the generation equipment at the primary energy side, to control the voltage in the network, in turns to meet the changeable energy need locally. When the load fluctuation is stabilized, the DLDRT will be disbanded and the MA will disappear. Otherwise, the upper layer MA is automatically established to form a wider range of DLDRTs until the DLDRT contains all the power supply devices at the primary energy side in the Energy Internet.

5.3 Distributed Coordinated Control for Energy Internet

Aiming at reasonably allocating the energy output of the energy supply devices at the primary energy side in the Energy Internet, as shown in Fig. 5.3, the coordination control strategy is proposed to stabilize load fluctuation, based on the multi-agent consensus algorithm.

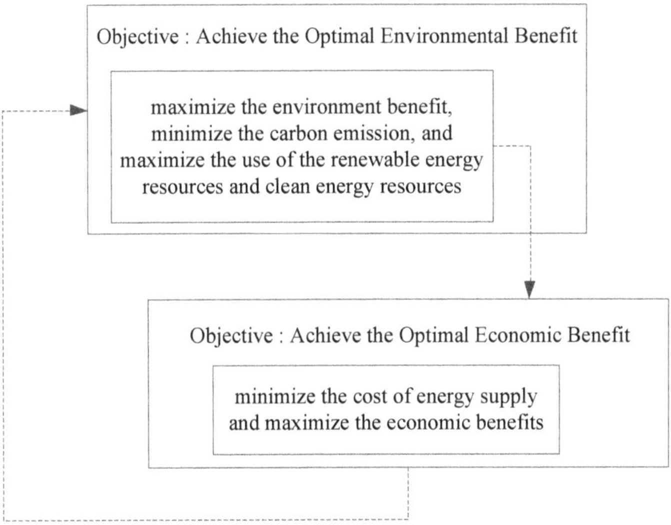

Fig. 5.4 Optimization objectives of primary energies

5.3.1 *Multi-objective Optimization Model*

In the Energy Internet, the secondary energy resources contain both electricity and heat. Therefore, to realize the optimization and address energy dispatch problem, it is necessary to consider the effect of heat. In addition, many other factors make the optimal dispatch of Energy Internet more difficult, such as the complexity of primary energy resources and diversity of the energy demand of the users. Accordingly, the multi-objective optimization model of Energy Internet is investigated to achieve the optimal dispatch, to realize the optimal utilization of distributed renewable energy resources and construct a low-carbon structure of the energy consumption, as shown in Fig. 5.4.

A. Objective Function

The aim of dispatching optimally energy suppliers is to utilize more renewable energy and less fossil energy, while minimize the environmental cost and the economic cost. The objective function is as follows:

$$f = \min(f_1, f_2), \tag{5.1}$$

where f_1 is the environmental benefit function of energy dispatch, f_2 is the economic benefit function of energy dispatch.

(1) Environmental Cost

To solve the energy crisis and environmental problems, the consumption of coal, natural gas and other fossil energy should be reduced. To maximize the environment benefit and minimize the carbon emission, the use of the renewable energy resources and clean energy resources should be maximized, which is expressed as

$$f_1 = \min \sum_{t=1}^{T} \sum_{j=1}^{N_G} \left(a_j P_{Gj,t}^2 + b_j P_{Gj,t} + c_j \right) + \sum_{t=1}^{T} \sum_{j=1}^{N_G} Q_{Grj,t} + \sum_{t=1}^{T} \sum_{j=1}^{N_F} Q_{Fj,t}, \quad (5.2)$$

where T is the number of dispatch intervals, N_G is total number of natural gas-fired CHP, jth is the generation parameters for the jth natural gas-fired CHP, $P_{Gj,t}$ is the active power of the jth natural gas-fired CHP. $Q_{Grj,t}$ is the heat supply of the jth natural gas-fired CHP at time t. N_F is the amount of the traditional coal-fired boiler unit. $Q_{Fj,t}$ is the heat supply of the jth traditional coal-fired boiler unit at time t.

(2) Economic Cost

Energy internet is a hybrid energy supply system, whose primary energy resources contain natural gas, fossil energy and renewable energy, such as wind, solar, biomass and so on. The optimal sub-objective is to minimize the cost of energy supply and maximize the economic benefits under the condition that the energy demand can be satisfied. The sub-objective function can be denoted as

$$f_2 = \min\left(F_{\cos t}^{W} + F_{res}^{W} + F_{\cos t}^{S} + F_{\cos t}^{B} + F_{\cos t}^{P} + F_{\cos t}^{G} + F_{\cos t}^{F} \right), \quad (5.3)$$

where $F_{\cos t}^{W}$ is the operating cost of wind turbine, $F_{\cos t}^{S}$ is the operating cost of the photovoltaic generator, $F_{\cos t}^{B}$ is the operating cost of the biomass generator. $F_{\cos t}^{W}$, $F_{\cos t}^{S}$ and $F_{\cos t}^{B}$ are all related to the capacity of active power and the feed-in tariff in electrical market. F_{res}^{W} is the compensation cost of standby capacity of wind turbines. $F_{\cos t}^{P}$ is the operating cost of the pumped storage unit. $F_{\cos t}^{G}$ is the operating cost of the natural gas-fired CHP.

B. Constraint Condition

Since heat is an energy transmission and consumption terminal in the energy internet, besides the conventional constraints in optimization problems of traditional power system, there are several conditions that should be considered such as the balance between the heat energy demand and heat energy supply, the maximum energy supply of heat producer and the capacity of heat transmission line.

5.3.2 Implementation of Distributed Coordinated Control

Energy Internet can be operated under two modes. The first mode is that Energy Internet has no responsibility to connect with Main-Grid and can be operated in the isolated mode. In this mode, one of the DGs should be chosen as the leader in the control system. In addition, the leader DG should be controlled to output rated power to guarantee its leader information with a constant and normal value.

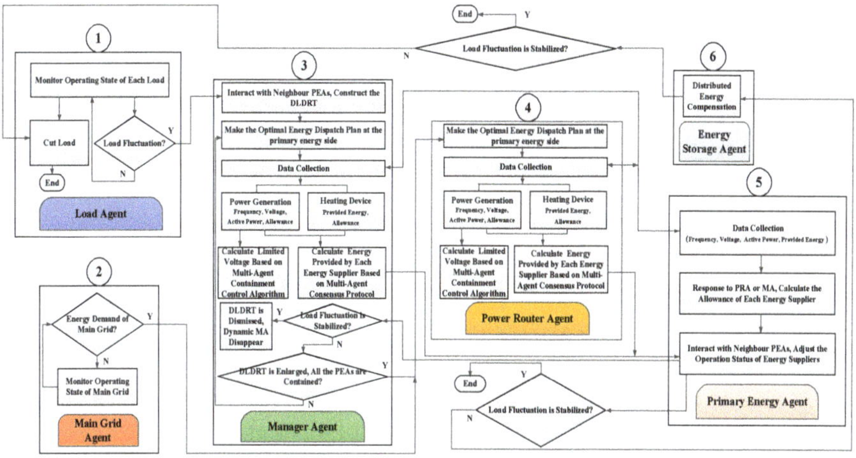

Fig. 5.5 Flow chart of energy coordinated control of energy internet

The second mode is that Energy Internet should be connected to the Main-Grid if necessary; the information from Main-Grid should be as the leader information. In this mode, Energy Internet can be operated under the spinning reserve condition with bidirectional power flow to compensate the power disturbance in both Main-Grid and Energy Internet. The main difference between the two modes is whether Energy Internet should relate to the Main-Grid. Except for that, other operations of the proposed controller of Energy Internet are same. The coordinated control strategies for the two modes are presented in the following subsections, as shown in Fig. 5.5.

5.3.3 Distributed Coordinated Control Strategy Responding to Users in Energy Internet

In order to stabilize the load fluctuation in the Energy Internet quickly, the global optimality can be ignored in the design of the energy coordination control strategy, the energy demand of the local user side is taken as the primary consideration to ensure the stability of the Energy Internet. Therefore, a dynamic multi-agent system is proposed, which includes a dynamic manager agent (MA) as a leader, to coordinate the energy resources in the Energy Internet, by the coordination with other agents in Energy Internet. The strategy can be further described as:

Step 1: If the LA detects the signal of the load fluctuation at the user side in the Energy Internet, the dynamic manager agent is generated. The MA communicates with the neighbor PEAs to establish the dynamic load demand response team (DLDRT). The mechanism is as shown in the Fig. 5.6.

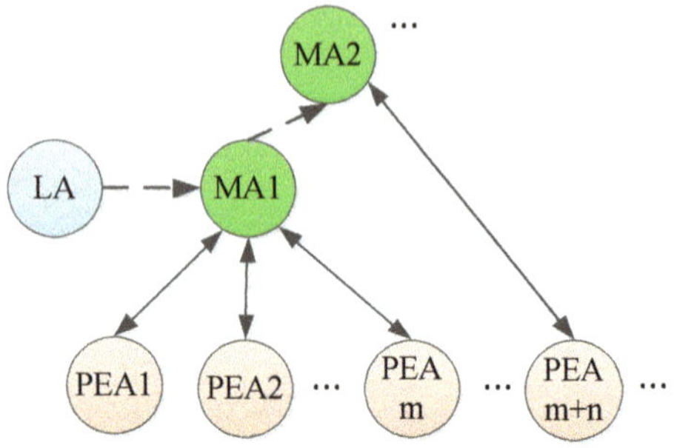

Fig. 5.6 Dynamic load demand response team forming mechanism

Step 2: Based on the energy dispatching optimization model at the primary energy side, MA optimizes the energy dispatch of PEAs at the primary energy side in DLDRT and selects PEAs to participate in the task.

Step 3: Collect the parameters of the PEAs participating in the load demand response, such as the frequency, voltage, active power and energy capacity.

Step 4: Based on the multi-agent consensus algorithm, MA calculates the power output that each energy supplier should supply, responding to the load fluctuation, and controls the voltage reasonably.

Step 5: According to the calculation results of MA, the PEA cooperates with its neighbor PEAs to adjust the operation status until the output of the active power of the power generator is $P_i + P_{ui}$, where P_i is the original output of power generator.

Step 6: As the load fluctuation is stabilized, the DLDRT will be disbanded and the MA will disappear. Otherwise, the upper-layer MA is generated and the scale of the DLDRT is expanded. And then repeat steps 2–6. If the DLDRT is extended to contain all the energy suppliers in the Energy Internet, the top-level MA becomes the PRA to execute the coordinated control strategy corresponding to the energy demand of the main grid.

5.3.4 Distributed Coordinated Control Strategy Responding to Main Grid

As the main grid needs energy sharing with the Energy Internet, to achieve the core of the Energy Internet content, which is the full use of renewable energy, a global optimization of the primary energy is needed. The coordinated control strategy can be further described as:

Step 1: If the PRA detects the signal of the energy request sent by the MGA, based on the energy dispatch optimization model at primary energy side, the PRA optimizes the energy dispatch of all PEAs at the primary energy side of the Energy Internet and selects PEAs participating in the energy demand response.

Step 2: Collect the parameters of the PEAs participating in the energy demand response of the main grid, such as the frequency, voltage, active power and the allowance of the energy capacity.

Step 3: Based on the multi-agent consensus algorithm, PRA calculates the electrical power that each energy supplier should provide and control the output voltage.

Step 4: According to the calculation results of MA, the PEA cooperates with its neighbor PEAs to adjust the operation status until the output of the active power of the power generator is $P_i + P_{ui}$, where P_i is the original output of power generator.

Step 5: As the energy demand of the user side in the Energy Internet is satisfied, the task will be over. Otherwise, the ESA will work and start the distributed energy storage device to make the local power compensation. If the ESA's work is finished, meanwhile the energy demand of the main grid or the user side in the Energy Internet has not yet been met, LA will shed load by the principle of minimizing the loss of operation benefits.

5.4 Analysis of Circulating Current and Design of the Primary Energy Agent Based on Nonlinear Model of Distributed Generator

The control loop, including the power calculation, voltage and current controllers, second order generalized integrator and droop control, is used to achieve local control on the amplitude and angle of output voltage produced by different DGs.

5.4.1 Circulating Current Analysis

Through the circulating current analysis between DGs, the circulating current between two parallel connected DG is divided into active circulating and reactive circulating current as

$$2\Delta \dot{I} = I_a - jI_r, \tag{5.4}$$

$$I_a = \frac{X_k^2 + 4R_o^2}{R_o}\left(E_1^{'} \sin\phi_1 - E_2^{'} \sin\phi_2\right)\frac{1}{A}, \tag{5.5}$$

$$I_r = \frac{X_k^2 + 4R_o^2}{R_o}\left(E_1^{'} \cos\phi_1 - E_2^{'} \cos\phi_2\right)\frac{1}{A}, \tag{5.6}$$

where I_a is the active circulating current, I_r is the reactive circulating current, ΔI is the total circulating current, E_1 is output voltage of the first DG, E_2 is output voltage of the second DG, X_{k1}, X_{k2} are line parameters between two DGs, R_o is the load between two DGs, $E_1^{'} = \frac{X_{k2}}{R_o}E_1$, $E_2^{'} = \frac{X_{k1}}{R_o}E_2$, $A = (X_{k1} + X_{k2})^2 + (X_{k1}X_{k2}/R_o)^2$.

From Eqs. (5.5) and (5.6), it can be found that active and reactive circulating currents cannot be effectively eliminated only by controlling the amplitudes or phase angles of the output voltages. Thus the $P-\delta$ and $Q-V$ droops are used to control output voltages amplitudes and angles. In order to mimic the behavior of a synchronous generator, the $P-\delta$ droop control represents the linear relationships between active power and angle of output voltage and the $Q-V$ droop control represents the linear relationships between reactive power and amplitude of output voltage.

5.4.2 Design of Primary Energy Agent Combined with Nonlinear Distributed Generator Model

In this section, dc-bus dynamics can be safely neglected, since a dc-link feed forward loop can be used. The block diagram of a DG based on a single-phase inverter is presented in Fig. 5.7.

The nonlinear dynamics of DGs are controlled under $d-q$ reference frame. In order to mimic a synchronous generator, the d-axis represents direction of rotor magnetic flux linkage and the q-axis is of 90° ahead of d-axis. Thus $d-q$ reference usually forms a rotating orthogonal reference frame used by three-phase DG. However, single-phase inverters are considered in this section, so that the inverters output voltage and current should be divided first into $\alpha-\beta$ components through a second order generalized integrator, as shown in Fig. 5.8. To be more specific, the $\alpha-\beta$ components are the transition reference frame from single-phase reference frame to the $d-q$ reference frame.

The phase of output voltage of one DG δ_i can be expressed as

$$\dot{\delta}_i = \omega_i \tag{5.7}$$

where δ_i is the angle of voltage from DG, ω_i is the angular velocity of DG to represent the frequency, i is the number of distributed generators (DGs).

Because the $P-\delta$ droop control is used in the system, no frequency deviations would occur under load disturbances.

The power controller shown in Fig. 5.9 consists of the second order generalized integrator, park transformation, phase-locked loop (PLL), power calculation, low-pass filter and droop controller. The power controller provides the amplitude reference of voltage v_{odi}^* for the first stage bridge and the angle reference of output voltage δ_i^* for the second stage bridge. The differential equations of the active and reactive power can be expressed as

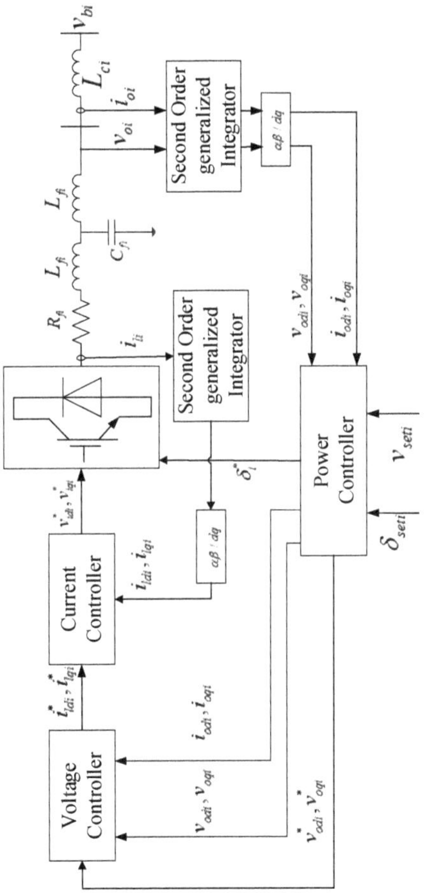

Fig. 5.7 Block diagram of a DG based on a single-phase inverter

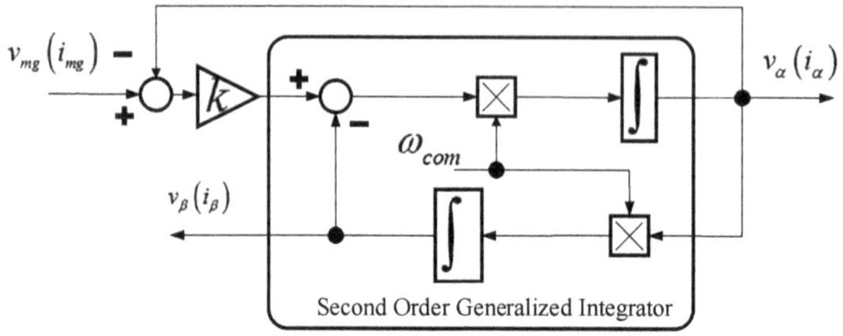

Fig. 5.8 Second order generalized integrator

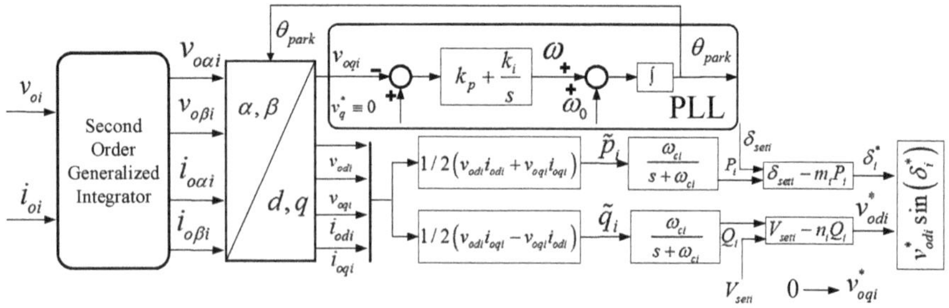

Fig. 5.9 The block diagram of power controller

$$\dot{P}_i = -\omega_{ci} P_i + \omega_{ci} \frac{\left(v_{odi} i_{odi} + v_{oqi} i_{oqi}\right)}{2}, \tag{5.8}$$

$$\dot{Q}_i = -\omega_{ci} Q_i + \omega_{ci} \frac{\left(v_{odi} i_{oqi} - v_{oqi} i_{odi}\right)}{2}, \tag{5.9}$$

where v_{odi} is the d-axis output voltage components for DG, v_{oqi} is the q-axis output voltage components for DG, i_{odi} is the d-axis output voltage components for DG, i_{oqi} is the q-axis output current components for DG, P_i is the average active power for DG, Q_i is the average reactive power for DG, ω_{ci} is the cutoff frequency of power filter.

The $P - \delta$ and $Q - V$ droop controls are presented as

$$\delta_i^* = \delta_{seti} - m_i P_i, \tag{5.10}$$

$$v_{odi}^* = v_{seti} - n_i Q_i, \tag{5.11}$$

$$v_{oqi}^* = 0, \tag{5.12}$$

where m_i is the angle-active droop parameter for DG, n_i is the voltage-reactive droop parameter for DG, δ_{seti} is the set value of angle-active droop control for DG, v_{seti} is the set value of voltage-reactive droop control for DG, δ_i^* is the reference of voltage angle for DG, v_{odi}^* is the d-axis reference of voltage amplitude in the voltage

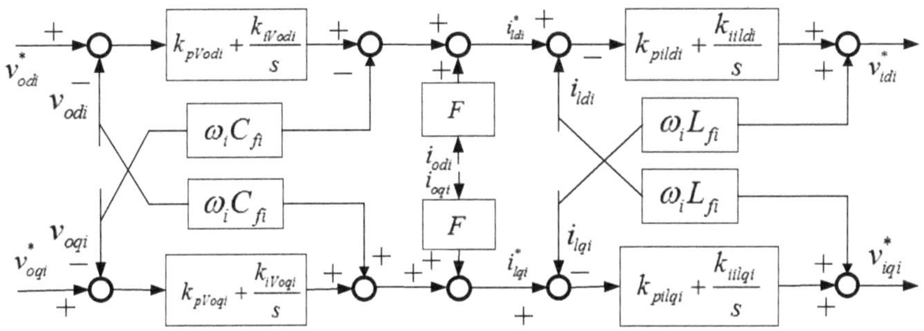

Fig. 5.10 The block diagram of the voltage and current controller

controller for DG, v_{oqi}^{*} is the q-axis reference of voltage amplitude in the voltage controller for DG.

By inserting the (5.10) and (5.11) into the (5.8) and (5.9), it yields to

$$\dot{P}_i = \frac{\omega_{ci}}{m_i}\left(\delta_i^* - \delta_{seti}\right) + \omega_{ci}\frac{\left(v_{odi}i_{odi} + v_{oqi}i_{oqi}\right)}{2}, \tag{5.13}$$

$$\dot{Q}_i = \frac{\omega_{ci}}{n_i}\left(v_{odi}^* - v_{seti}\right) + \omega_{ci}\frac{\left(v_{odi}i_{oqi} - v_{oqi}i_{odi}\right)}{2}. \tag{5.14}$$

During the process of our model, the normal control of voltage and current should be included as follows:

The block diagram of the voltage and current controller is shown in Fig. 5.10.

The differential equations of the voltage controller are presented as

$$\dot{\phi}_{odi} = v_{odi}^* - v_{odi}, \tag{5.15}$$

$$\dot{\phi}_{oqi} = v_{oqi}^* - v_{oqi}, \tag{5.16}$$

where ϕ_{odi} is the d-axis auxiliary state variables in the voltage controller for DG, ϕ_{oqi} is the q-axis auxiliary state variables in the voltage controller for DG.

The outputs of the voltage controller are presented as

$$i_{ldi}^* = Fi_{odi} - \omega_f C_f v_{oqi} + k_{pVodi}\left(v_{odi}^* - v_{odi}\right) + k_{iVodi}\phi_{odi}, \tag{5.17}$$

$$i_{lqi}^* = Fi_{oqi} - \omega_f C_f v_{odi} + k_{pVoqi}\left(v_{oqi}^* - v_{oqi}\right) + k_{iVoqi}\phi_{oqi}, \tag{5.18}$$

where F is the feedforward parameter in the voltage controller, C_f is the filter capacitor for DG.

In addition, inserting (5.11) into (5.15), we obtain

$$\dot{\phi}_{odi} = v_{seti} - n_i Q_i - v_{odi}. \tag{5.19}$$

The differential equations of the current controller are presented as

$$\dot{\gamma}_{di} = i_{ldi}^{*} - i_{ldi}, \tag{5.20}$$

$$\dot{\gamma}_{qi} = i_{lqi}^{*} - i_{lqi}, \tag{5.21}$$

where i_{ldi}^{*} is the d-axis current reference in the current controller for DG, i_{lqi}^{*} is the q-axis current reference in the current controller for DG, γ_{di} is the d-axis auxiliary state variables in the current controller for DG, γ_{qi} is the q-axis auxiliary state variables in the current controller for DG.

The outputs of the current controller are presented as

$$v_{idi}^{*} = -\omega_f L_f i_{lqi} + k_{pildi}\left(i_{ldi}^{*} - i_{ldi}\right) + k_{iildi}\gamma_{di}, \tag{5.22}$$

$$v_{iqi}^{*} = \omega_f L_f i_{ldi} + k_{pilqi}\left(i_{lqi}^{*} - i_{lqi}\right) + k_{iilqi}\gamma_{qi}, \tag{5.23}$$

where v_{idi}^{*} is the d-axis output of current controller for DG, v_{iqi}^{*} is the q-axis output of current controller for DG, i_{ldi} is the d-axis filter current components for DG, i_{lqi} is the q-axis filter current components for DG, L_{fi} is the filter inductance for DG, k_{pildi} is the d-axis proportion parameters in the current controller for DG, k_{pilqi} is the q-axis proportion parameters in the current controller for DG, k_{iildi} is the d-axis integral parameters in the current controller for DG, k_{iilqi} is the q-axis integral parameters in the current controller for DG.

As the connection between inverters are inductive, the differential equations for the output LCL filter L_{fi} and C_{fi} and the output inductor L_{Ci} are expressed as follows:

$$\dot{i}_{ldi} = -\frac{R_{fi}}{L_{fi}} i_{ldi} + \omega_i i_{lqi} + \frac{1}{L_{fi}}(v_{idi} - v_{odi}), \tag{5.24}$$

$$\dot{i}_{lqi} = -\frac{R_{fi}}{L_{fi}} i_{lqi} - \omega_i i_{ldi} + \frac{1}{L_{fi}}\left(v_{iqi} - v_{oqi}\right), \tag{5.25}$$

$$\dot{v}_{odi} = \omega_i v_{oqi} + \frac{1}{C_{fi}}(i_{ldi} - i_{odi}), \tag{5.26}$$

$$\dot{v}_{oqi} = -\omega_i v_{odi} + \frac{1}{C_{fi}}\left(i_{lqi} - i_{oqi}\right), \tag{5.27}$$

$$\dot{i}_{odi} = \omega_i i_{oqi} + \frac{1}{L_{ci}}(v_{odi} - v_{bdi}), \tag{5.28}$$

$$\dot{i}_{oqi} = -\omega_i i_{odi} + \frac{1}{L_{ci}}\left(v_{oqi} - v_{bqi}\right). \tag{5.29}$$

From the above equations, the model of ith DG can be rewritten into a matrix formation as

$$\begin{cases} \dot{x} = F(x) + gu \\ y = hx \end{cases}, \tag{5.30}$$

where the state vector is $x = \left[\delta_i, P_i, Q_i, \phi_{di}, \phi_{qi}, \gamma_{di}, \gamma_{qi}, i_{ldi}, i_{lqi}, v_{odi}, v_{oqi}, i_{odi}, i_{oqi}\right]$.

In order to keep the angle δ_i and the amplitude v_{oi} of the ith DG being synchronized with other DGs, the δ_{seti} and v_{seti} are selected to be the control inputs. Since v_{oqi} is kept to be zero, the amplitude of output voltage is

$$v_{oi} = v_{odi}. \tag{5.31}$$

Therefore, the inputs of the system are $u = \begin{bmatrix} \delta_{seti} & v_{seti} \end{bmatrix}^T$ and the outputs of the system are $y = \begin{bmatrix} \delta_i & v_{odi} \end{bmatrix}^T$. In the next Section, the feedback linearization method combined with the multi-agent consensus algorithm is used to design the distributed coordinated control strategy.

5.5 Design of Distributed Coordinated Control Strategy Based on Multi-agent Consensus Algorithm

In this Section, the design of the distributed coordinated control strategy based on the multi-agent consensus algorithm is proposed by using feedback linearization, which can transform the nonlinear model to the linear model. Each PEA has two control inputs and two outputs. The consensus problem about angles and amplitudes of output voltages among DGs is a synchronization tracking problem, which needs a leader in the system. The leader information is decided by the ERA. Each PEA only requires local and its neighbors' information, which can be used to produce the control decision signal. To be more specific, if the leader information in the ERA comes from Main-Grid, the Energy Internet can be operated under spinning reserve condition, which means it can be connected to Main-Grid and immediately provide power if necessary.

5.5.1 Graph Theory

The distributed coordinated controller should use a communication network called directed graph, which can be expressed as $G_r = (V_G, E_G, A_G)$. The set of nodes in the network can be expressed as $V_G = \{v_1, v_2, \ldots, v_n\}$, the set of edges can be expressed as $E_G \subseteq V_G \times V_G$ and the weighted adjacency matrix can be expressed as $A_G = \left[a_{ij}\right]_{n\times n}$ with nonnegative adjacency element a_{ij}. An edge rooted at node j and ended at node i is denoted by $\left(v_j, v_i\right)$, which means that information can flow from the node j to node i. For a graph with 0–1 adjacency elements, the in-degree and out-degree of node v_i are defined as follows:

$$\deg_{in}(v_i) = \sum_{j=1}^{n} a_{ji} \deg_{out}(v_i) = \sum_{j=1}^{n} a_{ij} \tag{5.32}$$

The degree matrix of diagraph G_r is a diagonal matrix $\Delta = \left[\Delta_{ij}\right]$ where $\Delta_{ij} = 0$ for all $i \neq j$ and $\Delta_{ii} = \deg_{out}(v_i)$. The Laplacian matrix associated with the digraph G_r is defined as

$$L(G_r) = \Delta - A. \tag{5.33}$$

5.5.2 Design of the Distributed Coordinated Control Strategy

From the discussion in Sect. 5.4.2, the δ_{seti} is used as the input to control angles of output voltages among different DGs to track the leader $y_{leader,1}$ and the v_{seti} is used as the input to control the amplitude of output voltage to track the voltage $y_{leader,2}$. The feedback linearization method can establish the relationship between the control outputs and inputs. The process of establishing the relationship in a MIMO nonlinear system is explained below:

Step 1: The relative degree $[r_1, r_2, \ldots, r_m]$ should be calculated.

Step 2: According to the relative degree, a decoupling matrix $A(x)$ which should be nonsingular should be calculated as given in (5.33), another vector $b(x)$ should be calculated at the same time as in (5.34). In (5.33) and (5.34), $L_f h_i(x)$ is the Lie derivative of $h_i(x)$ with respect to f and is defined as $L_f h_i(x) = (\partial(h_i)/\partial x_i) f$.

$$A(x) = \begin{pmatrix} L_{g_1} L_f^{r_1 - 1} h_1(x) & \ldots & L_{g_m} L_f^{r_1 - 1} h_1(x) \\ \vdots & & \vdots \\ L_{g_1} L_f^{r_1 - 1} h_m(x) & \cdots & L_{g_m} L_f^{r_1 - 1} h_m(x) \end{pmatrix}, \tag{5.34}$$

$$b(x) = \left[L_f^{r_1} h_1(x) \cdots L_f^{r_m} h_m(x)\right]^T. \tag{5.35}$$

Step 3: Based on the defined relative degree, the control law of a MIMO nonlinear system is defined as

$$u(x) = A^{-1}(x)[-b(x) + v(x)], \tag{5.36}$$

where $v(x) = [v_1 \cdots\cdots v_m]^T = \left[y_1^{(r_1)} \cdots\cdots y_m^{(r_m)}\right]^T$.

According to above processes, the relative degree should be calculated first. The relative degree of the nonlinear system is [1, 1]. The decoupling matrix is calculated as

$$A_i(x) = \begin{pmatrix} \omega_{ci}\delta_{seti} & 0 \\ 0 & \omega_{ci}V_{seti} \end{pmatrix}. \tag{5.37}$$

The matrix $A_i(x)$ is nonsingular. The matrix $b_i(x)$ is calculated as

$$b_i(x) = \left[L_f h_1(x)\ L_f h_2(x) \right]^T, \tag{5.38}$$

where $L_f h_1(x) = \omega_i + \omega_{ci}\left(\delta_{seti} - \delta_i^*\right) - m_i\omega_{ci}\frac{\left(v_{odi}i_{odi}+v_{oqi}i_{oqi}\right)}{2}$,

$$L_f h_2(x) = \omega_{ci}\left(v_{seti} - v_{odi}^*\right) + n_i\omega_{ci}\frac{\left(v_{odi}i_{oqi} - v_{oqi}i_{odi}\right)}{2} + \omega_i v_{oqi} + \frac{1}{C_{fi}}(i_{ldi} - i_{odi}).$$

From (5.37) and (5.38), the control law $u_i(x)$ can be calculated as in (5.36). The $v_i(x)$ can be calculated as follows to achieve the synchronization of y_i.

$$v_i(x) = A_i(x)u_i(x) + b_i(x), \tag{5.39}$$

where $v_i = \left[v_{i,1}\ v_{i,2} \right]^T$.

The nonlinear system can be transformed into the following linear system

$$\begin{pmatrix} \dot{y}_{i,1} \\ \dot{y}_{i,2} \end{pmatrix} = \begin{pmatrix} 1 & 0 \\ 0 & 1 \end{pmatrix}\begin{pmatrix} v_{i,1} \\ v_{i,2} \end{pmatrix}, \tag{5.40}$$

where $y_i = \left[y_{i,1}\ y_{i,2} \right]^T$, $y_{i,1} = \delta_i$, $y_{i,2} = v_{odi}$, $\forall i$, $B = \begin{pmatrix} 1 & 0 \\ 0 & 1 \end{pmatrix}$.

The nonlinear system (5.30) is transformed to the linear system (5.20) by using the well-known feedback linearization method. Note that (5.40) presents the ith DG linear dynamics. The control of angles of output voltage is the tracking synchronization problem with a leader in the system. The aim is to make all the angles be equal to the leader. The tracking error is given by

$$e_{i,1} = \sum_{j\in N_i} a_{ij}\left(y_{i,1} - y_{j,1}\right) + \varepsilon_{i,1}\left(y_{i,1} - y_{leader,1}\right), \tag{5.41}$$

where a_{ij} is the element of adjacency matrix, $e_{i,1}$ represents the error of angles of output voltage about the ith DG. If the node i can receive the leader's angle information, and edge $\left(v_{leader,1}, v_i\right)$ is said to exist with weighting gain $\varepsilon_{i,1}$. The node i with $\varepsilon_{i,1} = 1$ is used as a pinned and controlled node about angle. The $\varepsilon_{i,1}$ can be written into matrix as $E_1 = diag\{\varepsilon_{i,1}\} \in R^{N\times N}$.

The control of amplitude of the output voltage is also a tracking synchronization problem with the system leader. Thus, the tracking error is given as

$$e_{i,2} = \sum_{j\in N_i} a_{ij}\left(y_{i,2} - y_{j,2}\right) + \varepsilon_{i,2}\left(y_{i,2} - y_{leader,2}\right), \tag{5.42}$$

where $e_{i,2}$ represents the error of amplitudes of output voltage about the ith DG. If the node i can receive the leader's amplitude information, and edge $\left(v_{leader,2}, v_i\right)$ is said to exist with weighting gain $\varepsilon_{i,2}$. The node i with $\varepsilon_{i,2} = 1$ is used as a pinned and controlled node about amplitude. The $\varepsilon_{i,2}$ can be written into matrix as $E_2 = diag\{\varepsilon_{i,2}\} \in R^{N\times N}$.

The errors can be rewritten by using matrixes as following

$$e_1 = (L+E_1)\left(Y_1 - Y_{leader,1}\right), \tag{5.43}$$

$$e_2 = (L+E_2)\left(Y_2 - Y_{leader,2}\right), \tag{5.44}$$

where $e_1 = \begin{bmatrix} e_{1,1} & e_{2,1} & \ldots & e_{n,1} \end{bmatrix}^T$, $e_2 = \begin{bmatrix} e_{1,2} & e_{2,2} & \ldots & e_{n,2} \end{bmatrix}^T$, $Y_{leader,1} = 1_N y_{leader,1}$, $Y_{leader,2} = 1_N y_{leader,2}$, $Y_1 = \begin{bmatrix} y_{1,1} & y_{2,1} & \cdots & y_{n,1} \end{bmatrix}^T$, $Y_2 = \begin{bmatrix} y_{1,2} & y_{2,2} & \cdots & y_{n,2} \end{bmatrix}^T$.

The system (5.40) can be rewritten as

$$\begin{cases} \dot{Y}_1 = I_N v_1 \\ \dot{Y}_2 = I_N v_2 \end{cases}, \tag{5.45}$$

where $v_1 = \begin{bmatrix} v_{1,1} & v_{2,1} & \cdots & v_{n,1} \end{bmatrix}^T$, $v_2 = \begin{bmatrix} v_{1,2} & v_{2,2} & \cdots & v_{n,2} \end{bmatrix}^T$.

The relationship between e_1, e_2 and v_1, v_2 should be designed in order to keep the system stable with the control input. Let the auxiliary control $v_{i,1}$ and $v_{i,2}$ be defined as follows:

$$\begin{cases} v_{i,1} = -c_1 k_1 e_{i,1} \\ v_{i,2} = -c_2 k_2 e_{i,2} \end{cases}, \tag{5.46}$$

where $c_1, c_2 \in R$ is the coupling gain and $k_1, k_2 \in R$ is the feedback control gain.

According to the IEEE 1547–2003, if the amplitudes and angles of output voltages can be kept tracking with the Main-Grid, the system can be connected with Main-Grid if necessary and provide power, which is called spinning reserved condition. Under this condition, the system can compensate the power unbalance with the help from Main-Grid and can also provide the surplus power for Main-Grid. Along with the novel distributed coordinated controller (5.43)–(5.46), Energy Internet can be

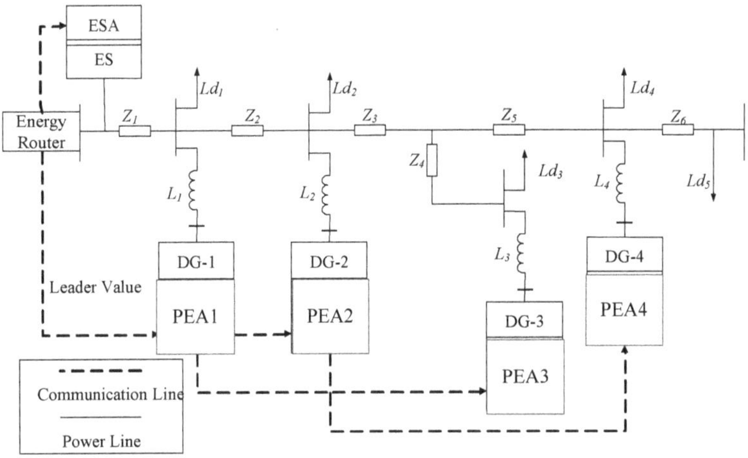

Fig. 5.11 Configuration of the energy internet test system

integrated into the Main-Grid seamlessly to provide reserve services at any time to ensure the reliability of the power system as long as the leader information comes from Main-Grid. To be specific, from the perspective of power management, the bidirectional power flow between Main-Grid and Energy Internet can be achieved to compensate power unbalance in both sides. Then Energy Internet can be operated under spinning reserve condition. From the perspective of control method, the angle and amplitude deviations between Main-Grid and Energy Internet are within a small range all the time, thus Energy Internet can be connected to the Main-Grid at any time.

5.6 Simulation and Results

In order to test the proposed approach, a number of simulations under different scenarios have been performed through MATLAB/SIMULINK. The electrical part of the simulation is established through the SIMULINK, meanwhile the control method including the feedback linearization and multi-agent consensus algorithm is programmed through the S-function. Different configurations of loads and DGs are considered in this section. Figure 5.11 shows the configuration of the Energy Internet test system. In addition, the communication structure between agents is also shown in Fig. 5.11 by dotted lines. In the system, the DG-1 and DG-2 are connected in series and the DG-3 and DG-4 are connected in parallel. This configuration consists of series and parallel connected DGs, which are common in Energy Internet, thus the simulation results can be more convincing. The system data used for simulation are listed in Table 5.1.

Table 5.1 System parameters

System quantities			Values
System frequency			50 Hz
DC voltage			800 V
Leader information	Voltage		1p.u
	Angle		57.4°
Line impedance $Z_1 = Z_2 = Z_3 = Z_4 = Z_5 = Z_6$			$0.05 + j1.5 * 10^{-5}\Omega$
Load ratings	Ld_1		10.1 KW and 5 kVAr
	Ld_2		10.1 KW and 5 kVAr
	Ld_3		10.1 KW and 5 kVAr
	Ld_4		10 KW and 6 kVAr
	Ld_5		11 KW and 7 kVAr
DG ratings	DG-1		10 KW and 5 kVAr
	DG-2		20 KW and 10 kVAr
	DG-3		13.3 KW and 6.6 kVAr
	DG -4		13.3 KW and 6.6 kVAr
Droop coefficients	Active power-angle	m_1	$1 * 10^{-4}$ rad/w
		m_2	$5 * 10^{-5}$ rad/w
		m_3	$7.5 * 10^{-5}$ rad/w
		m_4	$7.5 * 10^{-5}$ rad/w
	Reactive power-voltage	n_1	$4 * 10^{-4}$ rad/VAr
		n_2	$2 * 10^{-4}$ rad/VAr
		n_3	$3 * 10^{-4}$ rad/VAr
		n_4	$3 * 10^{-4}$ rad/VAr

5.6.1 Case 1: Conventional Controller

As the simulation starts, there are 4 distributed generators and 5 loads in the system. At 0.4 s, DG-4 cuts off, while the DG-1, DG-2 and DG-3 start the power supply. At 0.7 s, Ld_4 and Ld_5 are shed. And the results are shown in Figs. 5.12, 5.13, 5.14 and 5.15.

Moreover, the numerical values of power with conventional controller can be seen in Table 5.2.

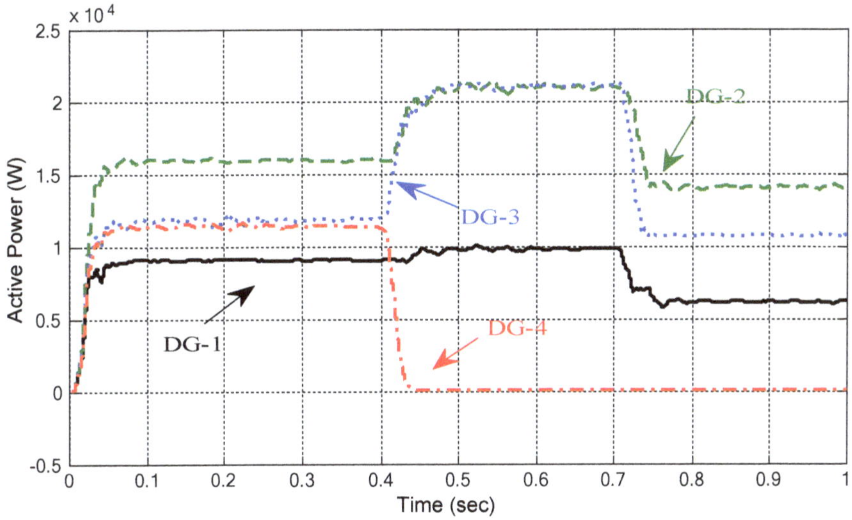

Fig. 5.12 Active power sharing with conventional controller

5.6.2 *Case 2: Proposed Controller*

To validate the performance of the proposed controller, the Energy Internet is operated as Case 1.

The proposed controller is started at 0.2 s. From the communication structure, only DG-1's PEA can receive the Main-Grid information from the Energy Router. The leader's information is shown in Table 5.1. The numerical values of power can be seen in Table 5.3. In Table 5.3, the output active power and reactive power from four DGs are shown in numerical results and the ratio between each DG output power to DG-1 output power are also shown in Table 5.3. The ratios in "()" are desired ratios between each DG to DG-1.

And it can be found that the output power-sharing ratios from four DGs are same with the desired ratios. The accuracies about power sharing can be verified. Compared with the conventional controller, DGs can share the active power and reactive power in the desired ratio as shown in Figs. 5.16 and 5.17. After 0.2 s, the amplitudes and angles of output voltages from DGs begin to reach a consensus with the leader shown in Figs. 5.18 and 5.19. And the current of the system is shown in Fig. 5.20. Thus the amplitude of output currents of four DGs are in desired ratio and the deviation angles between four output currents are very small as shown in Fig. 5.21. From the Eqs. (5.5) and (5.6), the circulating current between DGs can be minimized.

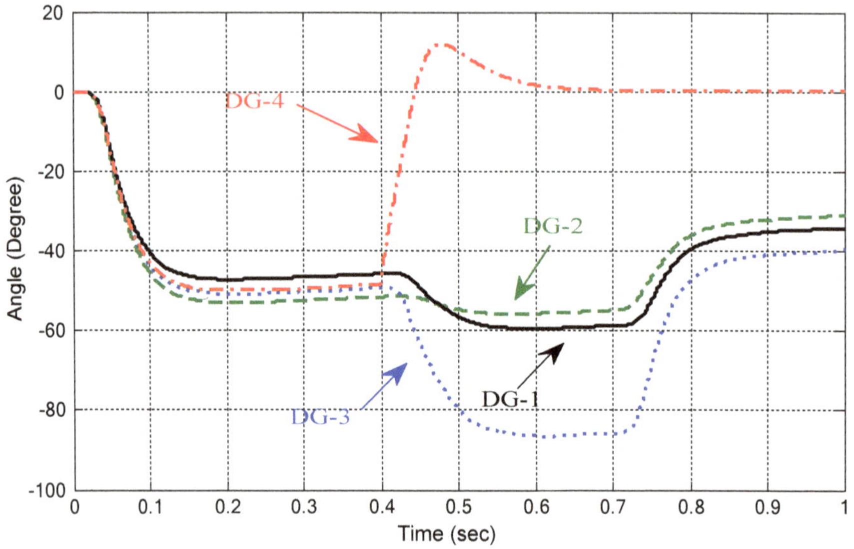

Fig. 5.13 Angles of output voltages from 4 DGs with conventional controller

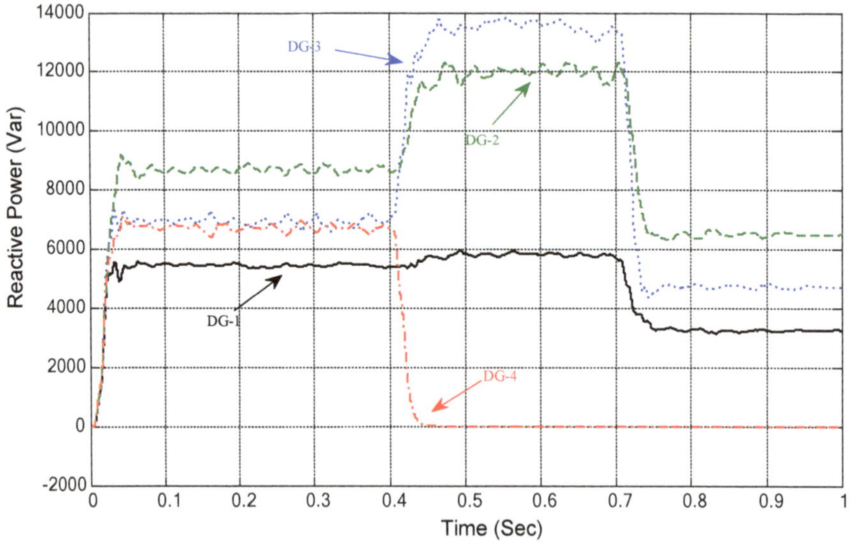

Fig. 5.14 Reactive power sharing with conventional controller

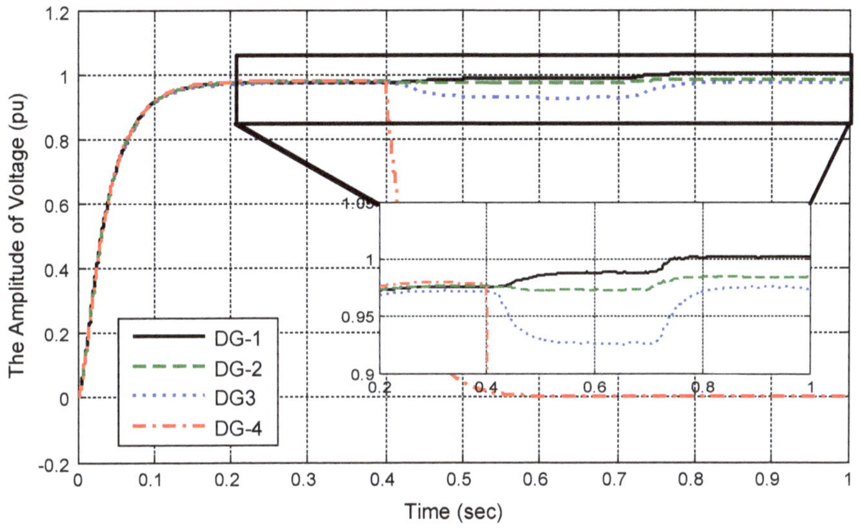

Fig. 5.15 Output voltages from 4 DGs with conventional controller

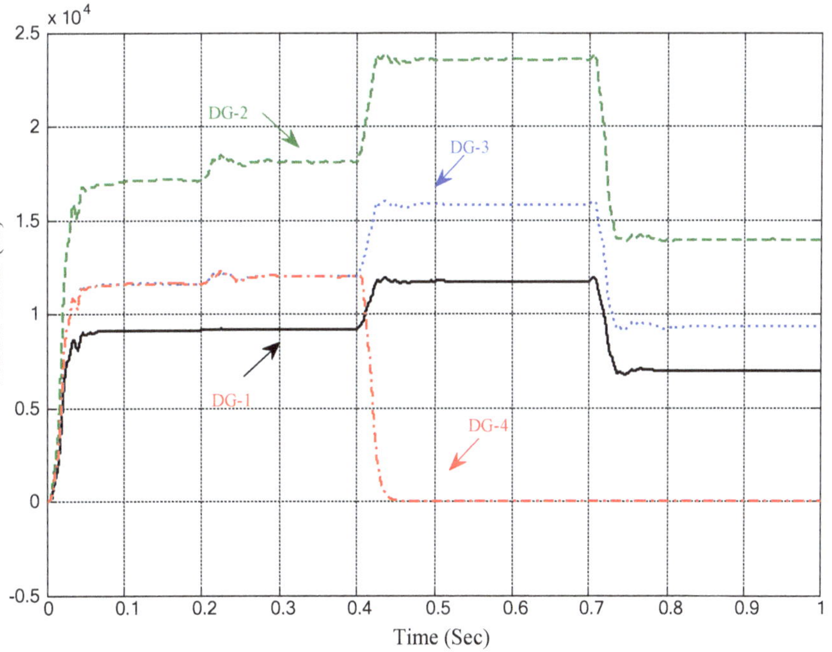

Fig. 5.16 Active power sharing

Table 5.2 Numerical results of the simulation with conventional controller

Active power	0 s–0.4 s	0.4 s–0.7 s	0.7 s–1 s
$P_{DG\text{-}1}$	9.1 kW	9.8 kW	6 kW
$P_{DG\text{-}2}$	17.1 kW	21 kW	14 kW
$P_{DG\text{-}3}$	11.6 kW	21 kW	11 kW
$P_{DG\text{-}4}$	11.6 kW	0 kW	0 kW
Sharing proportion	0 s–0.4 s	0.4 s–0.7 s	0.7 s–1 s
$P_{DG\text{-}1}$	1.0 (1.0)	1.0 (1.0)	1.0 (1.0)
$P_{DG\text{-}2}$	1.879 (2.0)	2.143 (2.0)	2.333 (2.0)
$P_{DG\text{-}3}$	1.26 (1.33)	2.142(1.33)	1.83 (1.33)
$P_{DG\text{-}4}$	1.26 (1.33)	0.0 (0.0)	0.0 (0.0)
Reactive power	0 s–0.4 s	0.4 s–0.7 s	0.7 s–1 s
$Q_{DG\text{-}1}$	5.4 kVAr	5.8 kVAr	3.2 kVAr
$Q_{DG\text{-}2}$	8.7 kVAr	12 kVAr	6.1 kVAr
$Q_{DG\text{-}3}$	6.8 kVAr	13 kVAr	4.7 kVAr
$Q_{DG\text{-}4}$	6.9 kVAr	0 kVAr	0 kVAr
Sharing proportion	0 s–0.4 s	0.4 s–0.7 s	0.7 s–1 s
$Q_{DG\text{-}1}$	1.0 (1.0)	1.0 (1.0)	1.0 (1.0)
$Q_{DG\text{-}2}$	1.61 (2.0)	2.06 (2.0)	1.9 (2.0)
$Q_{DG\text{-}3}$	1.29 (1.33)	2.24 (1.33)	1.469 (1.33)
$Q_{DG\text{-}4}$	1.27 (1.33)	0.0 (0.0)	0.0 (0.0)

Furthermore, since both angles and amplitudes of DGs output voltages in the Energy Internet can be controlled to reach consensus with the Main-Grid information due to the proposed method. With enough energy storage, the Energy Internet can be operated under spinning reserve condition.

Since large circulating currents in the Energy Internet can cause large unnecessary power losses, to reduce the circulating current can effectively decrease power losses in the Energy Internet. In addition, compared with the centralized approach:

Table 5.3 Numerical results of the simulation

Reduced system with proposed controller				
Active power	Initial value (0 s-0.2 s)	Intermediate value 1 (0.2 s–0.4 s)	Intermediate value 2 (0.4 s–0.7 s)	Final value (0.7 s–1 s)
P_{DG-1}	9.1 kW	9.2 kW	11.7 kW	7.0 kW
P_{DG-2}	17.1 kW	18.2 kW	23.6 kW	13.9 kW
Active power	Initial value (0 s–0.2 s)	Intermediate value 1 (0.2 s– 0.4 s)	Intermediate value 2 (0.4 s–0.7 s)	Final value (0.7 s–1 s)
P_{DG-3}	11.6 kW	12 kW	15.7 kW	9.3 kW
P_{DG-4}	11.6 kW	12 kW	0 kW	0 kW
Active power ratio of P_{DG-1}	Initial value (0 s–0.2 s)	Intermediate value 1 (0.2 s–0.4 s)	Intermediate value 2 (0.4 s–0.7 s)	Final value (0.7 s–1 s)
P_{DG-1}	1.0 (1.0)	1.0 (1.0)	1.0 (1.0)	1.0 (1.0)
P_{DG-2}	1.879 (2.0)	1.978 (2.0)	2.017 (2.0)	1.986 (2.0)
P_{DG-3}	1.26 (1.33)	1.304 (1.33)	1.341 (1.33)	1.328 (1.33)
P_{DG-4}	1.26 (1.33)	1.304 (1.33)	0.0 (0.0)	0.0 (0.0)
Reactive power	Initial value (0 s–0.2 s)	Intermediate value 1 (0.2 s–0.4 s)	Intermediate value 2 (0.4 s–0.7 s)	Final value (0.7 s–1 s)
P_{DG-1}	5.4 kVAr	4.9 kVAr	6.45 kVAr	3.5 kVAr
P_{DG-2}	8.7 kVAr	9.9 kVAr	12.95 kVAr	6.87 kVAr
P_{DG-3}	6.8 kVAr	6.6 kVAr	8.55 kVAr	4.63 kVAr
P_{DG-4}	6.7 kVAr	6.5 kVAr	0 kVAr	0 kVAr
Reactive power ratio of P_{DG-1}	Initial value (0 s– 0.2 s)	Intermediate value 1 (0.2 s–0.4 s)	Intermediate value 2 (0.4 s– 0.7 s)	Final value (0.7 s–1 s)
P_{DG-1}	1.0 (1.0)	1.0 (1.0)	1.0 (1.0)	1.0 (1.0)
P_{DG-2}	1.61 (2.0)	2.02 (2.0)	2.01 (2.0)	1.96 (2.0)
P_{DG-3}	1.26 (1.33)	1.34 (1.33)	1.33 (1.33)	1.32 (1.33)
P_{DG-4}	1.24 (1.33)	1.33 (1.33)	0.0 (0.0)	0.0 (0.0)

(1) Comparison about computational power

When a centralized controller is used, all the system information should be communicated to the central unit. With the increasing number of DGs and other energy sources, the communication burden may increase due to the additional variables and constraints. Meanwhile, the computation time of such a central controller is highly dependent on the number of devices in the control system. Thus large communication delays may occur in the system which may cause serious negative influence on the stability and system performances, thus reducing the system scalability drastically.

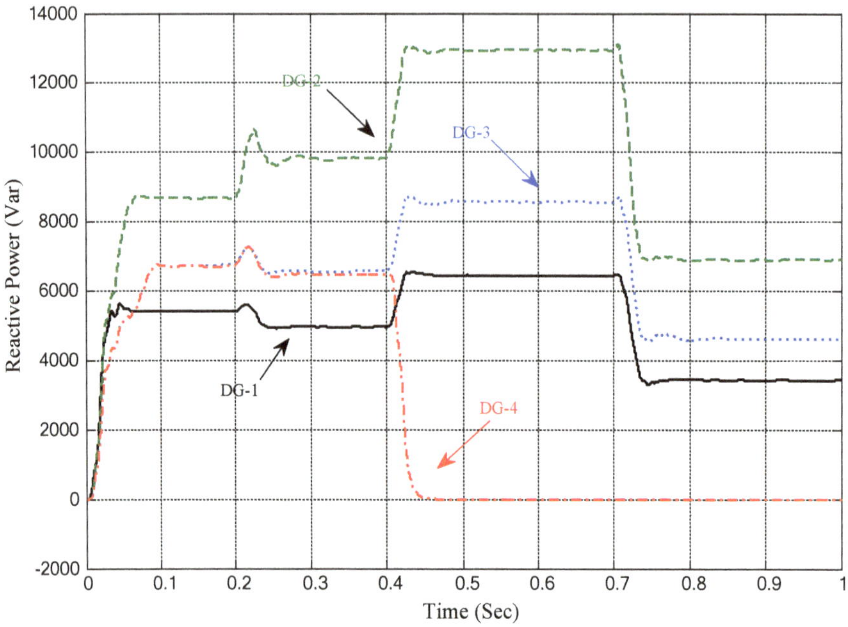

Fig. 5.17 Reactive power sharing

By comparison, in proposed controller, there is no centralized controller in the system and the communication load for each controller is uncorrelated with the increasing number of controllers in the system. Since the multi-agent consensus algorithm is applied in this paper, each controller only needs its local information and at least one of its neighbors' information. Thus each controller's computation load is nearly constant when controllers are added or removed from the control system.

(2) Comparison with the performance, resilience and scalability

When the system does not present faults, the performances between distributed or centralized controllers are almost equal. Indeed the centralized controller gives all the local DGs the same control signal, while the distributed controller make control decision of each DG according to its neighbors information. Thus the response speed of the distributed controller can be faster than that of centralized controller.

Under fault conditions, if the centralized controller fails down, the whole system will lose control and may become unstable. By comparison, once one distributed controller fails down, it cannot cause serious influence on the whole system. The reason of this is that each controller only needs its local and neighbors' information based on sparse communication structure. Furthermore, if this controller cannot be recovered in a short time, we can disconnect it to maintain the stability of the whole system. However, if the centralized controller cannot be recovered timely, it may not be possible to maintain stable operation. The stability about multi-agent consen-

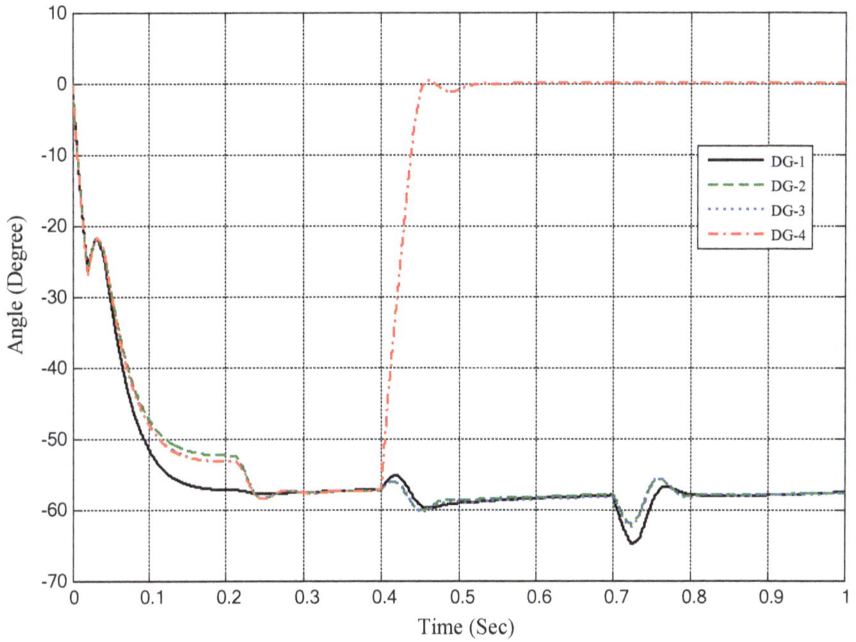

Fig. 5.18 Angles of output voltages from 4 DGs

sus algorithm considering switching topology has been proved in several previous works. In the future, this algorithm may be extended to the distributed controllers. Thus resilience of the distributed controller can be much better than the centralized controller one.

Based on the proposed algorithm combined with the multi-agent system, the distributed controller in this paper is still open for improvement, because several conditions are not included in this paper. Section 5.7 discusses some possibilities of improvement for reference to the interested reader. Meanwhile, the scale of the distributed system can be enlarged without communication limits, computation limits and algorithm limits. By comparison, the scale of the centralized controller is limited by its communication and computation ability. Thus the scalability of distributed controller is much higher than that of centralized controller.

5.7 Conclusion

The control issues of in Energy Internet were investigated in this paper. The multi-agent consensus algorithm and multi-agent system architecture are combined to design the control structure and control method applied in Energy Internet. The decomposed tasks, established and transformed model, information flow for each

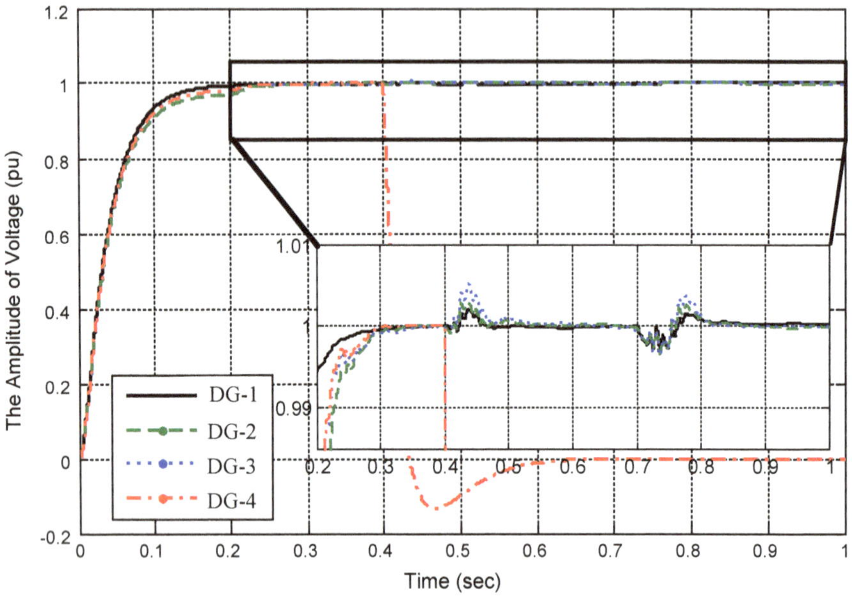

Fig. 5.19 The p.u. value of output voltages from 4 DGs

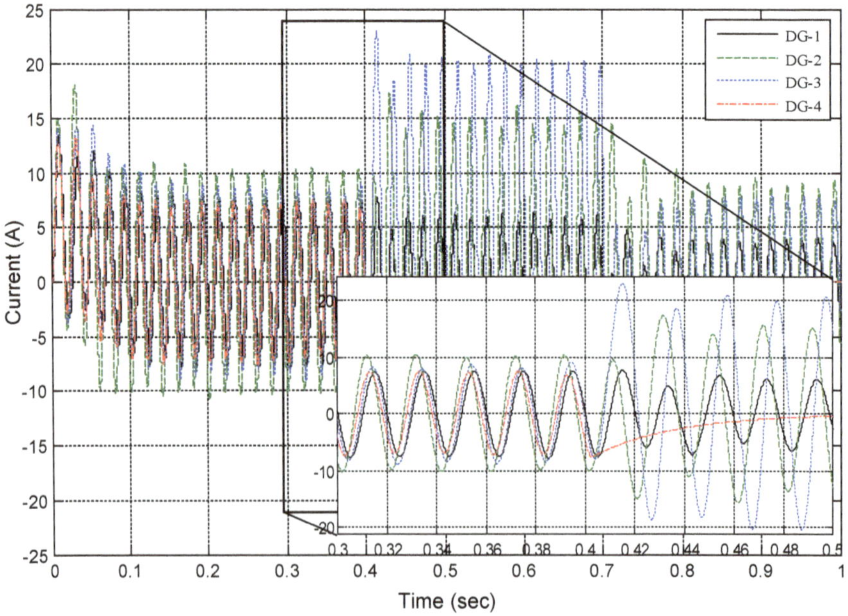

Fig. 5.20 The output current of distributed generators

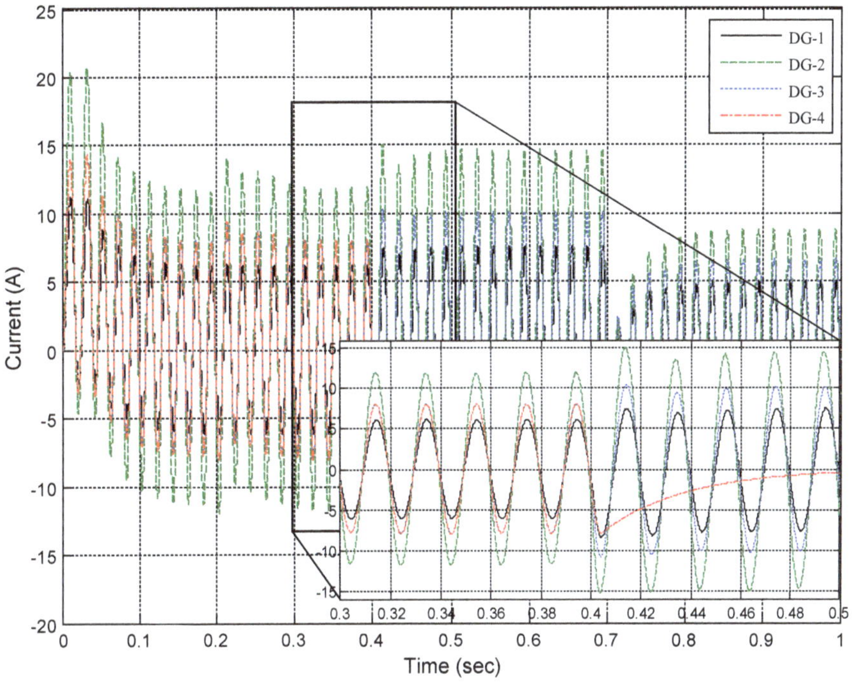

Fig. 5.21 The output current of system

agent are studied. The benefits of this paper are as follows: (1) The combination of multi-agent system and multi-agent consensus algorithm will provide an infrastructure for future research used in Energy Internet. (2) The circulating current reduction in Energy Internet and proportional power-sharing to the desired ratio among DGs can be guaranteed by combining with the $P - \delta$ and $Q - V$ controllers and multi-agent consensus algorithms. (3) The output voltages in Energy Internet can be recovered while being synchronized with the leader information from Main-Grid, which means the Energy Internet can operate as spinning reserve system. (4) For daily life, with sparse communication systems and bidirectional power flows, Energy Internet provides customers a reliable and efficient power supply, while minimizing energy costs and providing the different types of renewable energy resources the possibility to plug-in/out at any time.

Future work will consist of, but will not be limited to, the following aspects:

(1) The cost, location elements should be considered when the proposed algorithm decides the power sharing in Energy Internet. In this section, the power sharing between different DGs are only based on their rated power, but in practical the economic element is another important index. Meanwhile, the power market should be studied in Energy Internet in order to make the system operated more economical.

(2) The changes of communication topology should be considered in the future. In this section, the communication topology in the control system is kept constant. In the future, the switching topology multi-agent consensus algorithm should be applied in the control system. With this application, the comparison about resilience between distributed controller and centralized controller could be more obvious.
(3) The cost about hardware during the process of designation should be considered in the future. In this section, the cost about hardware between distributed controller and centralized controller are not compared which can illustrate benefits about distributed controller furthermore.
(4) In this section, we only introduce a little about the energy storage without its control strategy. Because the energy storage is very important under some extremely conditions, the control strategy for energy storage should be studied based on the proposed control architecture in the future.

References

1. M.C. Chandorkar, D.M. Divan, R. Adapa, Control of parallel connected inverters in standalone ac supply systems. IEEE Trans. Ind. Appl. **29**(1), 136–143 (1993)
2. Q. Sun, J. Zhou, J.M. Guerrero, H. Zhang, Hybrid three-phase/single-phase microgrid architecture with power management capabilities. IEEE Trans. Power Electron. **30**(10), 5964–5977 (2015)
3. Q. Sun, R. Han, H. Zhang, J. Zhou, J.M. Guerrero, A multiagent-based consensus algorithm for distributed coordinated control of distributed generators in the energy internet. IEEE Trans. Smart Grid **6**(6), 3006–3019 (2015)
4. D. Zhang, F. Wang, R. Burgos, R. Lai, D. Boroyevich, DC-link ripple current reduction for paralleled three-phase voltage-source converters with interleaving. IEEE Trans. Power Electron. **26**(6), 1741–1753 (2011)
5. J.M. Guerrero, J.C. Vasquez, J. Matas, J.G. de Vicuna, M. Castilla, Hierarchical control of droop-controlled AC and DC microgrids-a general approach toward standardization. IEEE Trans. Ind. Electron. **58**(1), 158–172 (2011)
6. Y.W. Li, C.-N. Kao, An accurate power control strategy for power-electronics-interfaced distributed generation units operating in a low-voltage multibus microgrid. IEEE Trans. Power Electron. **24**(12), 2977–2988 (2009)
7. A. Bidram, A. Davoudi, F.L. Lewis, S.S. Ge, Distributed adaptive voltage control of inverter-based microgrids. IEEE Trans. Energy Convers. **29**(4), 862–872 (2014)
8. J.M. Guerrero, L. Hang, J. Uceda, Control of distributed uninterruptible power supply systems. IEEE Trans Ind. Electron. **55**(8), 2845–2859 (2008)
9. X. Lu, J.M. Guerrero, K. Sun, J.C. Vasquez, R. Teodorescu, L. Huang, Hierarchical control of parallel ac-dc converter interfaces for hybrid microgrids. IEEE Trans. Smart Grid **5**(2), 683–692 (2014)
10. M. Savaghebi, A. Jalilian, J.C. Vasquez, J.M. Guerrero, Secondary control scheme for voltage unbalance compensation in an islanded droop-controlled microgrid. IEEE Trans. Smart Grid **3**(2), 797–807 (2014)
11. H. Zhang, T. Feng, G. Yang, H. Liang, Distributed cooperative optimal control for multiagent systems on directed graphs: an inverse optimal approach. IEEE Trans. Cybern. **45**(7), 1315–1326 (2014)

12. H. Zhang, J. Zhang, G. Yang, Y. Luo, Leader-based optimal coordination control for the consensus problem of multi-agent differential games via fuzzy adaptive dynamic programming. IEEE Trans. Fuzzy Syst. **23**(1), 152–163 (2015)
13. A.L. Dimeas, N.D. Hatziargyriou, Operation of a multiagent system for microgrid control. IEEE Trans. Power Syst. **20**(3), 1447–1455 (2005)
14. P. Papadopoulos, N. Jenkins, L.M. Cipcigan, I. Grau, E. Zabala, Coordination of the charging of electric vehicles using a multi-agent system. IEEE Trans. Smart Grid **4**(4), 1802–1809 (2013)
15. E.L. Karfopoulos, N.D. Hatziargyriou, A multi-agent system for controlled charging of a large population of electric vehicles. IEEE Trans. Power Syst. **28**(2), 1196–1204 (2013)
16. A. Bidram, A. Davoudi, F.L. Lwis, J.M. Guerrero, Distributed cooperative secondary control of microgrids using feedback linearization. IEEE Trans. Power Syst. **28**(3), 3462–3470 (2013)
17. W. Liu, W. Gu, W. Sheng, X. Meng, Z. Wu, W. Chen, Decentralized multi-agent system-based cooperative frequency control for autonomous microgrid with communication constraints. IEEE Trans. Sustain. Energy **5**(2), 446–456 (2014)
18. W. Yao, M. Chen, J. Matas, J.M. Guerrero, Z.M. Qian, Design and analysis of the droop control method for parallel inverters considering the impact of the complex impedance on the power sharing. IEEE Trans. Ind. Electron. **58**(2), 576–588 (2011)
19. J.E. Slotine, W. Li, *Applied Nonlinear Control* (Prentice-Hall, Upper Saddle River, 2009)
20. H. Zhang, D. Liu, Y. Luo, D. Wang, *Adaptive Dynamic Programming for Control-Algorithms and Stability* (Springer, London, 2013)
21. M.Q. Wang, H.B. Gooi, Spinning reserve estimation in microgrids. IEEE Trans. Power Syst. **26**(3), 1164–1174 (2011)
22. Y. Rebours, D.S. Kirschen, *What is Spinning Reserve*? (2005), http://eee.dev.ntweb.mcc.ac.uk/research/groups/eeps/publications/reportstheses/aoe/reboursetal_techrep_2005A.pdf
23. L. Rao, X. Liu, M.D. Ilic, J. Liu, Distributed coordination of internet data centers under multiregional electricity markets. Proc. IEEE **100**(1), 269–282 (2012)
24. T.L. Vandoorn, J.C. Vasquez, J.D. Kooning, J.M. Guerrero, L. Vandevelde, Microgrids: hierarchical control and overview of the control and reserve management strategies. IEEE Ind. Electron. Mag. **7**(4), 42–55 (2013)
25. C. Yuen, A. Oudalov, A. Timbus, The provision of frequency control reserves from multiple microgrids. IEEE Trans. Ind. Electron. **58**(1), 173–183 (2011)

Chapter 6
Control Strategy and Stability Analysis of Energy Router

Abstract This chapter introduces the structure of energy router in detail, deduction of the mathematical model of energy router, design of an energy management strategy and analyzation of the stability of energy router. By analyzing the mathematical model of the energy router, an optimal energy flow strategy is designed and the power exchange of each subsystem in the energy router is realized. The small signal model of the energy router is obtained and a new stability criterion is designed to judge the stability of the system.

6.1 Introduction

The Energy Router is an emerging device concept that is based on an advanced power electronic technique. It is able to realize flexible and dynamic electric power distribution in power systems analogous to the function of information routers in the internet [1]. In the energy internet vision, the energy router will become an important hardware foundation of the future distribution network. The analysis method and model of the energy router is of great significance to realize the optimal operation of distribution system.

At present, research on the energy router is mainly concentrated on its inner power electronic configuration [2]. The project on the Future Renewable Electric Energy Delivery and Management System (FREEDM) carried out by North Carolina State University is designing a solid-state-transformer-based version of the energy router for the purposes of developing a smart distribution network and micro grid [3]. The energy router is also equipped with intelligent communications that facilitate the smart control of both loads and generators in the distribution network or micro grid [4]. Researchers have reported different power electronic topologies for energy routers [5–7]. With the appropriate control strategies for power electronics, the power output/injection can be independently controlled to meet consumers' demands [8, 9]. Using the energy router as a basic element, the distribution system can achieve smarter energy management and improve the performance of the energy internet [10, 11].

Q. Sun, *Energy Internet and We-Energy*, Renewable Energy Sources & Energy Storage, https://doi.org/10.1007/978-981-13-0523-8_6

6.2 Mathematical Model of Energy Router

Figure 6.1 presents a new type of hybrid multi interface energy router architecture. This architecture expands the rich DC interface at the low voltage DC bus while also enabling multi-energy complementation in the energy internet with high new energy penetration rates. Power electronic transformer (PET) is the core of energy router. The power electronic transformer topology used in this book is a three-phase three-stage structure, which includes rectifier stage, Dual Active Bridge (DAB) converter stage and inverter stage.

Mathematical modeling of the system is the basis for its objective theoretical analysis as well as the cornerstone of improving the performance of the controller. Therefore, the system needs to be modeled first and then the controller design. In the actual modeling of single-phase inverter system, it is necessary to approximate for some conditions [12]:

(1) The switching devices required in the inverter are ideal switches, which is to say, they ignore the turn-on and turn-off delays.
(2) Since the system carrier frequency is much larger than the fundamental frequency, it can be approximated that the voltage and current in the inverter do not change during the adjacent switching cycles.
(3) The filter inductor used does not consider its magnetic saturation problem.
(4) Output filter capacitor is purely original, no resistance and parasitic inductance.

6.2.1 *Modeling of Rectifier Level Topology*

As shown in Fig. 6.2, this book is the study of three-phase three-wire two-level voltage rectifier circuit. It is composed of 6 IGBTs and 6 anti-parallel diodes. The AC side adopts the neutral symmetrical connection of three phases. It is important to

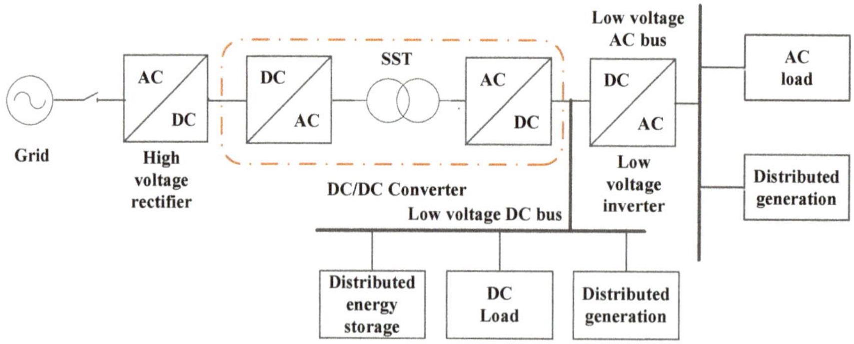

Fig. 6.1 Topology structure of SST system

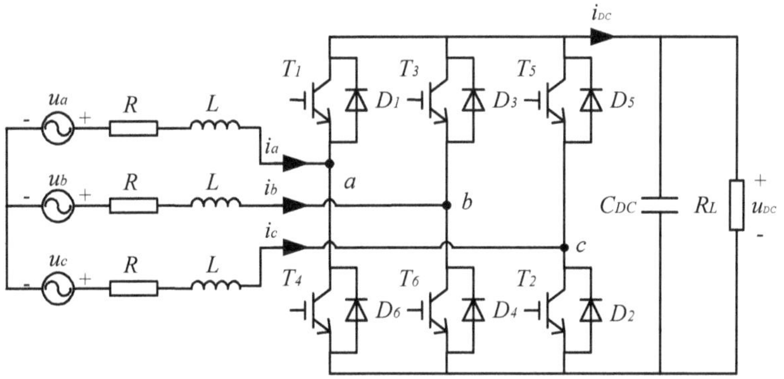

Fig. 6.2 The topology structure of rectifier for SST

note that the better the control performance of the voltage-fed bridge rectifier in the three-phase power grid, the more the unit-side power factor can be achieved [13].

The mathematical model of voltage source rectifier topology is established and the following assumptions are made:

(1) The AC side is a three-phase symmetrical voltage source, that is to say, $u_a + u_b + u_c = 0$, $i_a + i_b + i_c = 0$.
(2) The filter inductance of the network side is purely inductive, and the ideal inductance is not saturated [14].
(3) The power switch is represented by an ideal switch and a loss resistance (loss resistance included in the resistor).

For three phase symmetrical power supply, $u_{nN} = -1/3(S_a + S_b + S_c)u_{DC}$, so

$$\begin{cases} u_{aN} = (S_a - 1/3(S_a + S_b + S_c))u_{DC} \\ u_{bN} = (S_b - 1/3(S_a + S_b + S_c))u_{DC} \\ u_{cN} = (S_c - 1/3(S_a + S_b + S_c))u_{DC} \end{cases} \tag{6.1}$$

In the equation, S_j ($j = a, b, c$) is a unipolar logic switch function, $S_j = 1$ (upper bridge arm turn on, lower bridge arm turn off), $S_j = 0$ (upper bridge arm turn off, lower bridge arm turn on). The mathematical model of three-phase three-wire two-level voltage source PWM rectifier in three-phase coordinate system is shown in formula (6.2)

$$\begin{cases} L\frac{di_a}{dt} = u_a - Ri_a - d_a u_{DC} \\ L\frac{di_b}{dt} = u_b - Ri_b - d_b u_{DC} \\ L\frac{di_c}{dt} = u_c - Ri_c - d_c u_{DC} \\ C\frac{du_{DC}}{dt} = \sum\limits_{j=a,b,c} S_j i_j - i_L \end{cases} \tag{6.2}$$

In the equation, u_j, i_j ($j = a, b, c$) represent the three phase AC power supply voltage and three phase AC input current, L, R represent the total inductance and total resistance for AC side, C represent the DC side filter capacitor, R_L represent the load resistance, $i_L = u_{DC}/R_L$ represent the load current, u_{DC} represent the DC side voltage, $d_j = S_j - 1/3(S_a + S_b + S_c)$.

The mathematical model of PWM rectifier in three-phase *abc* coordinate system is transformed into the mathematical model in two-phase $\alpha\beta$ coordinate system. Equation 6.2 can be converted to Eq. 6.3 by an equal power transformation matrix $T_{Cabc/\alpha\beta}$.

$$T_{Cabc/\alpha\beta} = \frac{2}{3}\begin{bmatrix} 1 & -\frac{1}{2} & -\frac{1}{2} \\ 0 & \frac{\sqrt{3}}{2} & -\frac{\sqrt{3}}{2} \end{bmatrix}, T_{Pabc/\alpha\beta} = \sqrt{\frac{2}{3}}\begin{bmatrix} 1 & -\frac{1}{2} & -\frac{1}{2} \\ 0 & \frac{\sqrt{3}}{2} & -\frac{\sqrt{3}}{2} \end{bmatrix} \tag{6.3}$$

The PWM rectifier model with equivalent transformation is shown in Eq. (6.4). In the equation, $u_\alpha = U_m \cos\omega t$, $u_\beta = U_m \sin\omega t$, $S_\alpha = (2S_a - S_b - S_c)/3$, $S_\alpha = \sqrt{3}(S_b + S_c)$.

$$\begin{cases} L\frac{di_\alpha}{dt} = u_\alpha - Ri_\alpha - S_\alpha u_{DC} \\ L\frac{di_\beta}{dt} = u_\beta - Ri_\beta - S_\beta u_{DC} \\ C\frac{du_{DC}}{dt} = \frac{3}{2}(S_\alpha i_\alpha + S_\beta i_\beta) - i_L \end{cases} \tag{6.4}$$

Through the rotation transformation, Eq. (6.4) can be transfered into (6.5).

$$\begin{cases} L\frac{di_d}{dt} + \omega Li_q = u_d - Ri_d - S_d u_{DC} \\ L\frac{di_q}{dt} - \omega Li_d = u_\beta - Ri_q - S_q u_{DC} \\ C\frac{du_{DC}}{dt} = \frac{3}{2}(S_d i_d + S_q i_q - i_L). \end{cases} \tag{6.5}$$

6.2.2 Modeling of DAB Level Topology

The bi-directional full bridge converter with isolation transformer is shown in Fig. 6.3. The original side of the high-frequency transformer and secondary rectifier inverter unit are full-bridge structure. The original side is the voltage type and the low-voltage side is the current type.

Through the above analysis it can be drawn that DAB equivalents to high-frequency transformer circuit shown in Fig. 6.4. Further the modal analysis can draw leakage current waveform which is shown in Fig. 6.5. Due to the symmetry of the circuit, only half-cycle leakage current waveforms are considered. In the figure, d stands for phase shift duty ratio, V_{dc_in} stands for high voltage DC voltage, V_{dc_out}

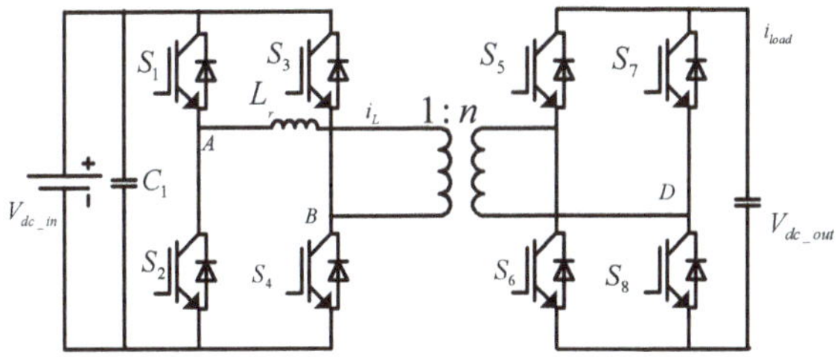

Fig. 6.3 Topology structure of DAB

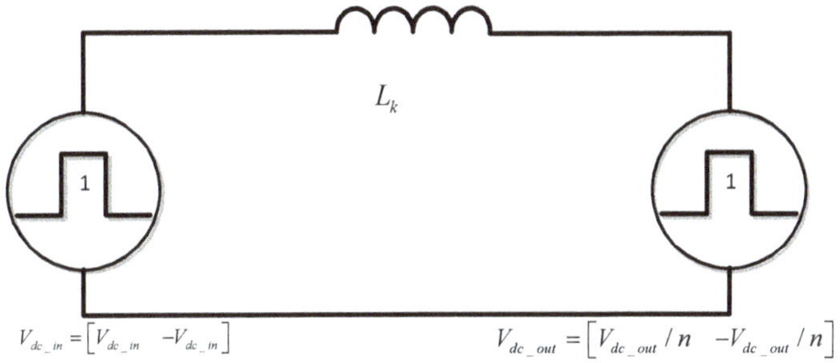

Fig. 6.4 Equivalent circuit of leakage inductance for DAB

Fig. 6.5 Equivalent analysis for input current of DAB

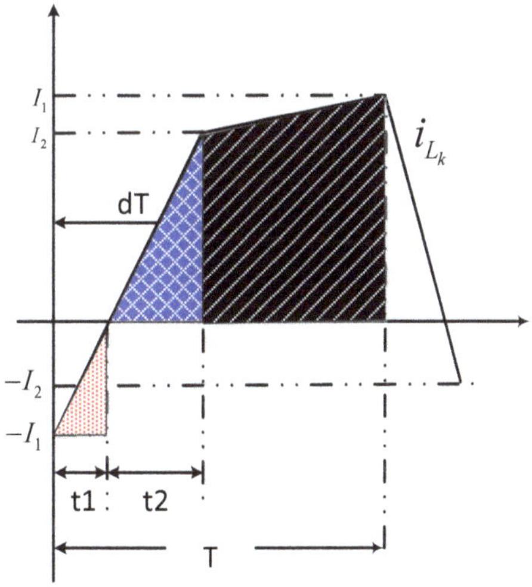

stands for low voltage DC voltage, V_1 stands for high voltage side square wave ($V_1 = V_{AB}$), V_2 stands for the low pressure side of square wave ($V_2 = V_{CD}$), I_1 and I_2 respectively represent the current value of the leakage current when the lead-bridge is opened and the lag-bridge is opened.

The leakage inductance formula can be deduced from Faraday's law (6.6):

$$\frac{di_{Lk}}{dt} = \frac{V_{AB} - V_{CD}}{L_k} = \frac{V_1 - V_2}{L_k} \tag{6.6}$$

The leakage current in accordance with phase shift duty cycle accounted for half the cycle time is divided into different states (6.7):

$$\begin{cases} V_{dc\_in} + \frac{V_{dc\_out}}{n} = L_k \frac{I_1 + I_2}{d \bullet T} & 0 < t < d \bullet T \\ V_{dc\_out} - \frac{V_{dc\_out}}{n} = L_k \frac{I_1 - I_2}{(1-d) \bullet T} & d \bullet T < t < T \end{cases} \tag{6.7}$$

$$\begin{cases} \frac{I_1}{t_1} = \frac{I_2}{t_2} \\ t_1 + t_2 = d \bullet T \end{cases} \tag{6.8}$$

The expression of current and its conduction time can be obtained by Eqs. 6.7 and 6.8:

$$\begin{gathered} I_1 = \frac{T}{2L_k}\left(2d\frac{V_{dc\_out}}{n} + V_{dc\_in} - \frac{V_{dc\_out}}{n}\right), I_2 = \frac{T}{2L_k}\left(2d\,V_{dc\_in} - V_{dc\_in} + \frac{V_{dc\_out}}{n}\right) \\ t_1 = T\left(\frac{2d\frac{V_{dc\_out}}{n} + V_{dc\_in} - \frac{V_{dc\_out}}{n}}{2\left(V_{dc\_in} + \frac{V_{dc\_out}}{n}\right)}\right), t_2 = T\left(\frac{2d\,V_{dc\_in} - V_{dc\_in} + \frac{V_{dc\_out}}{n}}{2\left(V_{dc\_in} + \frac{V_{dc\_out}}{n}\right)}\right) \end{gathered} \tag{6.9}$$

Since the DAB current flows from the high side to the low side, we can calculate the average current flowing to the low side capacitor and the load. The current equivalent area plotted in Fig. 6.5 shows the average output current (6.10):

$$\overline{i_o} = \frac{1}{nT}\left(\frac{1}{2}I_1 t_1 - \frac{1}{2}I_2 t_2 + (1-d)TI_2 + \frac{1}{2}(1-d)T(I_1 - I_2)\right) \tag{6.10}$$

The output average current Eq. 6.10 is further deduced from formula (6.11):

$$\overline{i_o} = \frac{(1-d)T\,V_{dc\_in}}{nL_k} \tag{6.11}$$

The area diagram of the input current can also be equivalent and the average equation of the input current is obtained (6.12):

$$\overline{i_l} = \frac{1}{T}\left(-\frac{1}{2}I_1 t_1 + \frac{1}{2}I_2 t_2 + (1-d)TI_2 + \frac{1}{2}(1-d)T(I_1 + I_2)\right) \tag{6.12}$$

Simultaneous Eqs. 6.11 and 6.12 can be obtained Eq. 6.13:

$$\overline{i_l} = \frac{(1-d)T\,V_{dc\_out}}{nL_k} \tag{6.13}$$

The average expression of the input and output current is obtained by Eqs. 6.11 and 6.13. In order to analysis the stability in the next section, the verification of the model's stability by small signal analysis method is necessary. The current expression of the disturbance in the specified location, can be deducted to following Eqs. (6.14) and (6.15):

$$\begin{cases} \hat{i}_l = \frac{\partial \bar{i}_l}{\partial d}\Big|_{\widehat{V_0}=0} \hat{d} + \frac{\partial \bar{i}_l}{\partial V_0}\Big|_{\hat{d}=0} \widehat{V_o} = G_{id}\hat{d} + G_{ivo}\widehat{V_0} \\ \hat{i}_o = \frac{\partial \bar{i}_l}{\partial d}\Big|_{\widehat{V_i}=0} \hat{d} + \frac{\partial \bar{i}_o}{\partial \widehat{V_i}}\Big|_{\hat{d}=0} \widehat{V_i} = G_{od}\hat{d} + G_{ovi}\widehat{V_i} \end{cases} \tag{6.14}$$

$$\begin{cases} G_{od} = \frac{V_o(1-2D)}{(1-D)DR} \\ G_{ov_i} = \frac{V_o}{V_i R} \\ G_{id} = \frac{V_o^2(1-2D)}{V_i(1-D)DR} = \frac{V_o}{V_i} G_{od}. \\ G_{iv_o} = \frac{V_o}{V_i R} \end{cases} \tag{6.15}$$

6.2.3 *Modeling of Inverter Level Topology*

Figure 6.6 is the topology of a single-phase inverter for bridge type, and its filtering uses LC filtering. V_{dc_in} represents DC bus interfaced voltage, L_1 is DC bus interfaced voltage, R_1 stands for equivalent impedance of filter inductance, C is output side filter capacitor, R_{load} is load. In the modeling process, consider the filter capacitor is purely original and the carrier frequency is much larger than the fundamental frequency.

Modeling and analysis of single-phase inverter are shown in Fig. 6.6, according to Kirchhoff's law of voltage and current, there is Eq. 6.16:

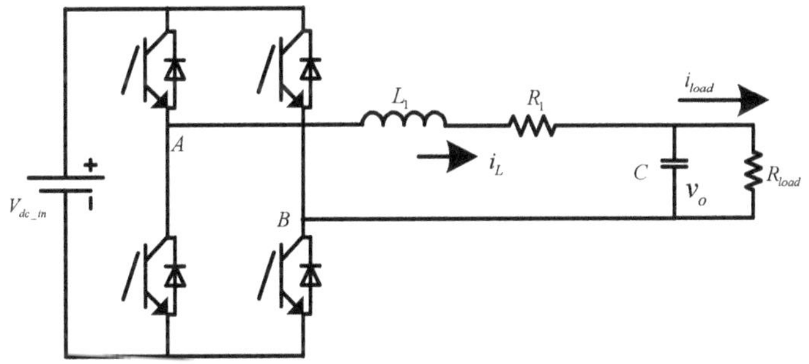

Fig. 6.6 The topology of the single-phase inverter in islanding

$$\begin{cases} L_1 \dfrac{di_L}{dt} = v_{AB} - v_o - R_1 i_L \\ C \dfrac{dv_o}{dt} = i_L - i_{load} \end{cases} \tag{6.16}$$

Write the state space equation from Eq. 6.16, there is Eq. 6.17:

$$\frac{d}{dt}\begin{bmatrix} i_L \\ v_o \end{bmatrix} = \begin{bmatrix} -\frac{R_1}{L_1} & -\frac{1}{L_1} \\ \frac{1}{C} & 0 \end{bmatrix}\begin{bmatrix} i_L \\ v_o \end{bmatrix} + \begin{bmatrix} 0 & \frac{1}{L_1} \\ -\frac{1}{C} & 0 \end{bmatrix}\begin{bmatrix} i_{load} \\ v_{AB} \end{bmatrix} \tag{6.17}$$

In the single-phase inverter system, the phase-shifted method is used to construct a quadrature signal with a 90° difference from the input current to obtain a two-phase $\alpha\beta$ signal. The model of the stationary two-phase coordinate system thus constructed is shown in Eq. 6.18. For single-phase inverter structure, the method of constructing a stationary coordinate system can make it have a similar approach with the three-phase structure.

$$\begin{cases} L_1 \dfrac{di_{L\alpha}}{dt} = v_{AB\alpha} - v_{o\alpha} - R_1 i_{L\alpha} \\ L_1 \dfrac{di_{L\beta}}{dt} = v_{AB\beta} - v_{o\beta} - R_1 i_{L\beta} \\ C \dfrac{dv_{o\alpha}}{dt} = i_{L\alpha} - i_{load\alpha} \\ C \dfrac{dv_{o\beta}}{dt} = i_{L\beta} - i_{load\beta} \end{cases} \tag{6.18}$$

Write the state space equation from Eq. 6.18, there is Eq. 6.19:

$$\frac{d}{dt}\begin{bmatrix} i_{L\alpha} \\ i_{L\beta} \\ v_{o\alpha} \\ v_{o\beta} \end{bmatrix} = \begin{bmatrix} -\frac{R_1}{L_1} & 0 & -\frac{1}{L_1} & 0 \\ 0 & -\frac{R_1}{L_1} & 0 & -\frac{1}{L_1} \\ \frac{1}{C} & 0 & 0 & 0 \\ 0 & \frac{1}{C} & 0 & 0 \end{bmatrix}\begin{bmatrix} i_{L\alpha} \\ i_{L\beta} \\ v_{o\alpha} \\ v_{o\beta} \end{bmatrix} + \begin{bmatrix} 0 & 0 & \frac{1}{L_1} & 0 \\ 0 & 0 & 0 & \frac{1}{L_1} \\ -\frac{1}{C} & 0 & 0 & 0 \\ 0 & -\frac{1}{C} & 0 & 0 \end{bmatrix}\begin{bmatrix} v_{AB\alpha} \\ v_{AB\beta} \end{bmatrix} \tag{6.19}$$

When the Park transformation is carried out, equal amplitude transformation matrix is Eq. 6.20:

$$T_{\alpha\beta-dq} = \begin{bmatrix} \sin(\omega t) & \cos(\omega t) \\ \cos(\omega t) & -\sin(\omega t) \end{bmatrix} \tag{6.20}$$

The inverse transformation matrix is:

$$T_{dq-\alpha\beta} = T_{\alpha\beta-dq} = \begin{bmatrix} \sin(\omega t) & \cos(\omega t) \\ \cos(\omega t) & -\sin(\omega t) \end{bmatrix} \tag{6.21}$$

Equations (6.19), (6.20) and (6.21) derive state space equation under d-q coordinate system:

$$\frac{d}{dt}\begin{bmatrix} i_{Ld} \\ i_{Lq} \\ v_{od} \\ v_{oq} \end{bmatrix} = \begin{bmatrix} -\frac{R_1}{L_1} & \omega & -\frac{1}{L_1} & 0 \\ -\omega & -\frac{R_1}{L_1} & 0 & -\frac{1}{L_1} \\ \frac{1}{C} & 0 & 0 & \omega \\ 0 & \frac{1}{C} & -\omega & 0 \end{bmatrix}\begin{bmatrix} i_{Ld} \\ i_{Lq} \\ v_{od} \\ v_{oq} \end{bmatrix} + \begin{bmatrix} 0 & 0 & \frac{1}{L_1} & 0 \\ 0 & 0 & 0 & \frac{1}{L_1} \\ -\frac{1}{C} & 0 & 0 & 0 \\ 0 & -\frac{1}{C} & 0 & 0 \end{bmatrix}\begin{bmatrix} v_{ABd} \\ v_{ABq} \end{bmatrix} \tag{6.22}$$

The Eq. 6.23 of the inner loop control of the amount of current differential equation is written as follows:

$$\begin{cases} \frac{di_{Ld}}{dt} = \frac{1}{L_1}(v_{ABd} - v_{od}) + \omega i_{Lq} - \frac{R_1}{L_1} i_{Ld} \\ \frac{di_{Lq}}{dt} = \frac{1}{L_1}(v_{ABq} - v_{oq}) - \omega i_{Ld} - \frac{R_1}{L_1} i_{Lq} \end{cases} \tag{6.23}$$

A mathematical model in frequency domain can be obtained by the Laplace transformation of Eq. 6.24:

$$\begin{cases} v_{ABd} - v_{od} + \omega L_1 i_{Lq} = (L_1 s + R_1) i_{Ld} \\ v_{ABq} - v_{oq} + \omega L_1 i_{Ld} = (L_1 s + R_1) i_{Lq} \end{cases} \tag{6.24}$$

The mathematical model in frequency domain is shown in Fig. 6.7.

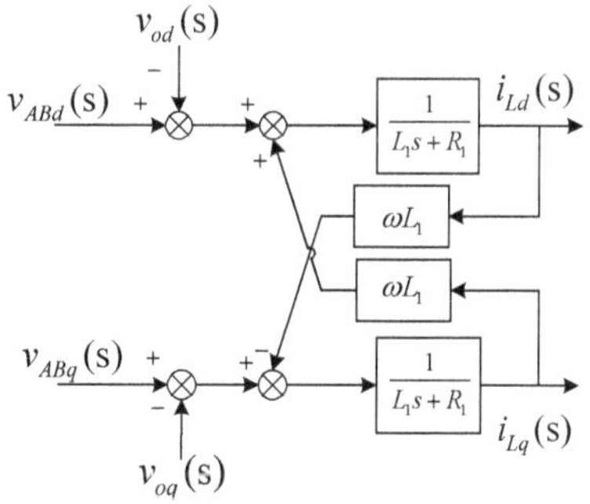

Fig. 6.7 The mathematical model of the inductor current in synchronous coordinate

6.3 Control and Power Optimization Strategy of Energy Router

6.3.1 Control Strategy of Energy Router

Rectifier in the SST undertake exchange function of the external power with grid power. When the power flows from the distribution network to SST, it is required that the rectifier can realize the unit power factor, inject less harmonic on the network side, protect the rapid action after the distribution network fault and then switch to the island mode smoothly. Obviously, when the grid is a three-phase symmetrical system, the grid-side reactive power is zero. u_d is constant, $u_q = 0$. According to the instantaneous power theory, the instantaneous active power and reactive power of the system are respectively Eq. (6.25):

$$\begin{cases} p = \frac{3}{2}u_d i_d \\ q = \frac{3}{2}u_q i_q \end{cases} \tag{6.25}$$

The system block diagram based on the grid voltage directional feedforward decoupling control is shown in Fig. 6.8. The control system consists of the DC voltage outer loop, the active current inner loop, the reactive current inner loop and the feedforward decoupling control loop. The DC voltage outer loop is designed to stabilize the DC side voltage. Obviously, direct current voltage feedback and a PI controller can be used to achieve DC voltage no-static control.

Since u_{DC} can be realized by the control of i_d, the output of the DC voltage outer loop PI controller is the current reference value of the active current inner loop. So that the active power of the PWM rectifier can be controlled.

The module DAB of SST is controlled by phase shift pulse width modulation (PWM). This control method is easy to be realized and can effectively eliminate

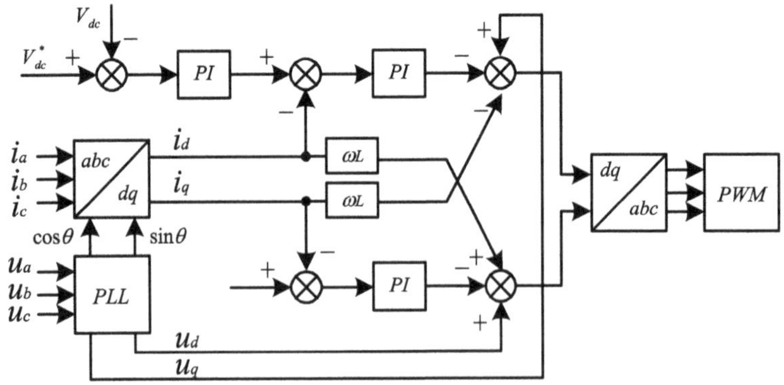

Fig. 6.8 Diagram of rectifier based VOC control

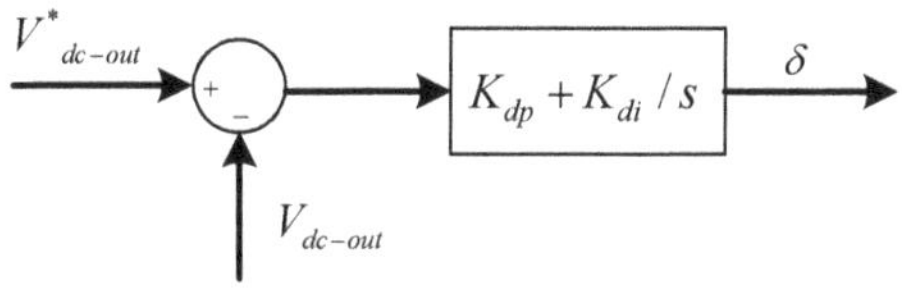

Fig. 6.9 The control diagram of DAB

low order harmonics. In the form of driving signal, the phase-shift control method adopted will not cause the output voltage distortion, and it is easy to realize soft switching control. For phase-shift control, there is a certain phase angle difference between the PWM waveform of the full bridge converter on the primary side and the second side of the high frequency transformer. The relation between the transmitted power and the phase shift angle is shown in Eq. 6.26.

$$P_o = \frac{V_{dc}V_{dc-low}}{2Lf_H}\delta(1-\delta) \tag{6.26}$$

In the equation, P_o is the active power transmitted for the DC-DC unit, V_{dc} is the input DC voltage to the high voltage side, f_H is the switching frequency, L is the leakage inductance, V_{dc-low} is the DC voltage output for the low-voltage side, and δ is the phase shift angle of the original side modulation signal. According to Eq. 6.26, we can get that the output voltage V_{dc-low} can be controlled by controlling the phase shift angle δ, the phase shift angle δ is directly controlled by PI. The control block diagram is shown in Fig. 6.9.

The inverter has two working modes of disconnecting the power grid (off-grid) and connecting the power grid (on-grid). In the off-grid mode, the design of the controller using the inner loop of the current and outer loop of the voltage can ensure the steady state performance and dynamic performance of the output voltage and current. To achieve this goal requires decoupling the inverter, and then control the amount of direct current. PI control in the d-q coordinate system can ensure the control of the target without static tracking.

In the previous sections, the mathematical model of single-phase inverters has been analyzed in detail and decoupled. In practical situations, the sampling delay caused by the transformer and the hardware delay caused by the modulation wave and the driver needs to be introduced into the controller. The current loop controller model with the inductance current as the control amount is shown in Fig. 6.10. The above two kinds of minute delays are merged. In Fig. 6.10, both d and q axis current control are implemented by PI controller.

According to Fig. 6.10, the open loop transfer function of the current loop is obtained:

$$G_i(s) = \frac{K_{PWM}(K_{ip}s + K_{ii})}{s(1.5T_s s + 1)(L_1 s + R_1)} \tag{6.27}$$

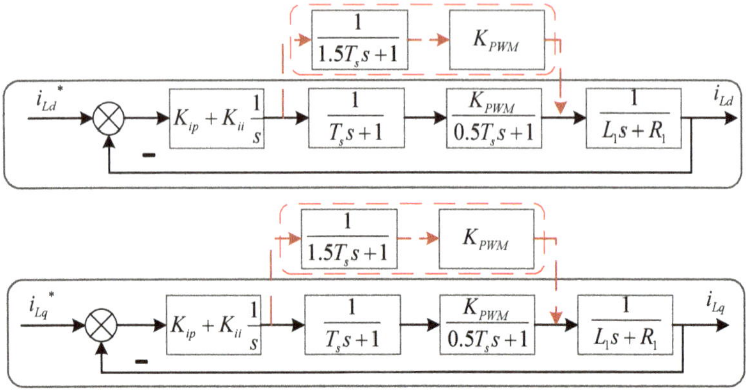

Fig. 6.10 The simplified block diagram of the current loop

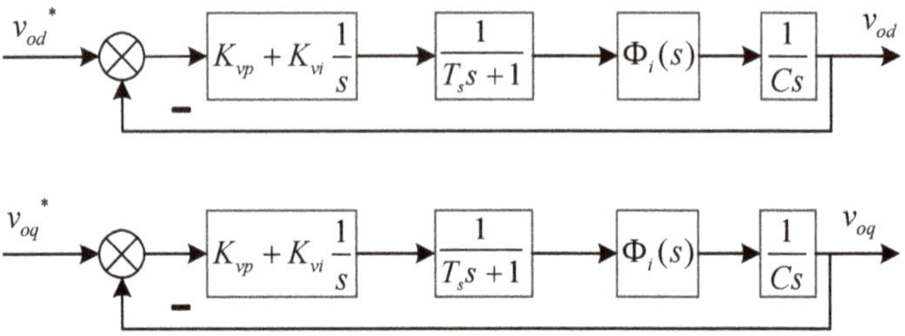

Fig. 6.11 The simplified block diagram of the voltage loop

Because of the decoupling of the external voltage loop of the inverter, the voltage loop of the inverter stage is simplified and the sampling delay of the voltage signal is taken into consideration. The block diagram is shown in Fig. 6.11.

The open loop transfer function of the external voltage loop can be obtained according to Fig. 6.11:

$$G_v(s) = \frac{(K_{vp}s + K_{vi})\Phi_i(s)}{Cs^2(T_s s + 1)} \tag{6.28}$$

Equations 6.27 and 6.28 respectively analyze the transfer function of the current inner loop and the voltage outer loop. The design of droop-based double closed-loop control requires the construction of $V - Q$, $P - f$ droop relations, which can be explained by the bottom equation:

$$\begin{cases} \omega = \omega_n - m(P - P_n) \\ V = V_n - n(Q - Q_n) \end{cases} \tag{6.29}$$

Fig. 6.12 The curve of droop control

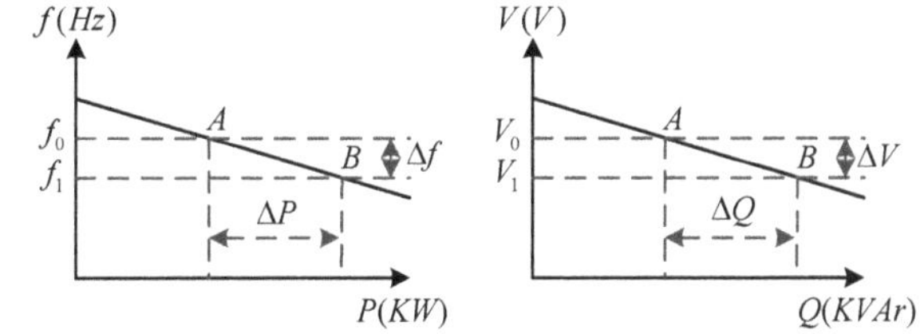

(a) P-f droop characteristics (b) Q-V droop characteristics

In the form of the inverter, ω_n is the rated angle frequency of the inverter, V_n is the output rated voltage amplitude of the inverter, m and n are the droop coefficient of the droop control.

The relation between $V - Q$ and $P - f$ can be drawn by Eq. 6.29 as shown in Fig. 6.12. In the diagram, the inverter adjusts its output power and reactive power by controlling the amplitude and frequency of output power.

6.3.2 Power Optimization Strategy of Energy Router

The new SST architecture designed in this paper extends the capabilities of the SST, with emphasis on incorporating DC loads, power generation and storage devices at the DC bus to make SST even more versatile:

(1) Easier access to a variety of DC source devices to build hybrid systems and system capacity can scale up gradually.
(2) Rich access types for a variety of new energy and AC and DC load distributed control.
(3) A variety of power flow ways to achieve the optimal power economy scheduling.

In the above section, the control strategies of SST are analyzed. The DC bus droop control strategy in this section will be applied to the design of SST to achieve control of distributed access devices and complete power management system. Depending on the battery state of charge (SOC) and the DC bus voltage to switch between energy storage and SST mode. So the low-voltage side access devices are divided into distributed renewable energy and distributed energy storage equipment.

The output power of the photovoltaic is greatly influenced by the environment, such as the intensity and temperature of light. The load of DC side is dynamic and not constant. The dynamic response of distributed renewable energy equipment and distributed energy storage equipment is inconsistent. The above points are determined by the need for proper power management at low voltage DC to provide reliable and high quality energy for the load. The SST can be equivalent to DC microgrid at low voltage DC bus, which is different from the frequency characteristic of AC microgrid. There is only the relationship between voltage and power in DC microgrid. When the system power is exceeded, the DC bus voltage will rise. When the system needs

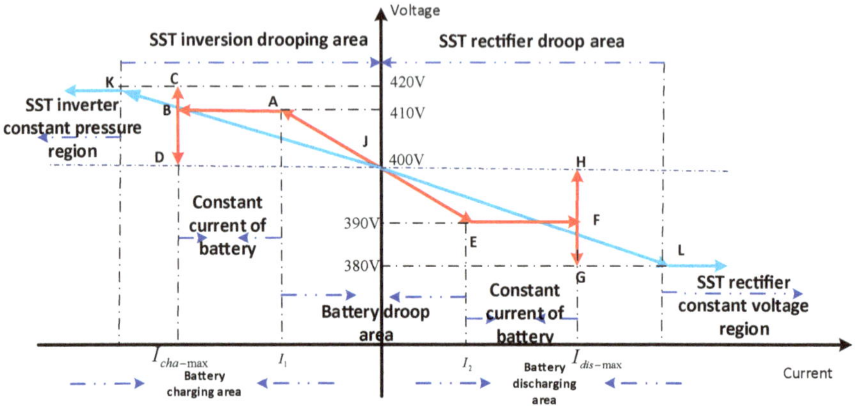

Fig. 6.13 The schematic diagram of integrative droop control for storage and SST

more power, the DC bus voltage will drop. Therefore, droop control can be applied to SST and battery. SST and energy storage devices can control the DC bus voltage in a specific range, and work in different modes according to the voltage range.

Figure 6.13 shows the proposed SST and energy storage integrated droop control strategy diagram. In the figure, the low voltage side DC bus voltage and output current are used as a reference for droop control. According to the droop control characteristics, the battery output power is only related to the operating mode of the system. For example, when the output of renewable energy is lower than the power required by the load, the energy is added to the load by the energy storage device. In the load peak, the grid power and frequency are still using droop control. In more extreme cases, the grid may collapse, and generate other serious accidents. On the other hand, energy storage does not supply its maximum output power to the load because the stored energy output is governed by droop control. Therefore, the strategy shown in the figure should consider three aspects: lack of grid power supply, maximum output power of energy storage and droop control. 4000 V is used as the boundary value between charging and discharging mode for SST and battery. When the DC bus voltage is higher than 400 V, the battery is in charging mode and the SST power flows from the DC bus to the distribution network. When the DC bus voltage is lower than 400 V, the battery is in discharging mode and the SST power flows from the distribution network to the DC bus. The droop slope for SST is chosen to be smaller than that of battery because the battery's power rating is smaller than SST.

The battery uses droop control when the DC link voltage is in the 390 V–410 V range, and is limited by the maximum value of the charge and discharge circuit of the battery. After the voltage exceeds the voltage range, the battery is transferred to the constant current charging control, and the current value corresponds to the maximum current of the battery charging and discharging. There will be two situations. (1) the output power of new energy is less than the power required by load, and the battery is in constant current and constant charging state. The power of SST will decrease,

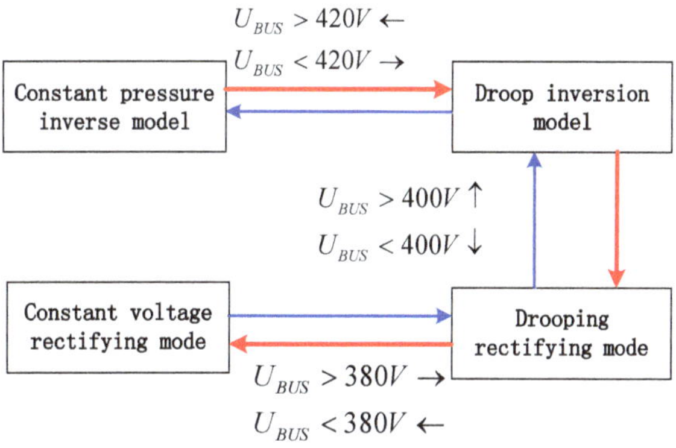

Fig. 6.14 Transfer principle of multi-mode operation for SST

and the DC bus voltage will also decrease until the system reaches the new power balance. For the battery running curve, it will drop vertically down to the B point. (2) the output power of the new energy is higher than the load demand and the maximum charge current of the battery. The DC bus voltage will rise and the battery running curve will rise vertically from the B point. When the DC bus voltage reaches 420 V, it means the battery will run from point B to point C. SST switches to constant voltage mode to maintain the DC link voltage at 420 V and reverse the excess power injected into the SST. In order to prevent the battery from oscillating between constant and droop modes, the bus voltage drops to 400 V for mode switching.

Under the same discharge mode, when the bus voltage reaches 390 V, SST and energy storage need to provide more power for the load. At this time, the battery is transferred to a constant discharge state, and the maximum discharge current is supplied to the load, and its running curve is from E to F. At this stage, with the matching of power system, the battery will move to H or G vertically. It depends on how well the total output power of the battery and SST match the power required by the load. When the load demand power increases, the DC bus voltage drops to 380 V. At the same time, SST will also be transferred from droop mode to constant voltage mode, so as to maintain DC bus voltage stability. When the bus voltage is restored to 400 V only, the battery can return to the droop mode.

The above control strategies based on droop control of the DC bus voltage respectively require the battery to be able to operate in the sagging charging mode and the constant charging/discharging mode. The SST is required to run in the droop control mode and the constant voltage mode. The battery's constant mode in the charge and discharge are two states of constant current state. In the droop control section, the droop control strategy is combined with the constant voltage—constant current two-stage charging method mentioned above, using the SOC of the battery as a determining amount to decide whether the battery is operating in the current droop or voltage droop mode. The control principle has conducted a detailed study in the third chapter, the two stage charging method by setting the constant voltage and current, the switching signal according to the battery state of SOC, to ensure that the battery is full at the same time with the characteristics of internal battery in a relatively short period of time, can protect the battery overcharge on the life of the damage. According to Fig. 6.13, further analysis of the operation mode of SST is shown in Fig. 6.14. The switching mode of SST is mainly changed according to the voltage value of the low voltage DC bus.

6.4 The Stability Analysis of the Energy Router

As previously mentioned, at the heart of the Energy Router are the power-electronic-enabled converters, the use of large amounts of the power-electronic-enabled converters are not only able to improve the renewable energy utilization efficiency are but also able to improve the power quality performance. However, many stability issues are introduced regarding this power-electronic-enabled system [12]. Interactions between power-electronic-enabled converters and passive components is a main reason for the instability of Energy Router. Thus, it is crucial for system integrators to analyze the Energy Router stability, design the power-electronic-enabled converter controller parameters and limit the load demanded active power and reactive power during Energy Router planning and maintenance periods to guarantee stable Energy Router operation [13–15].

The stability assessment of the Energy Router is a significant research topic to ensure the performance of a FREEDM system. An Energy Router consists of three stages: an active rectifier, a dual active bridge converter and an inverter. These three stages together can provide the key attractive features of the Energy Router. Meanwhile, the Energy Router is highly prone to instabilities due to the interaction among three stages. M. Khazraei was the first to propose Energy Router stability analysis strategy based on each subsystem model [16]. However, the control strategy of the each subsystem was so special that there was still a huge gap between theoretical analyses and actual utilizations. Furthermore, D. G. Shah analyzed the stability design criteria for distribution system with Energy Routers for the first time [17]. Nevertheless, the premise of those papers was that the each of the Energy Router was stability in the distribution system. Thus, the stability assessment of the Energy Router remains an open issue, which has rarely been researched.

The impedance criterion [18–33], which has been studied for ac-dc systems, dc-dc systems and dc-ac systems independently, was first established by Middlebrook for power electric converters [18]. The impedance criterion declares that the system stability can be predicted by dividing the overall system into the source subsystem and load subsystem, and next applying the generalized Nyquist criterion (GNC) to the ratio between the source subsystem output impedance $Z_s(s)$ and the load subsystem input impedance $Z_L(s)$ [19]. The main advantage of this impedance criterion is that the measured impedances intrinsically model overall circuit components, containing physical components and control systems. Thus, the impedance criterion can be applied to the stability assessment of the Energy Router well. Additionally, several researches in the literature have been proposed to improve the performance of the impedance criterion. To reduce the artificial conservativeness characteristics, the

Middlebrook Criterion, the Gain Margin and Phase Margin Criterion, the Energy Source Analysis Consortium Criterion, the Opposing Argument Criterion, and the Three-Step Impedance Criterion have been proposed one after another [19–23]. In order to real with disadvantage of the Nyquist criterion, the Inverse Nyquist Stability Criterion [24] and The Mikhailov Stability Criterion [25] have been investigated. Moreover, Sum Type Criterion has been proposed for the first time to predict the stability of the power electric converter with bidirectional power flow [32]. To decrease computational complexity using the GNC directly, several simplified stability criterions have been studied, such as singular-value criterion, G-norm criterion, infinity-one-norm criterion and infinity-norm criterion [33, 34].

Thus, the impedance-based stability analysis is always applied to analyze the stability of Energy Router. As shown in Sect. 6.1, there is two kinds of the structure of Energy Router: Single-Phase Energy Router and Three-Phase Energy Router. Thus, the stability of the Single-Phase Energy Router and Three-Phase Energy Router will be analyzed separately.

6.4.1 The Stability Analysis of the Single-Phase Energy Router

As is shown in Fig. 6.15, the single-phase SST structure model can be divided into three subsystems including ac-dc rectifier, DAB and dc-ac inverter. The ac-dc rectifier transforms the 240 V ac bus to a 400 V dc distribution bus. Furthermore, the DAB converter bidirectional power flow control is accustomed to step down this DC voltage to a regulated DC voltage. Eventually, the dc-ac inverter produces an AC voltage to integrate an AC load. Since this paper concentrates on the small signal instability of the single-phase SST due to the fast dynamics of inner control loop, therefore,slow control strategies such as energy management controls are ignored. Thus, the impedance criterion can be adopted to assess the overall stability of the single-phase SST. According to impedance criterion, the system is stability if and only if system satisfy following conditions: (1) the each of the subsystem ought to be stability; (2) the number of the counterclockwise encirclements of the $(-1+j0)$ point by the ratio between the output impedance of the source subsystem and the input impedance of the load subsystem locus should be equal to the number of the RHP poles of the minor loop gain. In practice, the second condition is usually judged only by checking whether the locus encircles the $(-1+j0)$ point.

Figure 6.15 shows the overall SST is divided into three subsystem containing rectifier [35], DAB with galvanic isolation [36] and inverter [37], and $Z_{out,rec}$, $Z_{in,DAB}$, $Z_{out,DAB}$, $Z_{in,inv}$ are the rectifier output impedance, DAB input impedance, DAB output impedance and inverter input impedance respectively. The small signal models of each subsystem in SST are investigated based on the converters' average models as follows.

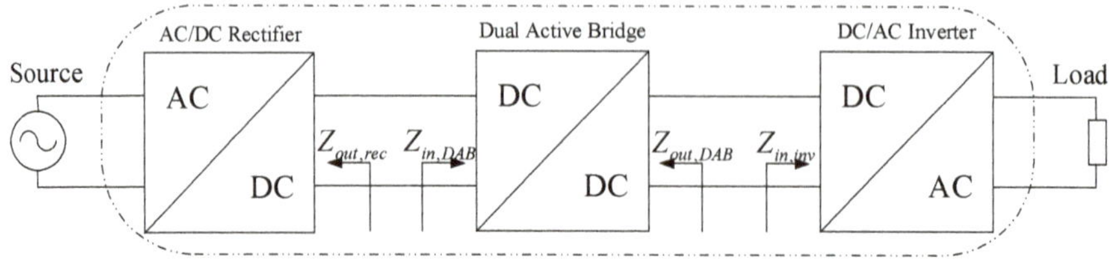

Fig. 6.15 The single-phase SST structure model with three subsystems

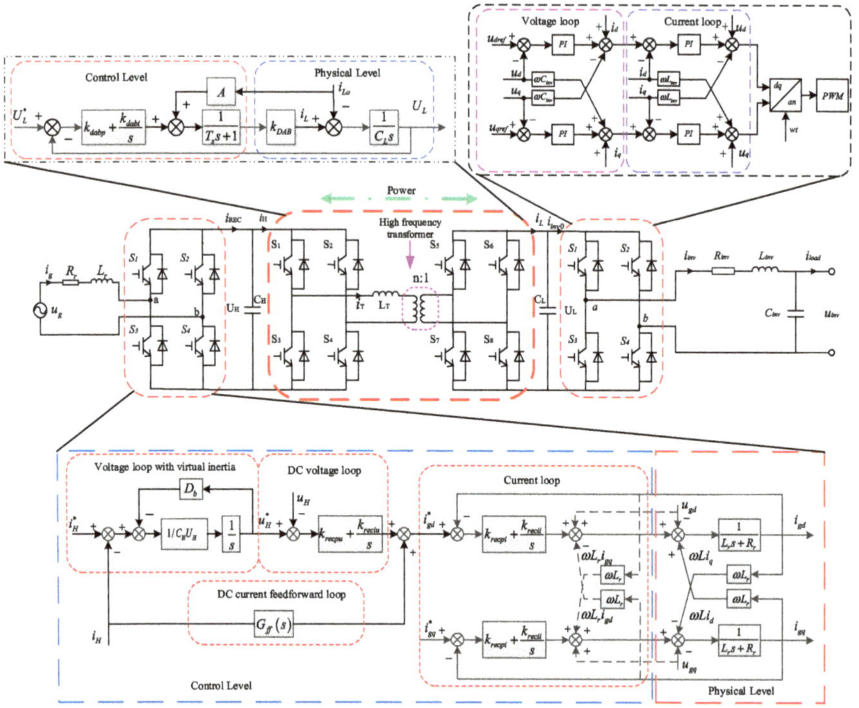

Fig. 6.16 The overall control strategy of the Energy Router

Rectifier

The control strategy with physical structure is shown in Fig. 6.16. The control strategy consists of voltage loop with virtual inertia, DC voltage loop, DC current feedforward loop and current loop, where C_H is the dc-link output capacitor, L_r is the ac-link input filter inductor, R_r is the ac-link equivalent series resistance, u_g is the utility grid voltage, i_g is the utility grid current, U_H is the dc-link voltage, D_b is the droop coefficient, U_n is the rated dc bus voltage, C_v is the introduced virtual capacitance, i_{REC} is the dc-link current and i_H is the dc-link output current.

From Fig. 6.16, the mathematical expression of the rectifier is shown as

$$\begin{cases} u_{gd} = u_{abd} + R_r i_{gd} + L_r \frac{di_{gd}}{dt} + \omega L_r i_{gq} \\ u_{gq} = u_{abq} + R_r i_{gq} + L_r \frac{di_{gq}}{dt} - \omega L_r i_{gd} \end{cases} \tag{6.30}$$

PI controller is utilized for the current control, which is described as $G_i(s) = k_{recpi} + k_{recii}/s$. Thus the Eq. (6.30) can be rewritten as follows:

$$\begin{cases} u_{gd} = -G_i(s)\left(i_{gd}^* - i_{gd}\right) + u_{abd} + \omega L_r i_{gq} \\ u_{gq} = -G_i(s)\left(i_{gq}^* - i_{gq}\right) + u_{abq} - \omega L_r i_{gd} \end{cases} \tag{6.31}$$

Furthermore, the small-signal model of current loop can be represented as:

$$\begin{cases} \triangle\, i_{gd}(s) = \left(\triangle\, i_{gd}^*(s) - \triangle\, i_{gd}(s)\right) G_i(s)/(R_r + L_r s) \\ \triangle\, i_{gq}(s) = \left(\triangle\, i_{gq}^*(s) - \triangle\, i_{gq}(s)\right) G_i(s)/(R_r + L_r s) \end{cases} \tag{6.32}$$

Neglecting the energy loss, the Eq. (6.33) can be gained in light of the power balance between two sides of the rectifier.

$$\frac{1}{2}\left(u_{gd} i_{gd} + u_{gq} i_{gq}\right) = u_H i_{REC} = u_H \left(C \frac{du_H}{dt} + i_H\right) \tag{6.33}$$

Furthermore, the small-perturbation Eq. (6.34) can be shown as:

$$\frac{1}{2}\left(U_{gd} \triangle\, i_{gd} + \Delta u_{gd} I_{gd}\right) = CU_H \frac{d \triangle\, u_H}{dt} + U_H \Delta i_H + \Delta u_H I_H \tag{6.34}$$

In light of the superposition theorem, the relation between $u_H(s)$ and $i_{gd}(s)$, $u_H(s)$ and $i_H(s)$, respectively, obtained as follows:

$$\begin{cases} G_1(s) = \triangle\, u_H(s)/ \triangle\, i_{gd}(s) = U_{gd}/2(CU_H s + I_H) \\ G_2(s) = \triangle\, u_H(s)/ \triangle\, i_H(s) = -U_H/(CU_H s + I_H) \end{cases} \tag{6.35}$$

Ignoring the influence of the utility grid voltage, and virtual inertia control, the dc output current feedforward control and PI controller of the voltage control loop are adopted, the small-signal closed-loop transfer function as follows:

$$TF(s) = \frac{\triangle\, u_H(s)}{\triangle\, i_H(s)} = \frac{G_{vir}(s) G_v(s) G_c(s) G_1(s) + G_2(s)}{1 + G_v(s) G_c(s) G_1(s)} \tag{6.36}$$

where,

$$G_v(s) = k_{recpu} + k_{reciu}/s,$$

$$G_c(s) = G_i(s)/(G_i(s) + (L_r s + R_r)),$$

$$G_{vir}(s) = -1/(D_b + G_v(s)U_n s).$$

DAB

The DAB is a bidirectional dc/dc converter with phase-shift control based on active bridges interfaced by a high frequency transformer, where, P_H is transmitted power, n is turn ratio of the transformer, U_L is the output dc-link voltage, L_T is the equivalent leakage inductance, d is the phase shift, T_s is half of the switching period, I_L is the output dc-link current, C_L is the output dc-link capacitor. The expression of the transmitted power can be illustrated as follows:

$$P_H = \frac{nU_H U_L}{4L_T} d(1-d)T_s = U_L I_{Lo} \tag{6.37}$$

Defining the relation between d and $1-d$ as $D_s = d(1-d)$. Thus the Eq. (6.37) can be rewritten using as:

$$\frac{nU_H}{4L_T} T_s \mathrm{D_s} = I_L \tag{6.38}$$

Defining the relation between D_s and dc-link current as $k_{DAB} = I_L/D_s$, where, $k_{DAB} = nU_H T_s/4L_T$.

Using Faraday's law, the capacitor voltage can be calculated as

$$I_L - I_{Lo} = C_L \frac{dU_L}{dt} \tag{6.39}$$

PI controller is utilized for the current control, which is described as $G_{dabpi}(s) = k_{dabp} + k_{dabi}/s$. The time delay which is shown as $G_d(s) = 1/(T_s s + 1)$ is considered. Especially, when a feedback loop (the feedback coefficient $A = 1/k_{DAB}$) is applied to the DAB, the small-signal closed-loop transfer function of the DAB as follows:

$$G_{DAB}(s) = \frac{G_d(s) - 1}{k_{DAB} G_{dabpi}(s) G_d(s) + C_L s}. \tag{6.40}$$

Inverter

As shown in Fig. 6.16, the inverter adopts the traditional voltage/current double closed control strategy, where, L_{inv} is the ac-link inductance, C_{inv} is the ac-link capacitor, R_{inv} is the equivalent ac-link resistance, i_{inv} is the output ac-link current, u_{inv} is the output ac-link voltage, u_{ab} is the dc-link voltage, K_{pwm} is the inverter gain and ω is the angular frequency. The current loop mathematical equations of inverter are expressed as (6.41).

$$\begin{cases} L_{inv}\frac{di_{invd}}{dt} + R_{inv}i_{invd} = u_{abd} - u_{invd} + \omega L_{inv}i_{invq} \\ L_{inv}\frac{di_{invq}}{dt} + R_{inv}i_{invq} = u_{abq} - u_{invq} - \omega L_{inv}i_{invd} \end{cases} \tag{6.41}$$

PI controller is used for the current control, and the Eq. (6.41) can be rewritten as follows:

$$\begin{cases} i_{invd}(s) = \left(i^*_{invd}(s) - i_{invd}(s)\right)\left(k_{invpi} + \frac{k_{invii}}{s}\right)/(R_{inv} + L_{inv}s) \\ i_{invq}(s) = \left(i^*_{invq}(s) - i_{invq}(s)\right)\left(k_{invpi} + \frac{k_{invii}}{s}\right)/(R_{inv} + L_{inv}s) \end{cases} \tag{6.42}$$

Furthermore, the time delay and symmetrical characteristic are considered, and the small-signal open-loop transfer function can be represented as:

$$G_{\text{oi}}(s) = \left(k_{invpi} + \frac{k_{invii}}{s}\right)\left(\frac{1}{1 + T_s s}\right)\left(\frac{K_{pwm}}{1 + 0.5T_s s}\right)/(R_{inv} + L_{inv}s) \tag{6.43}$$

Moreover, the small-signal closed-loop transfer function can be described as follows:

$$G_{invI}(s) = \frac{G_{\text{oi}}(s)}{1 + G_{\text{oi}}(s)} \tag{6.44}$$

For voltage control loop, a similar analysis is utilized to the one described previously to gain the voltage control loop closed-loop transfer function. The voltage loop mathematical equations of inverter are expressed as (6.45).

$$\begin{cases} i_{invd} = C_{inv}\frac{du_{invd}}{dt} + i_{loadd} - \omega C_{inv}u_{invq} \\ i_{invq} = C_{inv}\frac{du_{invq}}{dt} + i_{loadq} + \omega C_{inv}u_{invd} \end{cases} \tag{6.45}$$

PI controller is also used for the current control, and the Eq. (6.46) can be rewritten as follows:

$$\begin{cases} i^*_{invd} = \left(u^*_{invd}(s) - u_{invd}(s)\right)\left(k_{invpu} + \frac{k_{inviu}}{s}\right) + i_{loadd} - \omega C_{inv}u_{invq} \\ i^*_{invq} = \left(u^*_{invq}(s) - u_{invq}(s)\right)\left(k_{invpu} + \frac{k_{inviu}}{s}\right) + i_{loadq} + \omega C_{inv}u_{invd} \end{cases} \tag{6.46}$$

Eventually, the time delay and symmetrical characteristic are considered, and the open-loop transfer function and closed-loop transfer function of the overall inverter is being expressed as follows:

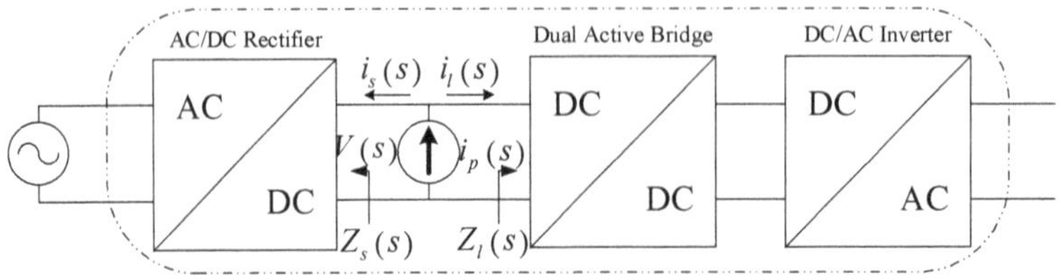

Fig. 6.17 The current-perturbation-based stability analysis model of the single-phase Energy Router

$$\begin{cases} G_{oinv}(s) = \left(k_{invpu} + \dfrac{k_{inviu}}{s}\right)\left(\dfrac{1}{1+T_s s}\right)\dfrac{G_{invI}(s)}{C_{inv}s} \\ G_{cinv}(s) = \dfrac{G_{oinv}(s)}{1+G_{oinv}(s)} \end{cases} \tag{6.47}$$

A. ***Impedance-based Energy Router stability analysis strategy based on measured data***

Compared with impedance-based stability analysis strategy based on system model, the impedance-based stability analysis strategy based on measured data has no difference between AC-AC Energy Router and AC-DC-AC Energy router. As shown in Fig. 6.17, the system stability can be predicted by dividing it into the source and load subsystems, and then applying the Nyquist criterion to the ratio between the source output impedance $Z_s(s)$ and the load input impedance $Z_l(s)$. However, this impedance criterion is complex, and not convenient for the design of the DPS. Then the impedance specification is defined for individual loads in the DPS so that the system stability can be guaranteed by design. Thus, Fred C. Lee has proposed a kind of stability margin via perturbation approaches, and this stability analysis can be applied in Energy Router stability analysis. Generally speaking, the source subsystem and load subsystem is designed stability respectively. It's known that the whole system is stable when the source output impedance $Z_s(s)$ is much smaller than the load input impedance $Z_l(s)$ within all frequency ranges. However, in many real Energy Router, it is impractical to have $|Z_s(s)| \ll |Z_l(s)|$ in all frequency ranges. To define a less conservative impedance specification, the forbidden region can be described as

$$\mathrm{Re}\left(\frac{Z_s(s)}{Z_l(s)}\right) \geq -\frac{1}{2} \tag{6.48}$$

To simplified analysis, the following variable is defined:

$$D(s) = \frac{|Z_s(s)/Z_l(s)|}{|1 + Z_s(s)/Z_l(s)|} \tag{6.49}$$

The physical meaning of D(s) is equal to the ratio of "the distance between point (0, 0) and $Z_s(s)/Z_l(s)$" and "the distance between point (−1, 0) and $Z_s(s)/Z_l(s)$" Thus, the forbidden region can be also shown as follow:

$$\mathrm{Re}\left(\frac{Z_s(s)}{Z_l(s)}\right) \geq -\frac{1}{2} \Leftrightarrow \mathrm{D}(s) < 1 \tag{6.50}$$

As shown in Fig. 6.17, there are

$$Z_s(s) = \frac{V(s)}{i_s(s)} \tag{6.51}$$

$$Z_l(s) = \frac{V(s)}{i_l(s)} \tag{6.52}$$

Substitution of (6.51) and (6.52) into (6.49) and (6.50) gives

$$\mathrm{D}(s) = \frac{\left|\frac{V(s)}{i_s(s)}\Big/\frac{V(s)}{i_l(s)}\right|}{\left|1+\frac{V(s)}{i_s(s)}\Big/\frac{V(s)}{i_l(s)}\right|} = \frac{|i_l(s)|}{|i_s(s)+i_l(s)|} = \frac{|i_l(s)|}{|i_p(s)|} < 1 \tag{6.53}$$

Thus, the impedance-based stability analysis strategy based on measured data is implemented simply with an impedance analyzer based on perturbation approaches. The sufficient condition of system stability is D(s) < 1. Furthermore, the measured data should not only be very high as it could impact the system operating point, but also could neither be low on account of the fact that noise can be dominating in the measured value, and calculate frequency responses that are meaningless. Fortunately, it is not complex to notice whether the measured system is under perturbed, as he transfer function that is measured and displayed on the monitor of the network analyzer will be very noisy. A good method can gradually increase the perturbation level (gain), and repeat the preceding step until the transfer functions achieve the acceptable form. As the perturbation level gain is higher, a modification would be noticed at the high frequency range and result with the accurate transfer functions in the low frequency range [38].

Similar to aforementioned method, the impedance-based stability analysis strategy based on measured data can also measure voltage perturbation to identify the stability of Energy Router. As shown in Fig. 6.18. Energy Router is stable when the following inequality is satisfied:

$$\left|\frac{V_p(s)}{V_l(s)}\right| \geq \frac{1}{2} \tag{6.54}$$

However, in order to reduce the conservatism of the stability analysis strategy, the unterminated small-signal behavioral model can be adopted to Energy Router. The two-port network can be directly used to build the small-signal linear model of the Energy Router around the particular operating point. As shown in Fig. 6.19, the four

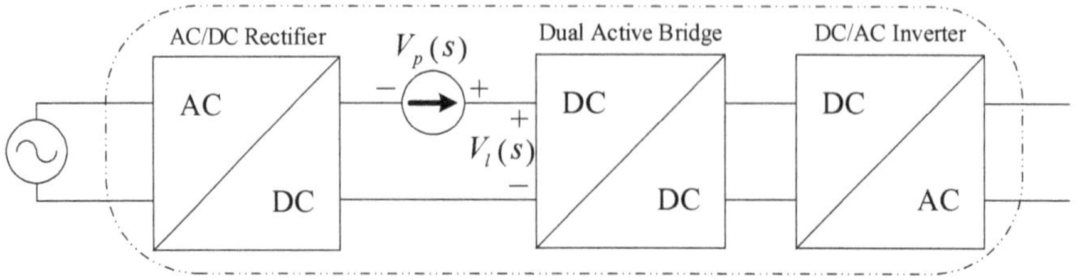

Fig. 6.18 The voltage-perturbation-based stability analysis model of the single-phase Energy Router

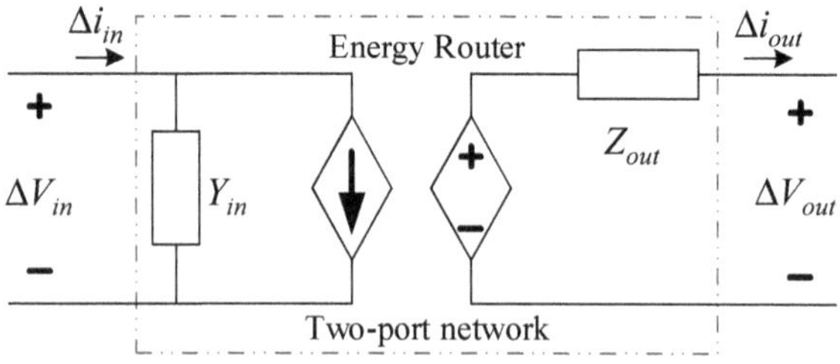

Fig. 6.19 The small-signal behavioral model of the Energy Router in two-port network

characteristic functions can be used to reflect stability of the Energy Router. K_V is voltage gain ratio, K_I is current gain ratio, Y_{in} is input admittance, and Z_{out} is output impedance, respectively.

$$\begin{cases} K_V(s) = \frac{\Delta V_{out}}{\Delta V_{in}}\left|_{i_{out}=0};\right. \\ K_V(s) = \frac{\Delta i_{in}}{\Delta i_{out}}\left|_{V_{in}=0};\right. \\ Y_{in}(s) = \frac{\Delta i_{in}}{\Delta V_{in}}\left|_{i_{out}=0};\right. \\ Z_{out}(s) = \frac{\Delta V_{out}}{\Delta i_{out}}\left|_{V_{in}=0};\right. \end{cases} \tag{6.55}$$

Thus, the small-signal two-port network model of the Energy Router can be rewritten.

$$\begin{bmatrix} \Delta V_{out}(s) \\ \Delta i_{in}(s) \end{bmatrix} = \begin{bmatrix} K_V(s) & -Z_{out}(s) \\ Y_{in}(s) & K_I(s) \end{bmatrix} \cdot \begin{bmatrix} \Delta V_{in}(s) \\ \Delta i_{out}(s) \end{bmatrix} \tag{6.56}$$

As shown in Fig. 6.4.5, the two individual sets of measurements on the Energy Router operating at rated state steady point is built in order to obtain the terminal-behavioral model. The aim is terminal identification of the inner dynamics represented via the four transfer functions (6.55). These transfer functions is "unterminated" to characterize dynamics of the Energy Router, which means that the source subsystem and load subsystem dynamics intrinsically reflect any of the transfer functions while measured online, should be decoupled from them insuring that the Energy

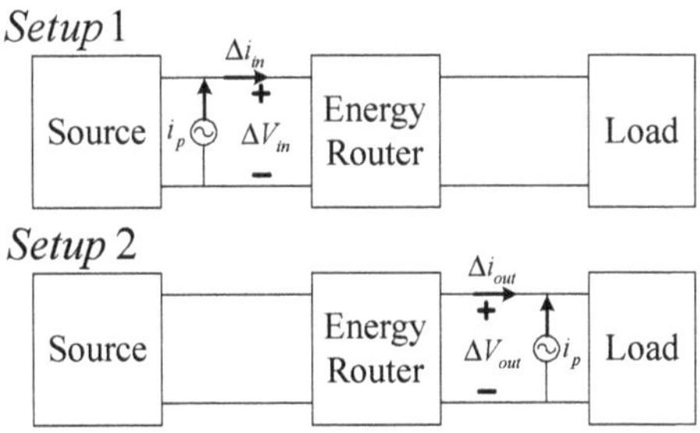

Fig. 6.20 Setup for terminal characterization of the Energy Router

Router, while is under different conditions in the simulations, would behave as it would in real [38].

The first setup of transfer functions (voltage gain ratio and input admittance) can be obtained by perturbing the input side of the converter (ac sweep—setup 1 in Fig. 6.20), and is shown in (6.57). The letter m in the index denotes measured quantities (terminated)

$$\begin{cases} K_{Vm}(s) = \frac{\Delta V_{out}}{\Delta V_{in}} \\ Y_{inm}(s) = \frac{\Delta i_{in}}{\Delta V_{in}} \end{cases} \tag{6.57}$$

The small-signal two-port network model that corresponds to Fig. 6.4.5 (setup 1) is

$$\begin{cases} \Delta V_{out}(s) = K_V(s) * \Delta V_{in}(s) - Z_{out}(s) * \Delta i_{out}(s) \\ \Delta i_{in}(s) = Y_{in}(s) * \Delta V_{in}(s) + K_I(s) * \Delta i_{out}(s) \end{cases} \tag{6.58}$$

From (6.58), it can be written

$$\begin{cases} \Delta V_{outm}(s) = K_{Vm}(s) \cdot \Delta V_{inm}(s) \\ \Delta i_{inm}(s) = Y_{inm}(s) \cdot \Delta V_{inm}(s) \end{cases} \tag{6.59}$$

Now combining (6.59) and (6.60), it is obtained

$$\begin{cases} K_{Vm}(s) \cdot \Delta V_{inm}(s) = K_V(s) \cdot \Delta V_{inm}(s) - Z_{out}(s) \cdot \Delta i_{outm}(s) \\ Y_{inm}(s) \cdot \Delta V_{inm}(s) = Y_{in}(s) \cdot \Delta V_{inm}(s) + K_i(s) \cdot \Delta i_{outm}(s) \end{cases} \tag{6.60}$$

where

$$\begin{cases} K_{Vm}(s) = K_V(s) - \frac{Z_{out}(s) \cdot \Delta i_{outm}(s)}{\Delta V_{inm}(s)} \\ Y_{inm}(s) = Y_{in}(s) + \frac{K_i(s) \cdot \Delta i_{outm}(s)}{\Delta V_{inm}(s)} \end{cases} \tag{6.61}$$

The last equations clearly show relationship between the terminated and unterminated transfer functions, as well as the common small-signal term $\Delta i_{outm}(s)/\Delta V_{inm}(s)$ that has dimension of conductance. The new term called trans-conductance $TK_{Vm}(s)$ can be now defined since it represents the input voltage to output current transfer function, so (6.61) becomes

$$\begin{cases} K_{Vm}(s) = K_V(s) - Z_{out}(s) \cdot TK_{Vm}(s) \\ Y_{inm}(s) = Y_{in}(s) + K_i(s) \cdot TK_{Vm}(s) \end{cases} \tag{6.62}$$

The second set of transfer functions could be given by perturbing the output subsystem side of the converter (setup two in Fig.6.20), and is represented in (6.63). Similarly, a letter m represents measured quantities [38]

$$\begin{cases} Z_{outm}(s) = -\frac{\Delta V_{outm}(s)}{\Delta i_{outm}(s)} \\ K_{im}(s) = \frac{\Delta i_{inm}(s)}{\Delta i_{outm}(s)} \end{cases} \tag{6.63}$$

The small-signal two-port network model that corresponds to Fig. 6.21 (setup 2) is

$$\begin{cases} \Delta V_{out}(s) = K_V(s) \cdot \Delta V_{in}(s) - Z_{ont}(s) \cdot \Delta i_{out}(s) \\ \Delta i_{in}(s) = Y_{in}(s) \cdot \Delta V_{in}(s) + K_i(s) \cdot \Delta i_{out}(s) \end{cases} \tag{6.64}$$

From (6.63), it can now be written

$$\begin{cases} \Delta V_{outm}(s) = -Z_{outm}(s) \cdot \Delta i_{outm}(s) \\ \Delta i_{inm}(s) = K_{im}(s) \cdot \Delta i_{outm}(s) \end{cases} \tag{6.65}$$

Now combining (6.64) and (6.65), it is obtained

$$\begin{cases} -Z_{outm}(s) \cdot \Delta i_{outm}(s) = K_V(s) \cdot \Delta V_{inm}(s) - Z_{out}(s) \cdot \Delta i_{outm}(s) \\ K_{Vm}(s) \cdot \Delta i_{outm}(s) = Y_{in}(s) \cdot \Delta V_{inm}(s) + K_V(s) \cdot \Delta i_{outm}(s) \end{cases} \tag{6.66}$$

where

$$\begin{cases} -Z_{outm}(s) = K_V(s) \cdot \frac{\Delta i_{outm}(s)}{\Delta V_{inm}(s)} - Z_{out}(s) \\ K_{Vm}(s) = Y_{in}(s) \cdot \frac{\Delta i_{outm}(s)}{\Delta V_{inm}(s)} + K_V(s) \end{cases} \tag{6.67}$$

The last equations again show relationship between the terminated and unterminated transfer functions and the common small-signal term $\Delta V_{inm}(s)/\Delta i_{outm}(s)$. The new term called trans-resistance $TR_m(s)$ can be now defined and represents the output

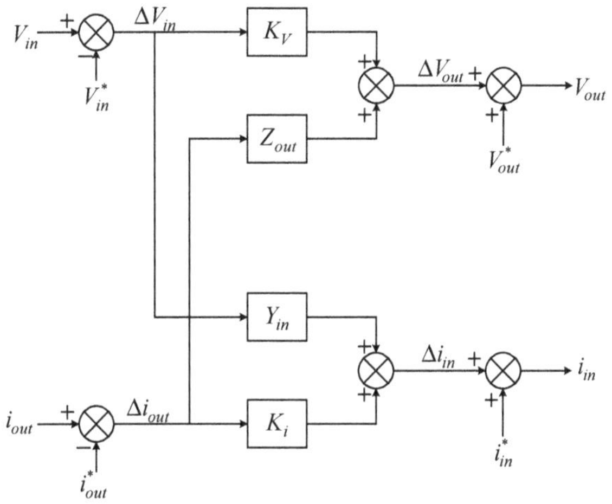

Fig. 6.21 Block diagram of a two-port network behavioral model of the Energy Router

current to input voltage transfer function. Equation (6.67) becomes

$$\begin{cases} -Z_{outm}(s) = K_V(s) \cdot TR_m(s) - Z_{out}(s) \\ K_{Vm}(s) = Y_{in}(s) \cdot TR_m(s) + K_V(s) \end{cases} \tag{6.68}$$

By representing (6.62) and (6.68) with the matrix form, it is obtained

$$\begin{bmatrix} K_V(s) & -Z_{out}(s) \\ Y_{in}(s) & K_i(s) \end{bmatrix} = \begin{bmatrix} K_{Vm}(s) & -Z_{outm}(s) \\ Y_{inm}(s) & K_{im}(s) \end{bmatrix} \cdot \begin{bmatrix} 1 & TR_m(s) \\ TK_{Vm}(s) & 1 \end{bmatrix} \tag{6.69}$$

The common algorithm that reflects steps of building the terminal behavioral model of the Energy Router via the measured frequency characteristics can be seen as follows,

Step1: Converter in steady state at the desired operating point. Record $V_{out}, V_{in}, i_{out}, i_{in}$;

Step2: By perturbing the input. Obtain $K_{Vm}(s), Y_{inm}(s), TK_{Vm}(s)$;

Step3: By perturbing the input. Obtain $K_{im}(s), Z_{outm}(s), TR_m(s)$;

Step4: Apply decoupling matrix to obtain un-terminated $K_V(s), -Z_{out}(s), Y_{in}(s), K_i(s)$;

Step5: Build the block diagram of the Energy Router as shown in Fig. 6.4.6.

In the end, the stability analysis of the Energy Router can be obtained by the block diagram of a two-port network behavioral model of the Energy Router. It should be noted that the criterion as already mentioned considers only magnitudes of impedances and it is soon realized that it is overly conservative since the forbidden region in the s-plane occupies much of the area, which is irrelevant from stability point of view. This kind of restriction can increase the cost of the design without improving system performance.

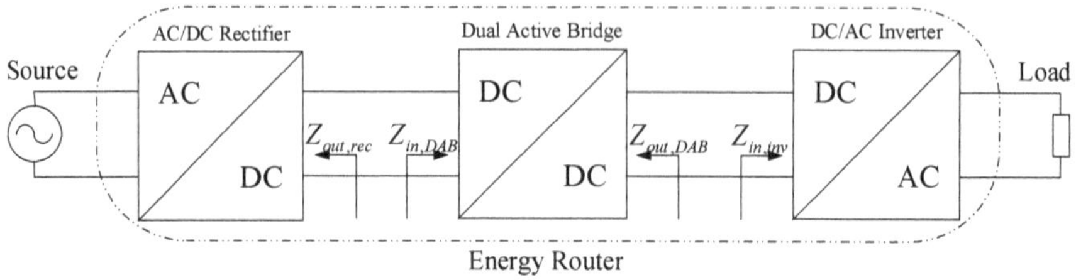

Fig. 6.22 Input and output transfer functions identification diagram

B. ***Impedance-based Energy Router stability analysis strategy based on system model***

Compared to the Energy Router stability analysis strategy based on measured date, the Energy Router stability analysis strategy based on system model has a lower conservatism and more computational complexity. As shown in the paper [16], the Energy Router stability analysis strategy can be reached by building system model. In order to analyze the stability, the source subsystem input impedance transfer functions and load subsystem output impedance transfer functions of the system are needed. As shown in Fig. 6.22, aforementioned transfer functions and the physical point that each of these transfer functions must be derived. $Z_{out,rec}$, $Z_{in,DAB}$, $Z_{out,DAB}$, $Z_{in,inv}$ are the rectifier output impedance, DAB input impedance, DAB output impedance, inverter input impedance and storage unit input impedance respectively. In this section, the small signal models of each subsystem in Energy Router are proposed based on the converters' average models. Then, input and output impedance transfer functions will be derived based on the small signal models. In the next sections, these transfer functions will be used to research the Energy Router's stability based on the Middlebrook stability criteria.

AC-DC Rectifier

The first subsystem is the ac-dc rectifier. Figure 6.23 reflects the average model of the rectifier. In real, it has been seen that in unity power factor networks such as Energy Routers and so on, the q-axis parameters and their impacts on the d-axis can be ignored. The closed loop output subsystem impedance of Energy Router can be obtained as follows [16]:

$$Z_{out,rec} = \frac{G_{v_{HV}i_0}\left(1 + G_{i_d,rec}G_{i_d d_d}\right) - G_{i_d i_0}G_{i_d,rec}G_{v_{HV}d_d}}{1 + G_{i_d,rec}G_{i_d d_d} + G_{v,rec}G_{i_d,rec}G_{v_{HV}d_d}} \tag{6.70}$$

where, $G_{v,rec}$, $G_{i_d,rec}$ are VLC and the d-axis current loop controller (CLC), respectively.

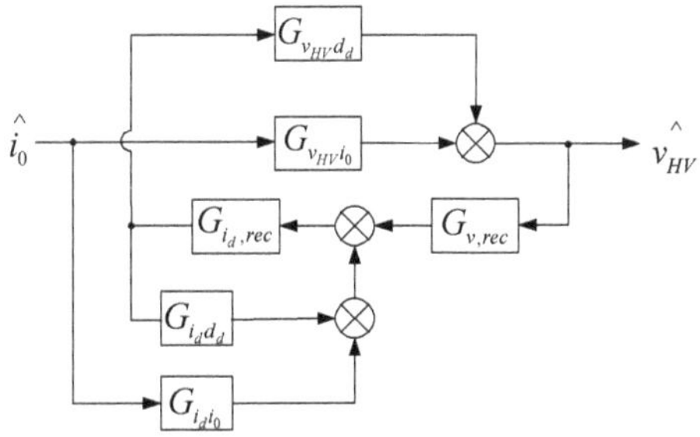

Fig. 6.23 Control block diagram model of the AC-DC Rectifier

$$
\begin{aligned}
G_{i_d i_0} &= -\frac{2}{D_d}\frac{1}{L_{in}C_{HV}/D_d^2 s^2+1} \\
G_{i_d d_d} &= -\frac{I_d}{D_d}\frac{2V_{HV}/(I_d D_d)s+1}{D_d^2/(L_{in}C_{HV})s^2+1} \\
G_{v_{HV} i_0} &= \frac{1}{D_d^2}\frac{L_{in}s}{L_{in}C_{HV}/D_d^2 s^2+1} \\
G_{v_{HV} d_d} &= \frac{V_{HV}}{D_d}\frac{5L_{in}I_d/(D_d V_{HV})s-1}{L_{in}C_{HV}/D_d^2 s^2+1}.
\end{aligned}
$$

Dual Active Bridge

The second subsystem is the DAB, of which characteristics and models have been researched in the past. The primary subsystem and secondary subsystem sides of the DAB both represent as an adaptive current source as follows:

$$
\begin{cases} I_{HV} = \alpha V_{LV}\varphi(1-|\varphi|) \\ I_{LV} = \alpha V_{HV}\varphi(1-|\varphi|) \end{cases} \tag{6.71}
$$

where,

$$
\alpha = \frac{1}{2fL_{leak}}
$$

$$
\varphi = \frac{\pm DAB\ Control\ Phase\ Shift}{\pi}
$$

Based on the average DAB model, one can propose a DAB small signal model where:

$$
\begin{aligned}
\widehat{i_{HV}} &= \beta\widehat{v_{LV}} + \lambda V_{LV}\hat{\varphi} \\
\widehat{i_{LV}} &= \beta\widehat{v_{HV}} + \lambda V_{HV}\hat{\varphi}
\end{aligned} \tag{6.72}
$$

where,

$$
\beta = \alpha\varphi(1-|\varphi|)
$$

$$\lambda = \alpha(1 - 2|\varphi|)$$

Consider that while the input subsystem impedance of the DAB is computed, the input subsystem impedance of any ones connected to the DAB output should be represented as the DAB load and be showed in the equation. Nevertheless, while the output subsystem impedance of the DAB is computed, the influence of the first subsystem could be ignored since the first subsystem only impacts the DAB small-signal model by applying the high voltage link perturbation. Whereas, a large high voltage capacitor value could make this variation ignorable. The input subsystem and output subsystem impedance of the DAB are given [16].

$$Z_{in,DAB} = \frac{1 + Z_1 G_{v.DAB} \lambda V_{HV}}{\beta^2 Z_1 - \beta Z_1 G_{v.DAB} \lambda V_{LV}}$$

$$Z_1(s) = \begin{cases} Z_{L.EQ}(s=0)//Z_{c.LV} & s \le Bw.DAB \\ Z_{c.LV} & s > Bw.DAB \end{cases}$$

$$Z_{o.DAB} = \frac{Z_{c.LV}}{1 + Z_{c.LV} G_{v.DAB} \lambda V_{HV}} \tag{6.73}$$

where, $Z_{c.LV}$ is the impedance of the DAB output capacitor, *Bw.DAB* is the bandwidth of the DAB, $Z_{L.EQ}$ is the equivalent DAB load impedance seen from the LV link and $G_{v.DAB}$ is the DAB VLC.

DC-AC Inverter

The third subsystem is the dc-ac inverter. The average model of the inverter can be shown via thinking it as a dc-dc buck converter [16]. Similarly, the voltage reference of the inverter voltage loop controller is a sinusoidal voltage waveform. At present, in order to achieve the inverter subsystem input impedance, equation (6.75) can be obtained as follows [16]:

$$Z_{in,inv} = \frac{1 + G_{i.inv} G_{i_{L_{fil}}d} + G_{v.inv} G_{v_{out}d} G_{v_{out}d}}{G_{i_{in}v_{LV}}\left(1 + G_{i.inv} G_{i_{L_{fil}}d} + G_{v.inv} G_{v_{out}d}\right) - G_{v_{out}} G_{v.inv} G_{v_{out}d} G_{i_{in}d}} \tag{6.74}$$

where,

$$G_{i_{L_{fil}}d} = \frac{V_{LV}}{R} \frac{RC_{fil}s+1}{L_{fil}C_{fil}s^2+L_{fil}s/R+1} \qquad G_{i_{L_{fil}}v_{LV}} = \frac{D}{R} \frac{RC_{fil}s+1}{L_{fil}C_{fil}s^2+L_{fil}s/R+1}$$

$$G_{v_{out}v_{LV}} = \frac{D}{L_{fil}C_{fil}s^2+L_{fil}s/R+1} \qquad G_{i_{in}v_{LV}} = \frac{D^2}{R} \frac{RC_{fil}s+1}{L_{fil}C_{fil}s^2+L_{fil}s/R+1}$$

$$G_{i_{in}d} = I_{fil} + \frac{V_{LV}D}{R} \frac{RC_{fil}s+1}{L_{fil}C_{fil}s^2+L_{fil}s/R+1} \qquad G_{v_{out}d} = \frac{V_{LV}}{L_{fil}C_{fil}s^2+L_{fil}s/R+1}$$

According to the paper [33], the whole Energy Router stability analysis can be researched by Sum Type Criterion. The cascade of two individually stable subsystems is shown in Fig. 6.24. The total input-to-output transfer function is

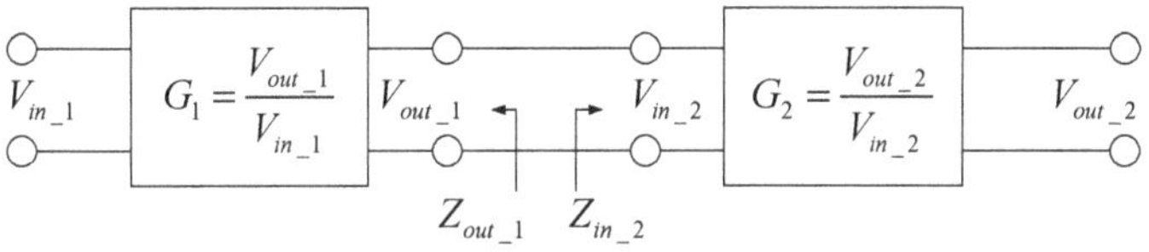

Fig. 6.24 Interconnection of two stable independent systems

Fig. 6.25 Z+Z parallel type connection

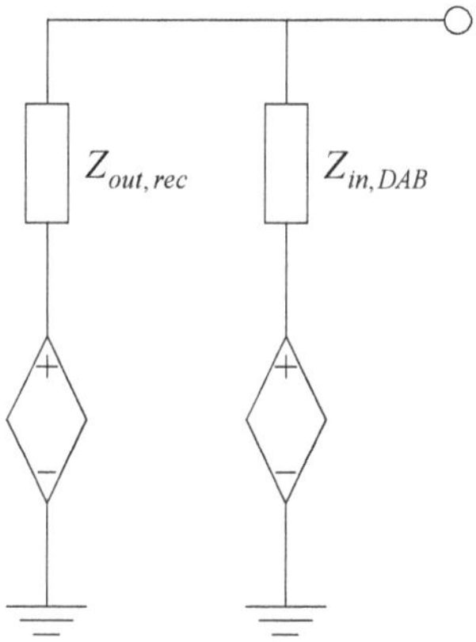

$$G_{12} = \frac{V_{out\_2}}{V_{in\_1}} = G_1 \cdot \frac{Z_{in\_2}}{Z_{in\_2} + Z_{out\_1}} \cdot G_2 \tag{6.75}$$

Lemma *As shown in Fig. 6.25, the closed minor loop gain* $T_{clm} = \frac{1}{1+T_m} = \frac{Z_{in,DAB}}{Z_{in,DAB}+Z_{out,rec}}$ *meets the Impedance-Sum-Type Criterion, and the equivalent impedance is* $Z_{eqv} = \frac{Z_{in,DAB}Z_{out,rec}}{Z_{in,DAB}+Z_{out,rec}}$.

Theorem *The sufficient condition of the Energy Router can be defined as: the number of the* $Z_{in,DAB} + Z_{out,rec}$ *and* $Z_{eqv} + Z_{in,inv}$ *counterclockwise encirclements of the* $(0+j0)$ *point is zero.*

Proof Draw the Nyquist plot of $Z_{in,DAB} + Z_{out,rec}$ and $Z_{eqv} + Z_{in,inv}$. Since there is no RHP in $Z_{in,DAB}$, $Z_{out,rec}$, Z_{eqv} and $Z_{in,inv}$, the number of RHZ in $Z_{in,DAB} + Z_{out,rec}$ could be judged by counting the number of times the trajectory encircles (0, 0) point in clockwise direction. The number of RHZ in $Z_{eqv} + Z_{in,inv}$ could be judged by counting the number of times the trajectory encircles (0, 0) point in clockwise direction. The system will be stable only if each number of times the trajectory encircles (0, 0) point, is zero.

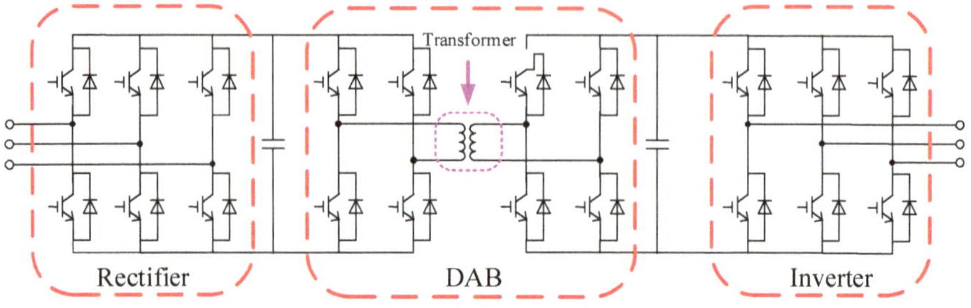

Fig. 6.26 Three-phase Energy Router system

6.4.2 *The Stability Analysis of the Three-Phase Energy Router*

Similar to the single-phase Energy Router, the three-phase Energy Router's stability analysis can also be divided into impedance-based stability analysis strategy based on measured data and impedance-based stability analysis strategy based on system model. In terms of the impedance-based stability analysis strategy based on measured data, there is no difference between the three-phase Energy Router and the single-phase Energy Router, not tired in words here. As shown in Fig. 6.26, the three-phase Energy Router can also be divided into three parts such as three-phase rectifier, dual active bridge and three-phase inverter. The three-phase Energy Router stability analysis is able to be shown as follows: Firstly, the each of subsystem model should be built. Secondly, the number of the $Z_{in,DAB}+Z_{out,rec}$ counterclockwise encirclements of the $(0+j0)$ point is zero. In the end, the number of the $Z_{eqv}+Z_{in,inv}$ counterclockwise encirclements of the $(0+j0)$ point is zero. If the appeal conditions are met, the three-phase Energy Router is stable.

AC-DC Rectifier

The first subsystem is the three-phase rectifier under Clark/Park and d-q reference frame current/voltage mode control [35]. And, Fig. 6.27 reflects the average model of the rectifier in the d-q reference frame, and it has been seen that in unity power factor systems such as Energy Routers and so on, the q-axis parameters and their influences on the d-axis can be ignored. The closed loop output subsystem impedance of Energy Router can be represented as [16]

$$Z_{out,recd} = \frac{G_{v_{HV}i_0}\left(1 + G_{i_d,rec}G_{i_d d_d}\right) - G_{i_d i_0}G_{i_d,rec}G_{v_{HV}d_d}}{1 + G_{i_d,rec}G_{i_d d_d} + G_{v,rec}G_{i_d,rec}G_{v_{HV}d_d}} \tag{6.76}$$

where, $G_{v,rec}$, $G_{i_d,rec}$ are VLC and the d-axis current loop controller (CLC), respectively.

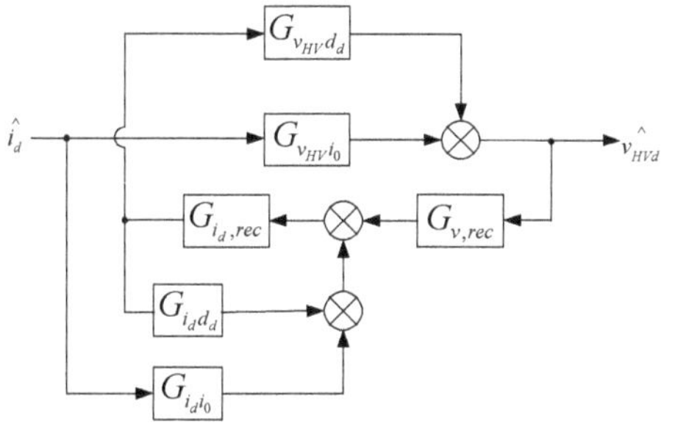

Fig. 6.27 Control block diagram model of the three-phase rectifier

$$G_{i_d i_0} = -\frac{2}{D_d}\frac{1}{L_{in}C_{HV}/D_d^2 s^2+1}$$

$$G_{i_d d_d} = -\frac{I_d}{D_d}\frac{2V_{HV}/(I_dD_d)s+1}{D_d^2/(L_{in}C_{HV})s^2+1}$$

$$G_{v_{HV} i_0} = \frac{1}{D_d^2}\frac{L_{in}s}{L_{in}C_{HV}/D_d^2 s^2+1}$$

$$G_{v_{HV} d_d} = \frac{V_{HV}}{D_d}\frac{5L_{in}I_d/(D_dV_{HV})s-1}{L_{in}C_{HV}/D_d^2 s^2+1}.$$

Dual Active Bridge

The second stage is the DAB, whose characteristics and models are same with DAB of the single-phase Energy Router [38]. Thus, there is not tired in words here.

DC-AC Inverter

The third subsystem is the three-phase dc-ac inverter [37]. The average model under Clark/Park of such an inverter can be shown by thinking it also as a dc-dc buck converter under average current/voltage mode control. Furthermore, the q-axis parameters and their influences on the d-axis can be ignored. And, the voltage reference of the inverter voltage loop controller is a sinusoidal voltage waveform. At present, to achieve the inverter input subsystem impedance, equation (6.77) can be seen as follows [16]:

$$Z_{in,invd} = \frac{1 + G_{i.inv}G_{i_{L_{fil}}d} + G_{v.inv}G_{v_{out}d}G_{v_{out}d}}{G_{i_{in}v_{LV}}\left(1 + G_{i.inv}G_{i_{L_{fil}}d} + G_{v.inv}G_{v_{out}d}\right) - G_{v_{out}}G_{v.inv}G_{v_{out}d}G_{i_{in}d}} \tag{6.77}$$

where,

$$G_{i_{L_{fil}}d} = \frac{V_{LV}}{R}\frac{RC_{fil}s+1}{L_{fil}C_{fil}s^2+L_{fil}s/R+1} \qquad G_{i_{L_{fil}}v_{LV}} = \frac{D}{R}\frac{RC_{fil}s+1}{L_{fil}C_{fil}s^2+L_{fil}s/R+1}$$

$$G_{v_{out}v_{LV}} = \frac{D}{L_{fil}C_{fil}s^2+L_{fil}s/R+1} \qquad G_{i_{in}v_{LV}} = \frac{D^2}{R}\frac{RC_{fil}s+1}{L_{fil}C_{fil}s^2+L_{fil}s/R+1}$$

$$G_{i_{in}d} = I_{fil} + \frac{V_{LV}D}{R}\frac{RC_{fil}s+1}{L_{fil}C_{fil}s^2+L_{fil}s/R+1} \qquad G_{v_{out}d} = \frac{V_{LV}}{L_{fil}C_{fil}s^2+L_{fil}s/R+1}$$

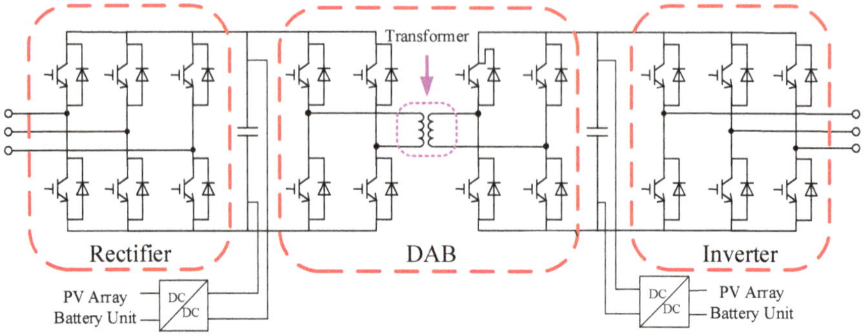

Fig. 6.28 The complex Energy Router system

As previously mentioned, the whole three-phase Energy Router stability analysis is same with the single-phase Energy Router.

6.4.3 The Stability Analysis of the Complex Energy Router

Due to the bidirectional power flow capability of Energy Router, the distributed renewable energy resources and the distributed energy storage devices, such as storage units and Photovoltaics (PVs), can connect to the Energy Router using dc-dc converter interfaces. Thus, the complex Energy Router stability analysis should be studied. As shown in Fig. 6.28, the paper in [18] provides a sort of effective method to resolve this problem. The majority of the connections among these passive and active components in Energy Router can be categorized into the following groups. Type 1: Y + Y parallel type connection. Type 2: Y + Z parallel type connection. Type 3: Z + Z series type connection. Type 4: Z + Z parallel type connection. Type 5: Y + Z series type connection. The stability criteria and total equivalent terminal characteristics of all the connection types are summarized in Table 6.1.

As previously mentioned, the whole complex Energy Router stability analysis is same with the single-phase Energy Router and the three-phase Energy Router.

6.5 Conclusion

This chapter introduces the mathematical model of the energy router in detail. Based on the analysis of the functions of each submodule, the energy flow model of the energy router is analyzed. Eventually, according to *Middlebrook* criterion, the stability of Energy Routers (Single-Phase Energy Router, Three-Phase Energy Router and Complex Energy Router) have been researched.

Table 6.1 The stability criteria and total equivalent terminal characteristics of all the connection types

Connection type	Stability criterion	Equivalent impedance or admittance
Type 1	Naturally stable	$Y_{eqc} = Y_1 + Y_2$
Type 2	The impedance ratio $T_m = Y_1 Z_2$ meet the Nyquist stability criterion	$Y_{eqc} = \frac{Y_1}{1+Y_1 Z_2}$
Type 3	Naturally stable	$Z_{eqc} = Z_1 + Z_2$
Type 4	The closed minor loop gain $T_{clm} = 1/(1+T_m) = Z_2/(Z_1+Z_2)$ meets the Impendance-Sum-Type Criterion	$Z_{eqc} = \frac{Z_1 Z_2}{Z_1+Z_2}$
Type 5	The impedance ratio $T_m = Y_1 Z_2$ meet the Nyquist stability criterion	$Z_{eqc} = \frac{Z_1 Z_2}{1+Y_1 Z_2}$

References

1. Y. Xu, J. Zhang, W. Wang, A. Juneja, S. Bhattacharys, Energy router: architectures and functionalities toward energy internet. IEEE Trans. Smart Grid 31–36 (2016)
2. P. Yi, T. Zhu, B. Jiang, R. Jin, B. Wang, Deploying energy routers in an energy Inter-net based on electric vehicles. IEEE Trans. Veh. Technol. **65**(6), 4714–4725 (2016)
3. T. Zhao, J. Zeng, S. Bhattacharya, M.E. Baran, A.Q. Huang, An average model of solid state transformer for dynamic system simulation, in *Proceedings of the IEEE Power and Energy Society General Meeting*, pp. 1–8
4. F. Gao, Z. Li, P. Wang, F. Xu, Z. Chu, Z. Sun, Y. Li, Prototype of smart energy router for distribution DC grid, in *Proceedings of 2015 Power Electronics and Applications* (*EPE'15 ECCE-Europe*), pp. 1–9
5. S. Lu, Z. Zhao, J. Ge, L. Yuan, T. Lu, A new power circuit topology for energy router, in *Proceedings of 2014 IEEE International Conference on Electrical Machines and Systems* (*ICEMS*), pp. 1921–1925
6. A. Sanchez-Squella, R. Ortega, R. Grino, S. Malo, Dynamic energy router. IEEE Con-trol Syst. **30**(6), 72–80 (2010)
7. X. Guo, H. Wang, Z. Lu, New inverter topology for ground current suppression in transformerless photovoltaic system application. J. Mod. Power Syst. Clean Energy **2**(2), 191–194 (2014)
8. F. Wang, A. Huang, G. Wang, X. She, R. Burgos, Feed-forward control of solid state transformer, in *Proceedings of 2012 IEEE, Applied Power Electronics Conference and Exposition* (*APEC*), pp. 1153–1158
9. Q. Duan, C. Ma, W. Sheng, C. Shi, Research on power quality control in distribution grid based on energy router, in *Proceedings of 2014 Power System Technology* (*POWERCON*), pp. 2115–2121
10. J. Cao, M. Yang, Energy internet - towards smart grid 2.0, in *Proceedings of 2013 Fourth International Conference on Networking and Distributed Computing* (*ICNDC*), pp. 105–110
11. L. Chen, Q. Sun, L. Zhao, Q. Cheng, Design of a novel energy router and its application in energy internet, in *Proceedings of 2015, Chinese Automation Congress* (*CAC*), pp. 1462–1467
12. B. Wen, D. Dong, D. Boroyevich, R. Burgos, P. Mattavelli, Z. Shen, Impedance-based analysis of grid-synchronization stability for three-phase paralleled converters. IEEE Trans. Power Electron. **31**(1), 26–38 (2016)

13. J.Z. Zhou, H. Ding, S. Fan, Y. Zhang, A.M. Gole, Impact of Short-Circuit Ratio and Phase-Locked-Loop Parameters on the Small-Signal Behavior of a VSC-HVDC Convert-er. IEEE Trans. Power Deliv. **29**(5), 2287–2296 (2014)
14. H. Xin, L. Huang, L. Zhang, Z. Wang, J. Hu, Synchronous instability mechanism of P-f droop-controlled voltage source converter caused by current saturation. IEEE Trans. Power Syst. **31**(6), 5206–5207 (2016)
15. H. Zhang, S. Kim, Q. Sun, J. Zhou, Distributed adaptive virtual impedance control for accurate reactive power sharing based on consensus control in microgrids. IEEE Trans. Smart Grid **PP**(99), 1–13
16. M. Khazraei, V.A.K. Prabhala, R. Ahmadi, M. Ferdowsi, Solid-state transformer stability and control considerations, in *2014 IEEE Applied Power Electronics Conference and Exposition - APEC 2014*, Fort Worth, TX (2014), pp. 2237–2244
17. D.G. Shah, M.L. Crow, Stability design criteria for distribution systems with solid-state transformers. IEEE Trans. Power Deliv. **29**(6), 2588–2595 (2014)
18. W. Cao, Y. Ma, F. Wang, Sequence-impedance-based harmonic stability analysis and controller parameter design of three-phase inverter-based multibus AC power systems. IEEE Trans. Power Electron. **32**(10), 7674–7693 (2017)
19. M. Amin; M. Molinas, Small-Signal stability assessment of power electronics based power systems: a discussion of impedance- and eigenvalue-based methods. IEEE Trans. Ind. Appl. **PP**(99), 1–1
20. A. Kahrobaeian, Y.A.R.I. Mohamed, Analysis and mitigation of low-frequency in-stabilities in autonomous medium-voltage converter-based microgrids with dynamic loads. IEEE Trans. Ind. Electron. **61**(4), 1643–1658 (2014)
21. Y. Wang, X. Wang, F. Blaabjerg, Z. Chen, Harmonic instability assessment using state-space modeling and participation analysis in inverter-fed power systems. IEEE Trans. Ind. Electron. **64**(1), 806–816 (2017)
22. L. Luo, S.V. Dhople, Spatiotemporal model reduction of inverter-based islanded microgrids. IEEE Trans. Energy Convers. **29**(4), 823–832 (2014)
23. F. Dorfler, F. Bullo, Kron reduction of graphs with applications to electrical net-works. IEEE Trans. Circuits Syst. I: Regul. Papers **60**(1), 150–163 (2013)
24. M. Rasheduzzaman, J.A. Mueller, J.W. Kimball, Reduced-order small-signal model of micro-grid systems. IEEE Trans. Sustain. Energy **6**(4), 1292–1305 (2015)
25. B. Wen, D. Boroyevich, R. Burgos, P. Mattavelli, Z. Shen, Inverse Nyquist stability criterion for grid-tied inverters. IEEE Trans. Power Electron. **32**(2), 1548–1556 (2017)
26. A.A.A. Radwan, Y.A.R.I. Mohamed, Stabilization of medium-frequency modes in isolated Microgrids supplying direct online induction motor loads. IEEE Trans. Smart Grid **5**(1), 358–370 (2014)
27. J. Zhou; P. Shi; D. Gan; Y. Xu; H. Xin; C. Jiang; H. Xie; W. Tao, Large-scale power sys-tem robust stability analysis based on value set approach. IEEE Trans. Power Syst. **PP**(99), 1–1
28. D. Dong, B. Wen, D. Boroyevich, P. Mattavelli, Y. Xue, Analysis of phase-locked loop low-frequency stability in three-phase grid-connected power converters considering impedance interactions. IEEE Trans. Ind. Electron. **62**(1), 310–321 (2015)
29. W. Cao, Y. Ma, L. Yang, F. Wang, L.M. Tolbert, D–Q impedance based stability analysis and parameter design of three-phase inverter-based AC power systems. IEEE Trans. Ind. Electron. **64**(7), 6017–6028 (2017)
30. B. Wen, D. Dong, D. Boroyevich, R. Burgos, P. Mattavelli, Z. Shen, Impedance-based analysis of grid-synchronization stability for three-phase paralleled converters. IEEE Trans. Power Electron. **31**(1), 26–38 (2016)
31. J. Sun, Impedance-based stability criterion for grid-connected inverters. IEEE Trans. Power Electron. **26**(11), 3075–3078 (2011)
32. S. Vesti, T. Suntio, J.A. Oliver, R. Prieto, J.A. Cobos, Impedance-based stability and transient-performance assessment applying maximum peak criteria. IEEE Trans. Power Electron. **28**(5), 2099–2104 (2013)

33. F. Liu, J. Liu, H. Zhang, D. Xue, Stability issues of $Z+Z$ type cascade system in hybrid energy storage system (HESS). IEEE Trans. Power Electron. **29**(11), 5846–5859 (2014)
34. Z. Liu, J. Liu, W. Bao, Y. Zhao, Infinity-Norm of impedance-based stability criterion for three-phase AC distributed power systems with constant power loads. IEEE Trans. Power Electron. **30**(6), 3030–3043 (2015)
35. W. Wu et al., A virtual inertia control strategy for DC microgrids analogized with virtual synchronous machines. IEEE Trans. Ind. Electron. **64**(7), 6005–6016 (2017)
36. A. Rodríguez, A. Vázquez, D.G. Lamar, M.M. Hernando, J. Sebastián, Different purpose design strategies and techniques to improve the performance of a dual active bridge with phase-shift control. IEEE Trans. Power Electron. **30**(2), 790–804 (2015)
37. J. Liu, Y. Miura, T. Ise, Comparison of dynamic characteristics between virtual synchronous generator and droop control in inverter-based distributed generators. IEEE Trans. Power Electron. **31**(5), 3600–3611 (2016)
38. I. Cvetkovic, D. Boroyevich, P. Mattavelli, F.C. Lee and D. Dong, Unterminated small-signal behavioral model of dc–dc converters. IEEE Trans. Power Electron. **28**(4), 1870–1879 (2013)

Chapter 7
The Model and Energy Measurement of Energy Hub

Abstract This chapter proposes a new method to evaluate integrated energy systems which take full consideration of the utilization effect of renewable energy. Exergy analysis of multiple energy system is introduced from the perspective of quality of energy. A new power flow expression of energy hub contain storage is given in order to overcome the disadvantage of traditional power flow expression. This chapter also presents a novel droop control method of energy hubs which contain thermal and electric droop control method in a multiple energy system. The proposed energy hub droop control method can proportional allocate different energy hubs thermal and electric outputs according to the corresponding energy maximum output power of energy hubs to allow independent operation of the energy hubs and ensure the stability of the multiple energy system. The three-dimensional diagrams of energy hub droop control method is presented to integrally represent the relationship between systems parameters and the inputs of energy hub meanwhile solely represent the relationship between systems parameters and each input of energy hub that using proposed droop control method. Numerical simulations demonstrate the effectiveness of the proposed droop control method based on energy hub.

7.1 Introduction

The energy hub concept has been recently introduced as a new paradigm for future multicarrier energy systems. An energy hub can be considered as a unit that offers the basic features conversion, and storage of different energy carriers and includes a variety of components, such as combined heat and power (CHP), transformers, boilers in order to meet energy demands [1]. The basic model is formulated in the literature. A new model of energy hub considering system efficiencies, storage losses and operating limits is presented in [2], but these three aspects are discussed separately which each aspect is also simplified. Thermal and electrical storage are simultaneously taken into account and a detailed evaluation about energy storage application benefit are given in [3]. All technical potential interconnections between

Q. Sun, *Energy Internet and We-Energy*, Renewable Energy Sources & Energy Storage, https://doi.org/10.1007/978-981-13-0523-8_7

different elements of energy hub and storage facility installation at both sides of converters are considered in [4].

Other than traditional energy hub concept, different kind of energy hub concept and multiple energy system structures are studied by some research. The energy hub concept at urban level is presented in [5], some different energy hub models at urban level has been presented and the influence of objection bring by different models has been discussed. A model of energy hub with combine distributed energy supply/combined cooling heating and power (DES/CCHP) has been proposed in [6]. The optimal operation of energy hubs connects with regional multi-energy prosumers which can sell electricity to grid by different price has been studied, the neglection of line impedance simplified the optimal operation problem.

Some research focus on the influence of different type electricity markets and the game in electricity markets to ensure the operation of energy hubs is optimally. The impact of time-of-use and dynamic pricing electricity markets for energy hubs optimal operation problem which is solved by a distributed arithmetic, is studied in [7]. Reference [8] has introduced the smart energy hub model and the interaction among smart energy hubs is formulated as a noncooperative game. Compare with [8], the game player has been changed in [9, 10] from hubs to energy companies and hubs.

There are several factors of uncertainty exist in multi–energy system which contains the energy hub. The uncertainty of wind and electricity market price which is forecasted by corresponding methods are considered in [11]. Two decision-making models are given and the risk aversion in management is taken into account. The uncertainty of wind, electricity market price and demand are considered in [12, 13]. Reference [14] has discussed the influence of the presence of data uncertainty which is calculated by affine arithmetic-based methodology. A stochastic bi-level model which has discussed three types of uncertainty in electricity market has been presented in [15]. The optimal operation problem based on the model is solved by using the kkt optimal condition.

The optimal operation problem of energy hubs and the optimal power flow problem of multiple energy system are the most studied directions. A Pareto optimal solution is determined for the multi objective optimization which uses two rival criteria of economics and environmental performance to solve optimized problem for smart energy network which contain multi energy hubs in [16]. Two long-term optimal planning are proposed in [17, 18] for energy hubs in multi-energy system. Reference [19] proposed a modified teaching–learning based optimization method to solve the optimal operation problem of energy power flow in multi–energy system, but the energy hub model and the framework of multicarrier system in the simulation of [19] is too simple. A multi-objective mixed integer linear programming model of multi–energy system with a district heating network is presented in [20], the optimal option of elements and respective capacities, district heating network distribution and optimal operation to meet demands can be determined by this model.

Security operation is important for energy system and many control methods were proposed to ensure it. The droop control method is one method that can achieve the reliability and stability of the system through decentralized control, reduce the impact of communication on system reliability. Hence, droop control method has been in deep studied. Reference [21] proposed a frequency and voltage droop control method to allow independent inverters to share the load in proportion to their capacities, but the line resistance impacts the effect of proportional share. Reference [22–25] focus on the effect of power sharing but overlooked the stability analysis. The model, stability analysis, and influence of parameters of the multi-inverter system are studied in [26].

From the perspective of quality of energy, energy analysis of multiple energy system is introduced. And the author proposed a new method to evaluate integrated energy systems which take full consideration of the utilization effect of renewable energy. To make the model of energy hub that contain storage, a new energy hub power flow expression is proposed. There is also little research focus on the security operation of multiple energy system with energy hubs. To enhance the system stability and decrease the influence of communication, we propose a droop control method of energy hub to carry out the proportional allocation of hubs and control the energy hub outputs properly based on the corresponding capacity. The droop control method of energy hub which contains thermal and electric droop control approaches can realize decentralized control of energy hubs and ensure the security operation of multiple energy system. The hub inputs, outputs and system parameters coupled together on account of using proposed method and hub own characteristic. In this regard, the three-dimensional diagram of droop control method of energy hub and two kinds of operation strategies are presented to solve the energy hub operation problem which is hard to dispose due to the intercoupling. The proposed methods are evaluated for their performance by the results of case studies.

7.2 Energy Hub Model

7.2.1 The Traditional Energy Hub Model

Energy hub is defined by a series of energy carriers, each energy hub includes multiple energy carriers as inputs and outputs. There are many elements such as connectors, conversion and storage facilities that can process multiple energy contained in Energy hub. Because of these elements, the input energy of hub can be converted to diversified forms in order to meet the energy demand at the output port of hub. The energy transfer from hub input energy to hub output energy can be expressed as:

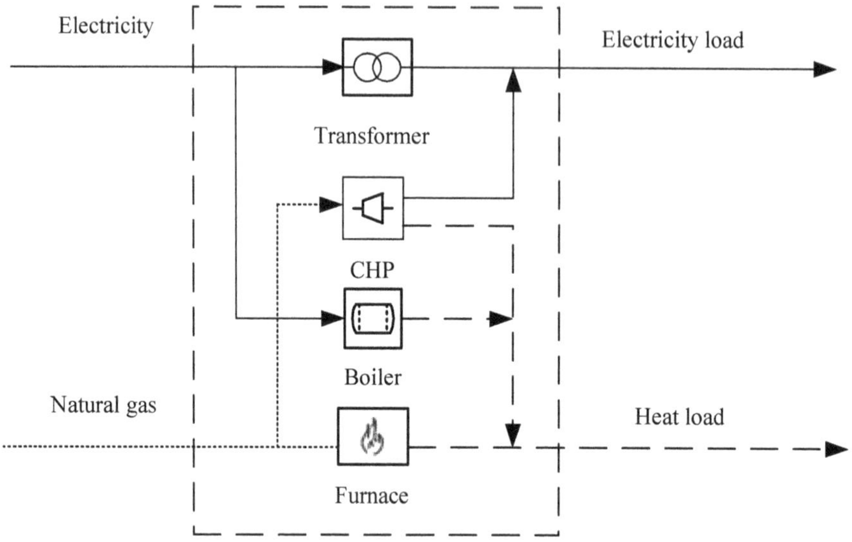

Fig. 7.1 Energy hub with electricity, natural gas, and heat systems

$$\underbrace{\begin{pmatrix} L_\alpha \\ L_\beta \\ \vdots \\ L_\gamma \end{pmatrix}}_{L} = \underbrace{\begin{pmatrix} C_{\alpha\alpha} & C_{\beta\alpha} & \cdots & C_{\gamma\alpha} \\ C_{\alpha\beta} & C_{\beta\beta} & \cdots & C_{\gamma\beta} \\ \vdots & \vdots & \ddots & \vdots \\ C_{\alpha\gamma} & C_{\beta\gamma} & \cdots & C_{\gamma\gamma} \end{pmatrix}}_{C} \underbrace{\begin{pmatrix} E_\alpha \\ E_\beta \\ \vdots \\ E_\gamma \end{pmatrix}}_{E} \tag{7.1}$$

where the various kinds of input energy and output energy are figured by $E = [E_\alpha, E_\beta, \ldots, E_\gamma]$ and $L = [L_\alpha, L_\beta, \ldots, L_\gamma]$, respectively. The matrix C is the forward coupling matrix which describes the conversion of energy from the input to the output. The elements of coupling matrix are coupling factors. For a single converter like gas furnace, transformer and so on, the coupling factor is decided by the converter efficiency. For a multiple input and output energy hub, the coupling factor is not only decided by each converter efficiencies, but also decided by dispatch factors which define how input energy be allocated to each convertor of energy hub. Figure 7.1 presents an energy hub with electricity, natural gas, and heat systems.

In Fig. 7.1, the energy hub is mainly constituted by four converter devices: transformer, CHP plant, electricity boiler and natural gas furnace. The specific energy input vector E formed by electricity and natural gas can be expressed as:

$$E = \begin{pmatrix} E_e \\ E_g \end{pmatrix} \tag{7.2}$$

The specific energy output vector L formed by electricity and heat can be expressed as

$$L=\begin{pmatrix} L_e \\ L_h \end{pmatrix} \tag{7.3}$$

The equation constrains between input E and output L can be expressed as

$$L_e = \kappa\eta_{ee}E_e + \nu\eta_{ge}^{CHP}E_g \tag{7.4}$$

$$L_h = (1-\nu)\eta_{gh}^{Fur}E_g + \nu\eta_{gh}^{CHP}E_g+(1-\kappa)\eta_{eh}E_e \tag{7.5}$$

where η_{ee} is the efficiency of electricity transformer, η_{eh} is the efficiency of electricity boiler, η_{ge}^{CHP} and η_{gh}^{CHP} are the gas-electric and gas-heat efficiencies of CHP plant, respectively. η_{gh}^{Fur} is the natural gas furnace efficiency. Moreover, ν is the dispatch factor for natural gas which can be consumed by both natural gas furnace and CHP. κ is the dispatch factor for electricity which can be consumed by both electricity boiler and transformer. Therefore, νE_g is the natural gas consumption which is converted to electricity and heat via CHP, and $(1-\nu)E_g$ is the natural gas consumption which used to generate heat by natural gas furnace. κE_e is the electricity consumption which is transferred by transformer, and $(1-\kappa)E_e$ is the electricity consumption which used to generate heat by electricity boiler.

The coupling factor $C_{\alpha\beta}$ is codetermined by the dispatch factors and the efficiency. The conversion matrix can be expressed as

$$\begin{pmatrix} L_e \\ L_h \end{pmatrix}=\begin{pmatrix} \kappa\eta_{ee} & \nu\eta_{ge}^{CHP} \\ (1-\kappa)\eta_{eh} & (1-\nu)\eta_{gh}^{Fur}+\nu\eta_{gh}^{CHP} \end{pmatrix}\begin{pmatrix} E_e \\ E_g \end{pmatrix} \tag{7.6}$$

In multi energy hubs system, energy is supplied by coupled energy carriers. Figure 7.2 shows a system with N energy hubs. Energy is transferred to the energy hub from the natural gas network and electricity network and then outputted to the demand side. In this system, energy hubs output parallel connecting to counter-PCC (Point of Common Coupling). All energy hubs satisfy the electricity and heat load together but not each energy hub meet the corresponding load.

7.2.2 A New Model of Energy Hub Contain Storage

When storages are contained by the energy hub, the elements of forward coupling matrix C can't be derived by physical meanings. The coupling factors only can be calculated by mathematical method when the inputs and outputs are exactly known. The effect of storage and other parts which contained by the energy hub during power

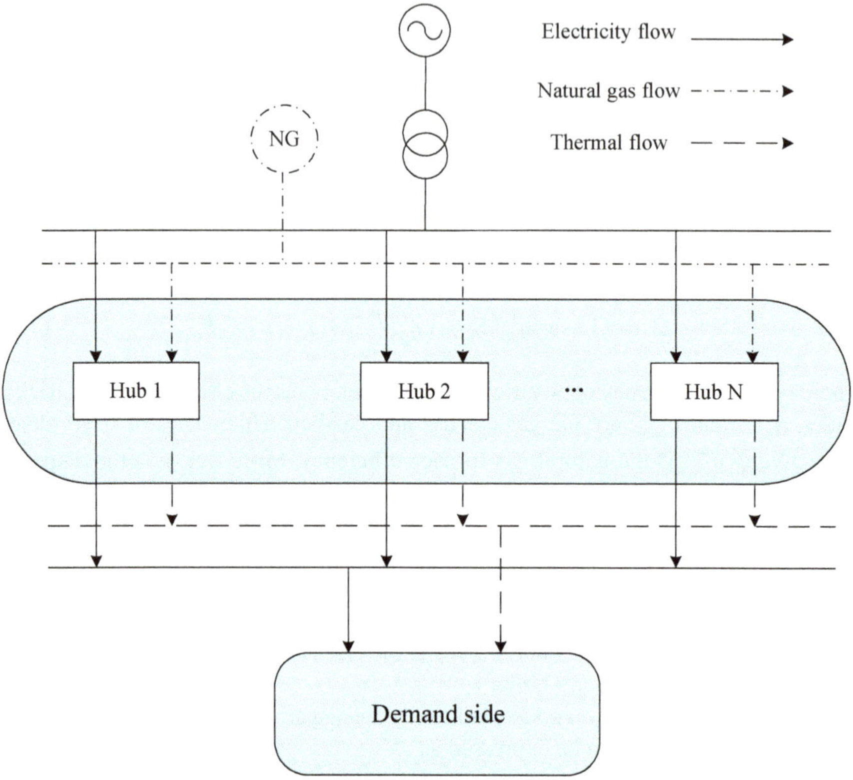

Fig. 7.2 Multi energy carriers system contain energy hubs

flow can't be expressed by the elements. To solve this problem, another power flow expression is given by some references. The expression is stated as:

$$L = CE - ST \tag{7.7}$$

where the matrix S is the storage coupling matrix. It describes how changes within the storage energies affect the output, i.e., how the storage energy derivatives are mapped into equivalent output-side flows. The matrix T describes storages change of multiple energy carriers. The disadvantage of this expression is the change of storages must be known.

Because of all these reasons, the traditional power flow expression is extended. The new power flow expression is stated as:

$$\underbrace{\begin{bmatrix} L_e \\ L_g \\ L_h \\ L_{e'} \\ L_{g'} \\ L_{h'} \end{bmatrix}}_{L_1} = \underbrace{\begin{bmatrix} a_{ee}\lambda_{e1} & a_{Ge}\lambda_{g1} & a_{he}\lambda_{h1} & a_{ee}\lambda_{e'1} & a_{Ge}\lambda_{g'1} & a_{he}\lambda_{h'1} \\ a_{eg}\lambda_{e2} & \lambda_{g2} & a_{hg}\lambda_{h2} & a_{eg}\lambda_{e'2} & \lambda_{g'2} & a_{hg}\lambda_{h'2} \\ a_{eh}\lambda_{e3} & a_{Gh}\lambda_{g1} + a_{gh}\lambda_{g3} & a_{hh}\lambda_{h3} & a_{eh}\lambda_{e'3} & a_{Gh}\lambda_{g'1} + a_{gh}\lambda_{g'3} & a_{hh}\lambda_{h'3} \\ \eta_e a_{ee}\lambda_{e4} & \eta_e a_{Ge}\lambda_{g4} & \eta_e a_{he}\lambda_{h4} & \lambda_{e'4} & \eta_e a_{Ge}\lambda_{g'4} & \eta_e a_{he}\lambda_{h'4} \\ \eta_g a_{eg}\lambda_{e5} & \eta_g \lambda_{g5} & \eta_g a_{hg}\lambda_{h5} & \eta_g a_{eg}\lambda_{e'5} & \lambda_{g'5} & \eta_g a_{hg}\lambda_{h'5} \\ \eta_h a_{eh}\lambda_{e6} & \eta_h a_{Gh}\lambda_{g4} + \eta_h a_{gh}\lambda_{g6} & \eta_h a_{hh}\lambda_{h6} & \eta_h a_{eh}\lambda_{e'6} & \eta_h a_{Gh}\lambda_{g'4} + \eta_h a_{gh}\lambda_{g'6} & \lambda_{h'2} \end{bmatrix}}_{C_1} \underbrace{\begin{bmatrix} T_e \\ T_g \\ T_h \\ T_{e'} \\ T_{g'} \\ T_{h'} \end{bmatrix}}_{T_1} \tag{7.8}$$

where variables $T_{e'}$, $T_{g'}$ and $T_{h'}$ are electricity, natural gas and heating stored in storages before power flow, respectively. Variables $L_{e'}$, $L_{g'}$ and $L_{h'}$ are of electricity, natural gas and heating stored in storages after power flow, respectively. Parameters a_{ee}, a_{eh} and a_{gh} denote the efficiencies of the transformer, the electric boiler and the auxiliary boiler, respectively. Parameters η_e, η_g and η_h are the efficiencies of the electric power storage, the natural gas storage and the heat storage, respectively. Parameters a_{Gh} and a_{Ge} are the heating and the electrical efficiency of the CHP. Parameters λ_{ei}, .., λ_{hi} are the dispatch factors of P_e, P_g and P_h. In the same way, parameters $\lambda_{e'i}$, $\lambda_{g'i}$, $\lambda_{h'i}$ are the dispatch factors of $T_{e'}$, $T_{g'}$ and $T_{h'}$. In this energy hub power flow expression which is newly introduced, the input and output of storages do not simply be regarded as the energy exchange inner energy hub but be treated as input and output terms of the power flow expression. The forward coupling matrix C_1 is extended from original a third-order matrix to a sixth-order matrix. The coupling factors of the forward coupling matrix can be derived from the converter efficiencies and power dispatch. In this way, the details of power flow transformation can be known from the forward coupling matrix, include one storage output converted into other kinds of energy.

Limits on factor value are some inequality and equality constraints at time t stated as:

$$0 \le T_\alpha^t \le \bar{T}_\alpha \quad \alpha \in e, g, h \tag{7.9}$$

$$0 \le T_{\alpha'}^t \le Q_\alpha \quad \alpha \in e, g, h \tag{7.10}$$

$$\lambda_{\beta i} \ge 0 \quad \beta \in [e, g, h, e', g', h'], i \in [1, 2, 3, 4, 5, 6] \tag{7.11}$$

$$\sum_{i=1}^{6} \lambda_{\beta i} = 1 \quad \beta \in [e, g, h, e', g', h'] \tag{7.12}$$

where Variable Q_α is the maximum storage capacity of corresponding store. Variable α means all kind of inputs and outputs energy. Variable β means not only all kind of inputs and outputs energy but also all kind of storages.

7.3 The Droop Control Method of Energy Hub

For a network contain multi energy hubs, if there is no communication between energy hubs, it is hard to determine how much energy each hub output. To solve this problem and guarantee the multi energy hubs system stability, the heat droop control method that combine with electricity droop control method can determine the energy hub output is proposed.

7.3.1 Thermal Droop Control Method

Energy hub outputs thermal and electric power to satisfy the corresponding loads. For the thermal load which is supplied heat by energy hubs, we assume the environment temperature will not change suddenly and the environment temperature in an area is similar. So the thermal load change discussed is about the heat pipeline join in or break away from the network. Generally the heat pipelines at demand side connect parallelly. Each thermal load pipeline has the impedance of heat water flow. Pipeline impedance is only decided by pipeline parameters such as pipe diameter, length, absolute roughness of pipeline inner surface and so on. In other words, pipeline impedance is determined by the character of pipeline and do not change with water flow rate. The heat water flow rate of thermal load water inlet and outlet is identical, and the water flow rate is equivalent in one pipeline.

The whole heat pipeline impedance of thermal load S and the heat water flow rate of thermal load water inlet V can be expressed as follows

$$\frac{1}{\sqrt{S}}=\frac{1}{\sqrt{S_1}}+\frac{1}{\sqrt{S_2}}+\cdots\frac{1}{\sqrt{S_i}} \tag{7.13}$$

$$V=V_1+V_2+\cdots V_i \tag{7.14}$$

Equation (7.9) defines that the impedance of thermal load is inversely proportional with thermal demand. S_1, S_2 and S_i are impedance of each thermal load pipeline. V is the heat water flow rate of thermal load water inlet and outlet. V_1, V_2 and V_i are water flow rate of each thermal load pipeline. p_{in} and p_{out} are the intensity of pressure of thermal load water inlet and outlet, respectively. The relation expression of intensity of pressure and water flow rate is as follow

$$p_{in} - p_{out} = SV^2 \tag{7.15}$$

When the thermal load is determined, this equation shows that the pressure drop of thermal load is proportional to the square of total water flow rate.

The power balance equation of the district heating sub-network is expressed as follows

$$L_{hi} = Q_{D,i} + \sum_{j}^{N_{line}^{heat}} Q_{H,ij} \tag{7.16}$$

$$Q_{D,i} = cm_i \bigvee t_i \tag{7.17}$$

$$Q_{H,ij} = cm_{ij} \bigvee t_{ij} \tag{7.18}$$

$$m_i = \sum_{j}^{N_{line}^{heat}} m_{ij} \tag{7.19}$$

The output heat power of energy hub i (7.12) can be divided into the loss of transmission pipe and the power thermal load get. Where L_{hi} is the heat output of energy hub i, $Q_{D,i}$ is the heat loss at transmission pipe and $Q_{H,ij}$ is the heat that demand side get from each thermal load pipeline. In Eqs. (7.13) and (7.14), c is the specific heat capacity of water. m_{ij} is the mass flow rate of each pipeline, m_i is the output mass flow rate of energy hub i and m_i is equal to the sum total of m_{ij}.Similarly, $\triangle\ t_i$ and $\triangle\ t_{ij}$ is the temperature different of the transmission pipe and the each thermal load pipeline. Commonly, there is a thermal-protective coating at the outside of the transmission pipe to reduce the heat loss. So $\triangle\ t_i$ is much lower than $\triangle\ t_{ij}$ and the $\triangle\ t_i$ can be ignored. In this way, the output heat power of energy hub i is equal to the power demand side get from hub i and can be expressed as follow

$$L_{hi} = \sum_{j}^{N_{line}^{heat}} Q_{H,ij} = cm_i \triangle t \tag{7.20}$$

In Eq. (7.16), $\triangle\ t$ means the temperature differences between thermal load inlet and outlet.

There are two energy regulation ways for the thermodynamic system: flow control and temperature control. Because the temperature variation is a slowly changing process. To improve the system response speed of the system, flow control is chosen as the control method of energy hubs. According to the assumption at Sect. 7.2 beginning, the environment temperature is stable. Because the flow control response fast, so the temperature difference change little during the variation process. Though these reason, the change of $\triangle\ t$ can be ignored when the flow control method is implemented.

The relationship between water flow rate V and mass flow rate m can be expressed as

$$m = \rho V \tag{7.21}$$

where ρ is a constant which stands for the water density. It defines that the mass flow rate is proportional to the water flow rate.

The way that energy hub transfer thermal power to the heat load can be described as follow: Hot water outputted from energy hub, flow though outlet pipe into thermal load and traverse inlet pipe back to energy hub. Because the water flow rate is equivalent in one pipeline, the flow rate in inlet and outlet pipe is equality. Equations (7.11) and (7.17) shows that the mass flow rate depends on the pressure drop for certain pipelines. In order to easy the analysis, the impedance of inlet pipe is integrated into the impedance of outlet pipe. Because of the integration, the pressure of thermal load water outlet is equal to the energy hub water inlet pressure. The thermal load water outlet pressure is set as the basic value. According to the analysis of the energy hub heat transfer, Eqs. (7.11), (7.14) and (7.17), the following expression can be derivated as

$$\left(\frac{L_h}{c\rho \Delta t}\right)^2 S = p_2 - p_1 \tag{7.22}$$

where p_1 and p_2 are the pressure of energy hub outlet and thermal load inlet, respectively. S is the transmission pipeline impedance which integrate the impedance of energy hub inlet pipe and outlet pipe. m is the output mass flow rate of energy hub. For stable operation the thermal power flow from the energy hubs to the load should be properly controlled. Equation (7.18) defines the relationship between the transmission pipeline pressure drop and the energy hub output heat power. Based on the analysis preceding part, all the parameters can be considered as constant except for p_1, p_2 and L_h.So the hub output heat power are mainly influenced by transmission pipeline pressure drop, the relationship between the two is positively correlated. The impendence of thermal load will be changed when the demand of thermal changes. For example, if the heat demand increases, the thermal load impendence will decrease. Since the mass flow rate can't mutate suddenly, p_2 will decline and the output power can be varied by adjusting hub outlet pressure p_1 .The same conclusion can be easily derived when the heat demand decrease. To avoid overloading on energy hub we want each energy hub to respond to the load change so as to automatically take a share proportional to its power rating. In analogy with the $Q - V$ droop control of electric system, the $L_h^2 - p$ droop control formula can be expressed as follow

$$p - p_N = k_p(L_{h,N}^2 - L_h^2) \tag{7.23}$$

where k_p is the droop control coefficient of the thermal droop control method. p_N is the rating pressure of energy hub outlet and $L_{h,N}^2$ is the square of rating energy hub heat output power.

In order to reduce the heat loss, generally the impedance of transmission pipeline is far less than the impedance of load. So the outlet pressure of energy hubs which supply the identical load is same. So as to enhance the stability of the multi energy hub system, the thermal droop control method can be utilized to make the heat power output is proportional to heat capacity of energy hub. Under this circumstance, the relationship between the maximal output heat power $L_{h,\max}$ and the droop control coefficient k_p can be expressed as

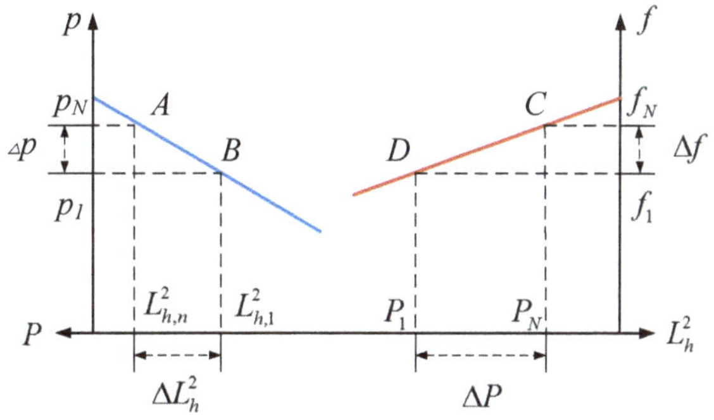

Fig. 7.3 The curve of energy hub droop control

$$\frac{L_{hn,\max}^2}{L_{hm,\max}^2} = \frac{k_{m,p}}{k_{n,p}} \tag{7.24}$$

where $L_{hn,\max}^2$ and $L_{hm,\max}^2$ are the square of energy hub n and energy hub m maximal output heat power, respectively. $k_{n,p}$ and $k_{m,p}$ are the thermal droop control coefficient of energy hub n and energy hub m respectively.

7.3.2 Electricity Droop Control Method

The $P - f$ droop control method is used to determine the electric output of energy hubs. The $P - f$ droop control formula can be expressed as follow

$$f - f_N = k_q(P_N - P) \tag{7.25}$$

where k_q is the droop control coefficient of the electric droop control method. f_N is the rating frequency of electricity sub-network and P_N is the rating energy hub electric output power.

According to Eqs. (7.19) and (7.21), the curve of energy hub droop control method can be drawn up like Fig. 7.3.

In Fig. 7.3, point a represents the rating condition of energy hub thermal output, point b represents another condition of energy hub thermal output, point c represents the rating condition of energy hub electric output and point d represents another condition of energy hub electric output. Parameter p_1, $L_{h,1}^2$ denote the pressure and thermal output power of energy hub at the condition of point b, respectively. The pressure and thermal output power of energy hub at the condition of point d, is denoted by f_1 and P_1, respectively. $\triangle p$ is the pressure difference between the operating point a and b. ΔL_h^2 is the difference of energy hub output thermal power square. The frequency different between the operating point c and d is represented by $\triangle f$. $\triangle P$ is the energy hub output active power difference between the operating condition c and d.

According to the proposed energy hub droop control method, energy hubs can proportionally output power by their power rating.

7.3.3 The Three-Dimensional Diagram of Energy Hub Droop Control Method

The utilization of energy hub droop control method can determine the output of energy hub. But in fact, the operating state of energy hub is decided by the inputs of energy hub. To find out the relationship between the system parameters and energy hub inputs, the proposed energy hub droop control method is divided into thermal droop control method and electric droop control method to analysis.

Combining with Eqs. (7.7) and (7.19), a new thermal droop control formula can be deduced as follow:

$$p - p_N = k_p\left[L_{h,N}^2 - \left((1-\kappa)\eta_{eh}E_e + (1-\nu)\eta_{gh}^{Fur}E_g + \nu\eta_{gh}^{CHP}E_g\right)^2\right] \quad (7.26)$$

It defines that the electricity and natural gas input of energy hub have the equality relationship with energy hub outlet pressure. According to Eq. (7.22), a three-dimensional diagram of thermal droop control method can be drawn like Fig. 7.6. Figure 7.6 describes a feasible region of energy hub inputs and outlet pressure. Each point of the region stands for a working position of energy hub. It also shows that each outlet pressure value of energy hub is towards to multiple statuses of energy hub electricity and natural gas input. All of these statuses are possible energy hub inputs of one certain outlet pressure.

For Fig. 7.4, three kinds of sections which respectively parallel coordinate surfaces can be sectioned. Figure 7.5 shows one section that parallel $p-E_g$ coordinate surface. It defines the relationship between outlet pressure p and energy hub natural gas input E_g when the electricity input E_e is determined. Similarly, a plane parallel $p - E_e$ coordinate surface defines the relationship between outlet pressure p and energy hub electricity input E_e when the natural gas input E_g is determined. A plane parallel $E_e - E_g$ coordinate surface defines the relationship between energy hub electricity input E_e and energy hub natural gas input E_g when the outlet pressure p is determined. According to Figs. 7.4, 7.5 and other two kinds sections, the relationship between the outlet pressure and energy hub inputs can be integrally represented and the relationship between the outlet pressure and each energy hub input can be solely represented.

Combining with Eqs. (7.6) and (7.21), a new electric droop control formula can be deduced as follow similarly:

$$f - f_N = k_q(P_N - \kappa\eta_{ee}E_e - \nu\eta_{ge}^{CHP}E_g) \quad (7.27)$$

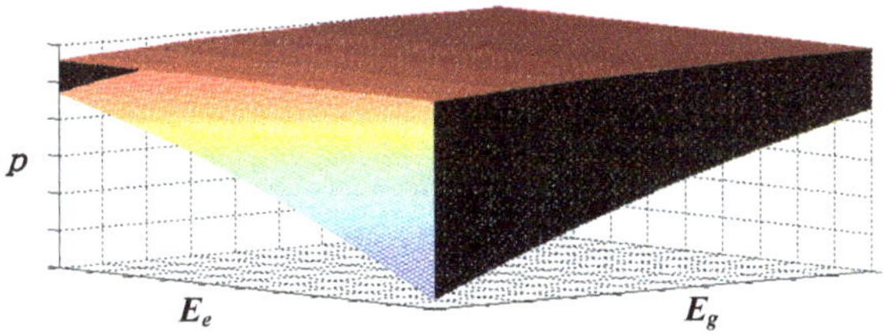

Fig. 7.4 The three-dimensional diagram of thermal droop

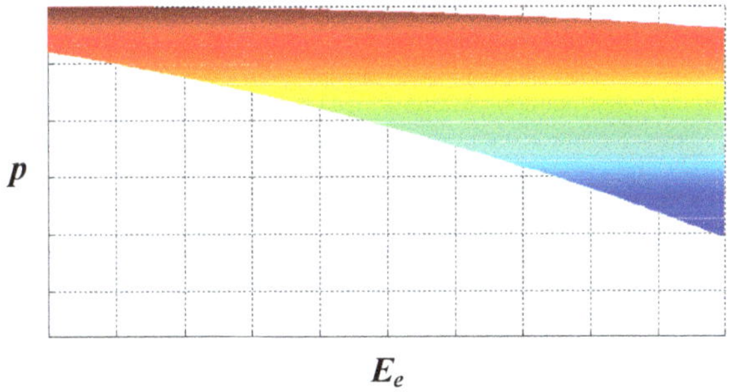

Fig. 7.5 The section parallel $p - E_e$ coordinate surface

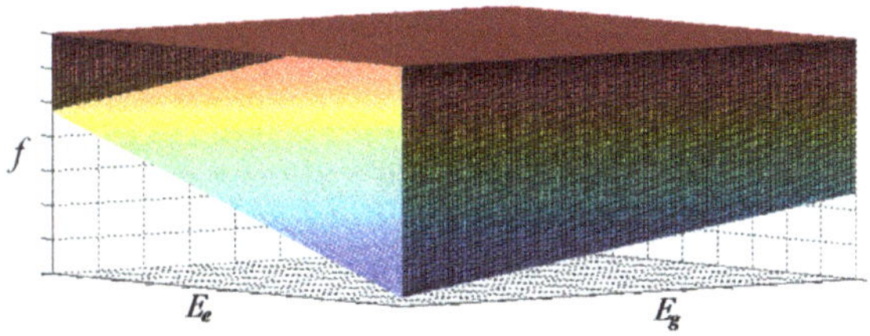

Fig. 7.6 The three-dimensional diagram of electricity droop

It defines that the electricity and natural gas input of energy hub have the equality relationship with the frequency of system. According to Eq. (7.23), a three-dimensional diagram of electricity droop control method can be drawn like Fig. 7.6. Figure 7.6 describes a feasible region of energy hub inputs and system frequency. Each frequency value of energy hub is towards to multiple statuses of energy hub electricity and natural gas input. All of these statuses are possible energy hub inputs of one certain frequency value.

For Fig. 7.6, three kinds of sections which respectively parallel coordinate surfaces can be sectioned. Figure 7.7 shows one section that parallel $f - E_e$ coordinate surface. It defines the relationship between frequency f and energy hub electricity input E_e when the natural gas input E_g is determined. Similarly, a plane parallel $f - E_g$ coordinate surface defines the relationship between frequency p and energy hub natural gas input E_g when the electricity input E_e is determined. A plane parallel $E_e - E_g$ coordinate surface defines the relationship between energy hub electricity input E_e and energy hub natural gas input E_g when the frequency f is determined. According to Figs. 7.8, 7.9 and other two kinds sections, the relationship between the

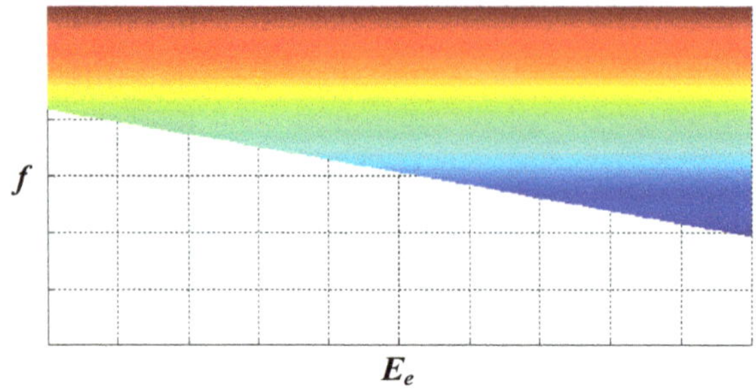

Fig. 7.7 The section parallel $f - E_e$ coordinate surface

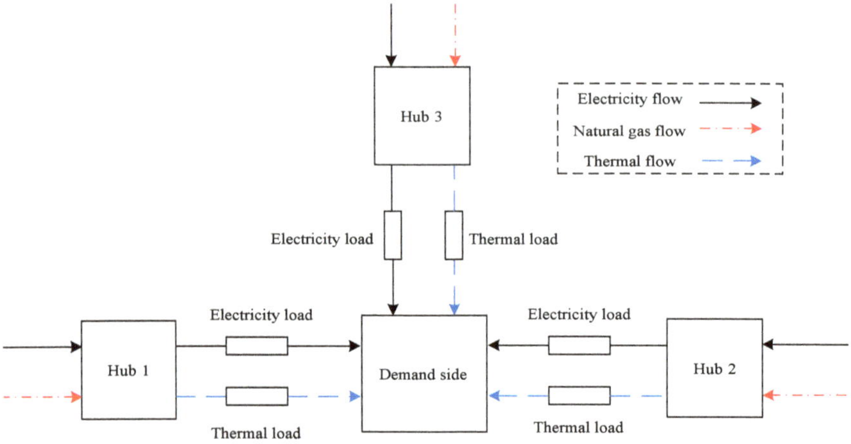

Fig. 7.8 Energy hubs supply energy to demand

frequency and energy hub inputs can be integrally represented and the relationship between the frequency and each energy hub input can be solely represented.

7.3.4 The Operation Strategies of Energy Hub

The energy hub outputs can be determined by using the electric and thermal droop control method to ensure the stability of the multi energy carriers system contains energy hubs. The different control objectives may obtain different results of energy hub inputs for a certain energy hub outputs. We provide two different operation strategies of energy hub for different purposes. The specific content of the two operation strategies are as follow:

(1) The energy hub proportionally allocates the energy outputs to inner facilities on the basis of their capacities;

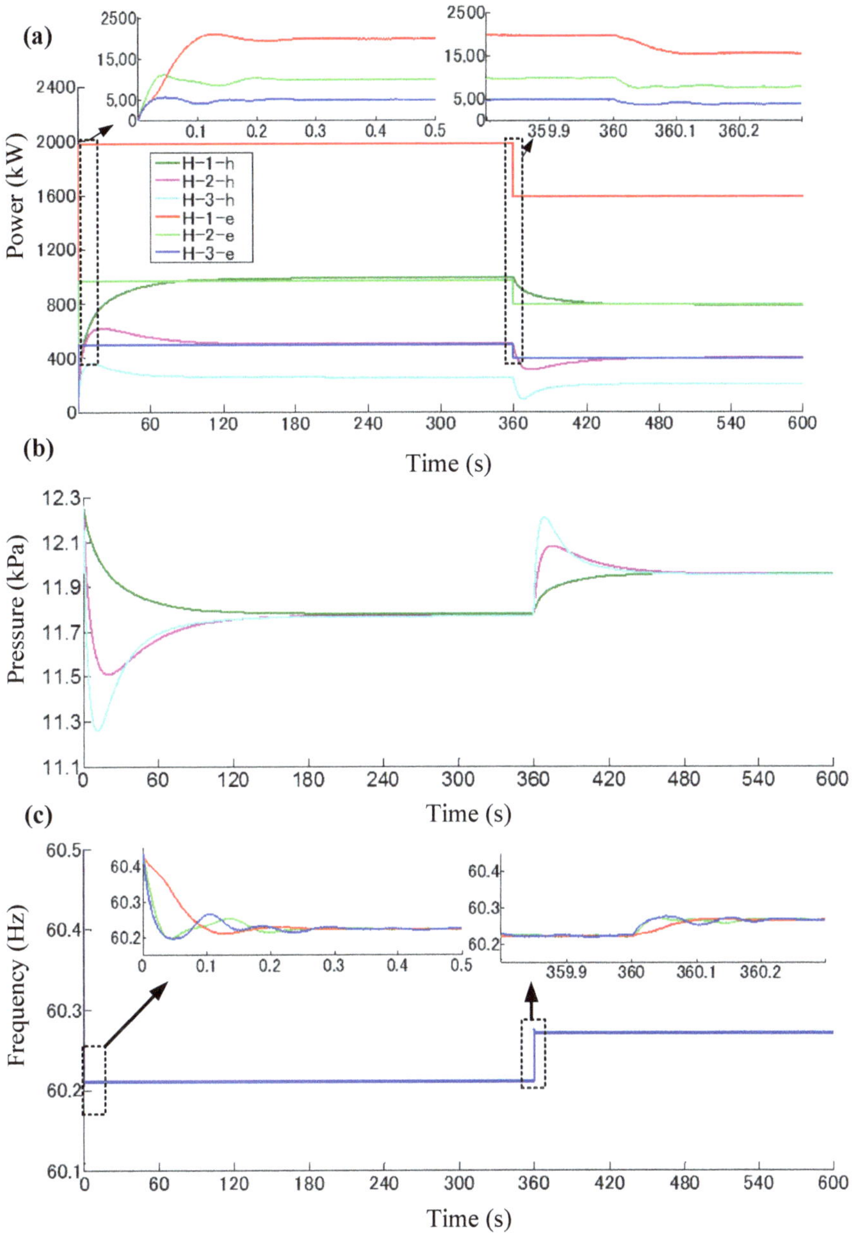

Fig. 7.9 The simulation result of case 1

(2) The energy hub operates on the basis of optimal result which is obtained by a certain optimal objective such as lowest the energy hub operation cost.

The purpose of first operation strategy is to maximize the stability of the entire system and the second operation strategy is designed to achieve the optimal operation of energy hubs on the basis of certain energy hub outputs. For the first operation strategy, because the operation of CHP which is contained in energy hub produce the electric and thermal power according to the operational efficiencies of CHP. It causes the outputs proportional allocation of energy hub inner facilities cannot realize. To solve this problem, we subdivide the first operation strategy into two implementation approaches: The first approach is regarding CHP as electric power output facility. After the proportional allocation of energy hub electric power output, other heat power output facilities proportional allocate the remainder thermal output which equal to the thermal output of energy hub remove the thermal output of CHP; The second approach is regarding CHP as thermal power output facility. CHP participate the proportional allocation of energy hub thermal power output at this point. After get rid of the thermal output of CHP, the rest electric output of energy is proportional allocated to the electric output facilities. These two implementation approaches generally are chosen depend on the comprehensive consideration of demand side and the facilities of energy hub.

7.3.5 Simulation and Results

The proposed energy hubs supply energy model with three energy hubs is illustrated in Fig. 7.8. The energy hub structure is the same as Fig. 7.1 which contains transformer, CHP, electricity boiler and gas furnace. Each energy hub is connected to electricity and natural gas networks to satisfy the requirement of inner facilities. The energy form of demand side is electric and thermal. Both thermal and electricity transmission line exist impedance.

The efficiencies and capacity of energy hub inner devices such as transformer, CHP are showed in Table 7.1. The unlit of capacity in Table 7.1 is kW. The inputs of energy hub can be obtained by using the two operation strategies provided in the mentioned above. Because there only one electric output device in the energy hub model, the implementation approach of the first energy hub operation strategy is regarding CHP as electric power output facility. The optimal objective of the second energy hub operation strategy is the lowest cost of purchasing energy in these cases and the objective function and be expressed as follow.

$$pr_t = pr_e E_e + pr_g E_g \tag{7.28}$$

where parameter pr_t, pr_e and pr_g represent the total cost of purchasing energy from corresponding network, electricity price and natural gas price, respectively. The con-

Table 7.1 Efficiencies and capacity of devices

Devices	Efficiency	Capacity
The transformer in hub1	$\eta_{ee} = 0.98$	3000
The natural gas furnace in hub1	$\eta_{gh} = 0.95$	1000
The electric boiler in hub1	$\eta_{eh} = 0.95$	1000
The CHP in hub1	$\eta_{ge} = 0.35$, $\eta_{gh} = 0.45$	2000
The transformer in hub2	$\eta_{ee} = 0.98$	1500
The natural gas furnace in hub2	$\eta_{gh} = 0.9$	500
The electric boiler in hub2	$\eta_{eh} = 0.95$	500
The CHP in hub2	$\eta_{ge} = 0.35$, $\eta_{gh} = 0.45$	1000
The transformer in hub3	$\eta_{ee} = 0.98$	750
The natural gas furnace in hub3	$\eta_{gh} = 0.95$	250
The electric boiler in hub3	$\eta_{eh} = 0.95$	250
The CHP in hub3	$\eta_{ge} = 0.35$, $\eta_{gh} = 0.45$	500

Table 7.2 Price Of Multi Energy

Energy	Price
Electricity	0.4
Natural gas	0.2

crete data of energy prices are showed in Table 7.2. The unlit of capacity in Table 7.1 is $.

In these case studies, a stability parameter is proposed to assess the stability of energy hub. The stability parameter can be calculated according to the following equation.

$$F = \left(D_e^{CHP} - D_e^{trans}\right)^2 + \left(D_h^{CHP} - D_h^{boiler}\right)^2 + \left(D_h^{CHP} - D_h^{furnace}\right)^2 + \left(D_h^{boiler} - D_h^{furnace}\right)^2 \tag{7.29}$$

where F denotes the stability parameter of energy hub. D_e^{CHP}, D_e^{trans}, D_h^{CHP}, $D_h^{furnace}$ and D_h^{boiler} represent the ratio between the corresponding element actual output power and the corresponding element maximum output power. the stability parameter value is negatively related to stability.

The specific heat capacity of water c is equal to 4.2 kJ/ (kg · K) and the water density ρ is equal to 10^3 kg/ m^3. The temperature difference of thermal load inlet and outlet $\Delta\ t$ is set as 10 K. The magnitude of thermal impedance 1 is 0.05 Pa/ (m^3/h)^2, the magnitude of thermal impedance 1 is 0.08 Pa/ m^3/h) ^{2} and the magnitude of thermal impedance 1 is 0.08 Pa/ (m^3/h)^2. Electricity impedance 1, impedance 2 and impedance 3 are equal to 0.0175Ω+0.2mH, 0.0265Ω+0.3mH and 0.0315Ω + 0.3mH, respectively. Because the electric and thermal capacity ratio between energy hub 1 and 2 are all 2:1, the thermal and electric droop control coefficient of energy hub 1, 2 and 3 are set as 1×10^{-3}, 4×10^{-3}, 1.6×10^{-2}, 2×10^{-5}, 4×10^{-5} and $8 \times$

10^{-5}, respectively. The rated pressure of energy hub 1, 2 and 3 is 117.14 kPa and the rated thermal power are 1000, 500 and 250 kW, respectively. The rated frequency is 60 Hz and the active power of energy hub 1, 2 and 3 are 4000, 2000 and 1000 kW, respectively.

Four cases are presented to illustrate the effectiveness of the energy hub droop control method. The case study results are compared and analyzed based on the outputs of energy hub at the same demand condition.

Case 1: The load of heat and electricity both suddenly decrease into new values.
Case 2: The load of heat and electricity both suddenly increase into new values.
Case 3: The load of heat decrease and the load of electricity increase into new values.
Case 4: The load of electricity decrease and the load of heat increase into new values.

These cases are discussed as follows and the thermal and electric load are same for all cases before 360 s.

Case 1

There are no electric and thermal loads in the system at the beginning of this case. The 1500 kW thermal load and 3000 kW electric load are joined into the system at beginning. Thermal load decreases 350 kW and electric load decreases 700 kW at 360 s. Figure 7.9a shows the thermal and electric outputs of energy hub 1, 2 and 3. Figure 7.9b shows the pressure variation of energy hub 1, 2 and 3. Figure 7.9c shows the frequency variation of energy hub 1, 2 and 3. The time unit of adjusting thermal and electric outputs are second. When the thermal output of energy hub 1, 2 and 3 are stabilized before 360 s, energy hub 1 outputs 986 kW thermal power, energy hub 2 outputs 495 kW thermal power and energy hub 3 outputs 249 kW thermal power. The electric outputs are respective 1972, 981 and 491 kW of energy hub 1, 2 and 3 when the electric outputs are stabilized. For the first operation strategy, the input electricity and natural gas of energy hub 1 are respective 1887.6 and 1345.8 W, the input electricity and natural gas of energy hub 2 are respective 941.4 and 671.9 kW and the input electricity and natural gas of energy hub 3 are respective 471.8 and 336.9 kW. The cost which energy hub 1, 2 and 3 purchase energy are 1024.2, 510.9 and 256.1$, respectively. The stability parameter of energy hub 1, 2 and 3 are 0.156, 0.149 and 0.146, respectively. The total cost of the first operation cost is 1791.2$. For the second operation strategy, the input electricity and natural gas of energy hub 1 are respective 1298.0 and 2090.5 kW, the input electricity and natural gas of energy hub 2 are respective 643.8 and 1047.4 kW and the input electricity and natural gas of energy hub 3 are respective 322.4 and 525.3 kW. The cost which energy hub 1, 2 and 3 purchase energy are 937.3, 466.9 and 234.1$, respectively. The total cost of the second operation cost is 1638.3$. The stability parameters of energy hub 1, 2 and 3 are 2.157, 2.154 and 2.143. Compared with the first operation cost, the second operation can decrease 8.53% cost of purchasing energy.

When the thermal output of energy hub 1, 2 and 3 are stabilized after 360 s, energy hub 1 outputs 782 kW thermal power, energy hub 2 outputs 396 kW thermal power and energy hub 3 outputs 195 kW thermal power. The electric outputs are respectively 1579, 788 and 387 kW of energy hub 1, 2 and 3 when the electric outputs are stabilized. For the first operation strategy, the input electricity and natural gas of energy hub 1 are respectively 1507.5 and 1073.7 W, the input electricity and natural gas of energy hub 2 are respectively 755.3 and 538.3 kW and the input electricity and natural gas of energy hub 3 are respectively 371.2 and 264.9 kW. The cost which energy hub 1, 2 and 3 purchase energy are 817.7, 409.9 and 201.5$, respectively. The stability parameter of energy hub 1, 2 and 3 are 0.104, 0.098 and 0.094, respectively. The total cost of the first operation cost is 1429.1$. For the second operation strategy, the input electricity and natural gas of energy hub 1 are respectively 990.6 and 1737.8 kW, the input electricity and natural gas of energy hub 2 are respectively 489.8 and 880 kW and the input electricity and natural gas of energy hub 3 are respectively 240.1 and 433.3 kW. The cost which energy hub 1, 2 and 3 purchase energy are 743.8, 371.9 and 182.7$, respectively. The total cost of the second operation cost is 1298.4$. The stability parameter of energy hub 1, 2 and 3 are1.8, 1.855 and 1.801. Compare with the first operation cost, the second operation can decrease 9.15% cost of purchasing energy.

Case 2

The situation that thermal load and electric load increase at a certain time is discussed in case 2. Thermal load increase 600 kW and electric load increase 1000 kW at 360 s Fig. 7.10a shows the thermal and electric outputs of energy hub 1, 2 and 3. Figure 7.10b shows the pressure variation of energy hub 1, 2 and 3. Figure 7.10c shows the frequency variation of energy hub 1, 2 and 3. When the thermal output of energy hub 1, 2 and 3 are stabilized after load change, energy hub 1 outputs 1321 kW thermal power, energy hub 2 outputs 661 kW thermal power and energy hub 3 outputs 329 kW thermal power. The electric outputs are respectively 2533, 1258 and 633 kW of energy hub 1, 2 and 3 when the electric outputs are stabilized. For the first operation strategy, the input electricity and natural gas of energy hub 1 are respectively 2453.3 and 1757.4 W, the input electricity and natural gas of energy hub 2 are respectively 1221.1 and 875.4 kW and the input electricity and natural gas of energy hub 3 are respectively 612.5 and 438.6 kW. The cost which energy hub 1, 2 and 3 purchase energy are 1332.8, 663.5 and 332.7$, respectively. The stability parameters of energy hub 1, 2 and 3 are 0.218, 0.208 and 0.221, respectively. The total cost of the first operation cost is 2329$. For the second operation strategy, the input electricity and natural gas of energy hub 1 are respectively 1870.4 and 2443.2 kW, the input electricity and natural gas of energy hub 2 are respectively 926.5 and 1222.1 kW and the input electricity and natural gas of energy hub 3 are respectively 467.3 and 609.5 kW. The cost which energy hub 1, 2 and 3 purchase energy are 1236.8, 615.2 and 308.8$, respectively. The total cost of the second operation cost is 2160.8$. The stability parameters of energy hub 1, 2 and 3 are1.648, 1.652 and 1.649. Compared with the first operation cost, the second operation can decrease 11.56% cost of purchasing energy.

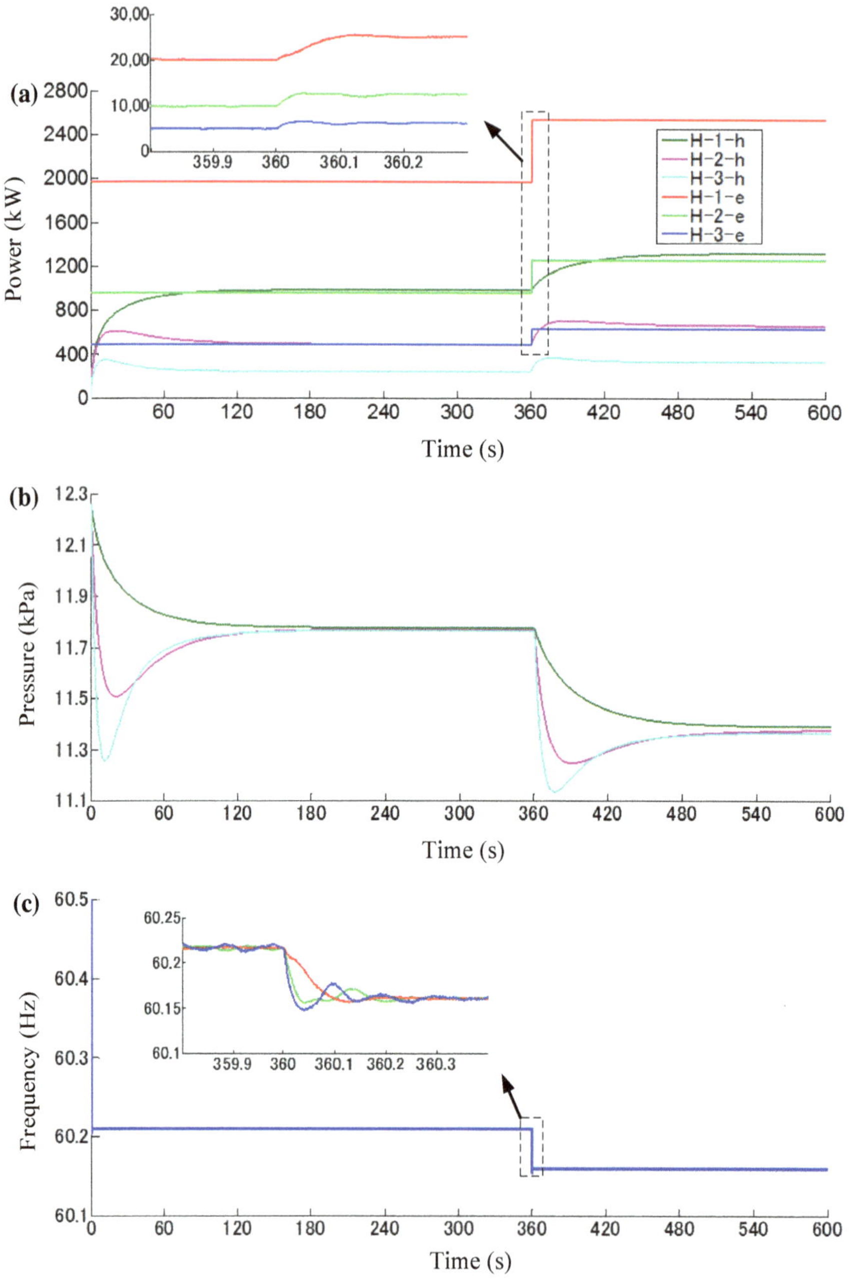

Fig. 7.10 The simulation result of case 2

Case 3

The situation that thermal load decreases and electric load increases at a certain time is discussed in case 3. Thermal load decreases 300 kW and electric load increases 700 kW at 360 s. Figure 7.11 (a) shows the thermal and electric outputs of energy hub 1, 2 and 3. Figure 7.11b shows the pressure variation of energy hub 1, 2 and 3. Figure 7.11c shows the frequency variation of energy hub 1, 2 and 3. When the thermal output of energy hub 1, 2 and 3 are stabilized after load change, energy hub 1 outputs 818 kW thermal power, energy hub 2 outputs 414 kW thermal power and energy hub 3 outputs 210 kW thermal power. The electric outputs are respectively 2361, 1183 and 589 kW of energy hub 1, 2 and 3 when the electric outputs are stabilized. For the first operation strategy, the input electricity and natural gas of energy hub 1 are respectively 2069.9 and 1421 kW, the input electricity and natural gas of energy hub 2 are respectively 1038.9 and 713.9 kW and the input electricity and natural gas of energy hub 3 are respectively 519.3 and 357.5 kW. The cost which energy hub 1, 2 and 3 purchase energy are 1112.1, 558.4 and 279.2$, respectively. The stability parameters of energy hub 1, 2 and 3 are 0.553, 0.545 and 0.524, respectively. The total cost of the first operation cost is 1949.7$.

For the second operation strategy, the input electricity and natural gas of energy hub 1 are respectively 1760 W and 1817.8 kW, the input electricity and natural gas of energy hub 2 are respectively 878.6 and 920 kW and the input electricity and natural gas of energy hub 3 are respectively 434.4 and 466.7 kW. The cost which energy hub 1, 2 and 3 purchase energy are 1067.6, 535.4 and 267.1$, respectively. The total cost of the second operation cost is 1870.1$. The stability parameter of energy hub 1, 2 and 3 are1.756, 1.805 and 1.868. Compared with the first operation cost, the second operation can decrease 4.08% cost of purchasing energy.

Case 4

The situation that thermal load increases and electric load decreases at a certain time is discussed in case 4. Thermal load increases 500 kW and electric load decreases 1000 kW at 360 s. Figure 7.12a shows the thermal and electric outputs of energy hub 1, 2 and 3. Figure 7.12b shows the pressure variation of energy hub 1, 2 and 3. Figure 7.12c shows the frequency variation of energy hub 1, 2 and 3. When the thermal output of energy hub 1, 2 and 3 are stabilized after load change, energy hub 1 outputs 1267 kW thermal power, energy hub 2 outputs 634 kW thermal power and energy hub 3 outputs 318 kW thermal power. The electric outputs are respectively 1417 kW, 709 and 359 kW of energy hub 1, 2 and 3 when the electric outputs are stabilized. For the first operation strategy, the input electricity and natural gas of energy hub 1 are respectively 1650.3 kW and 1261 W, the input electricity and natural gas of energy hub 2 are respectively 825.8 and 630.9 kW and the input electricity and natural gas of energy hub 3 are respectively 416.5 and 317.9 kW. The cost which energy hub 1, 2 and 3 purchase energy are 912.3, 456.5 and 230.2$, respectively. The stability parameters of energy hub 1, 2 and 3 are 0.017, 0.017 and 0.016, respectively. The total cost of the first operation cost is 1825.3$. For the second operation strategy, the input electricity and natural gas of energy hub 1 are

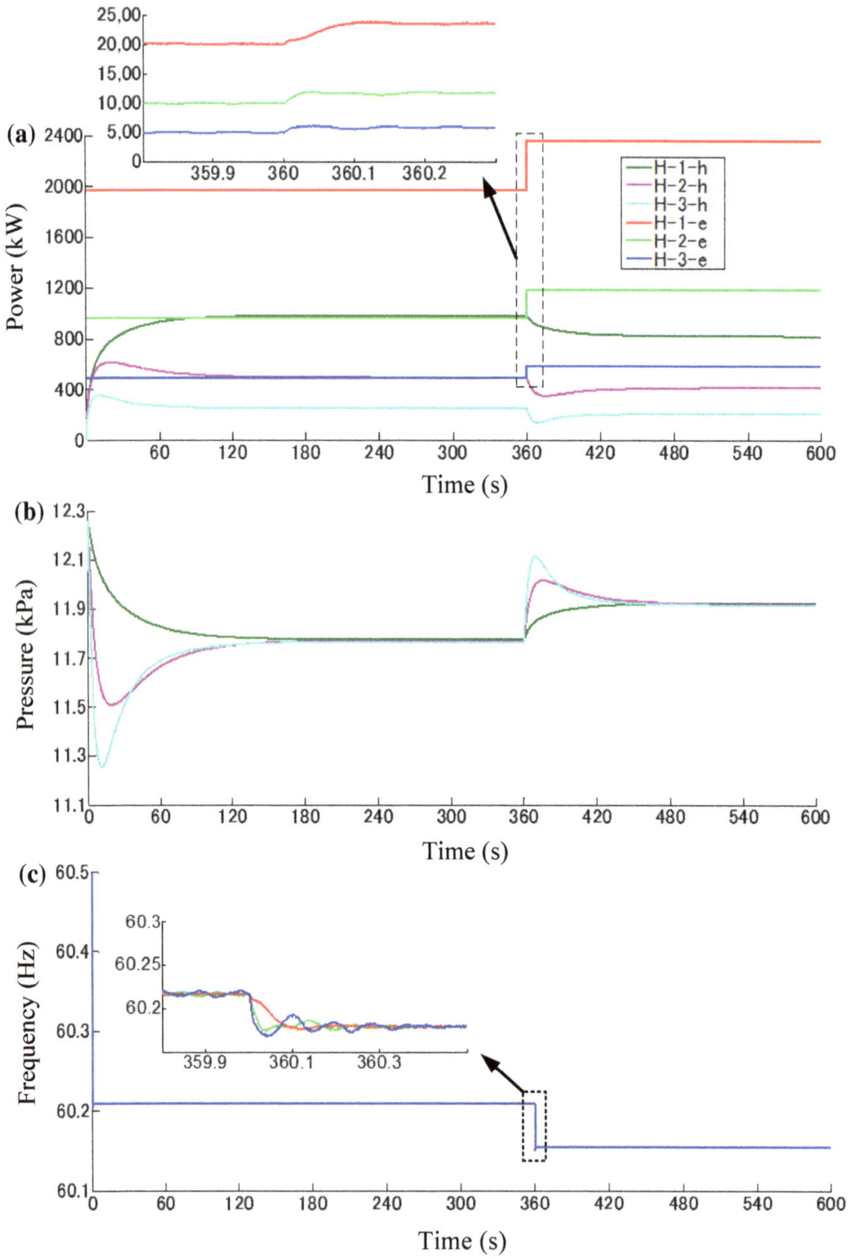

Fig. 7.11 The simulation result of case 3

respectively 731.6 W and 2386.3 kW, the input electricity and natural gas of energy hub 2 are respectively 366.3 and 1193.7 kW and the input electricity and natural gas of energy hub 3 are respectively 187.8 and 597.9 kW. The cost which energy hub 1, 2 and 3 purchase energy are 769.9, 385.3 and 194.7$, respectively. The total cost of the second operation cost is 1349.9$. The stability parameter of energy hub1, 2 and 3 are 2.098, 2.097 and 2.086. Compare with the first operation cost, the second operation can decrease 26.05% cost of purchasing energy.

From the result of case studies, the proportional allocation function of energy hub droop control method is proved and the results of different operation strategies are presented.

7.4 Exergy Analysis of Multiple Energy System

Energy is usually quantitatively measured by power as in the energy hub method. In this way, different forms of energy are equally treated.

However, for those energy with the same power their performance in practice differ a lot. For example, the waste heat has a great quantity of energy while few can be used. So we say that energy has not only quantity but quality.

We use the concept exergy to measure quality of energy. Exergy can be understood as the real usable part of energy. It describes how much of energy can be converted into work. Accordingly, the unusable part is called anergy. We have:

$$E = E_x + A_n \tag{7.30}$$

where E represents energy. E_x, A_n are exergy and anergy respectively.

7.4.1 Energy and Exergy General Equations

The ordinary energy equation is acquired from the first law of thermodynamics. Based on the law, the energy input of the system is equivalent to the energy output of the system. The equation is divided by mass m[kg] at both sides and can be expressed as:

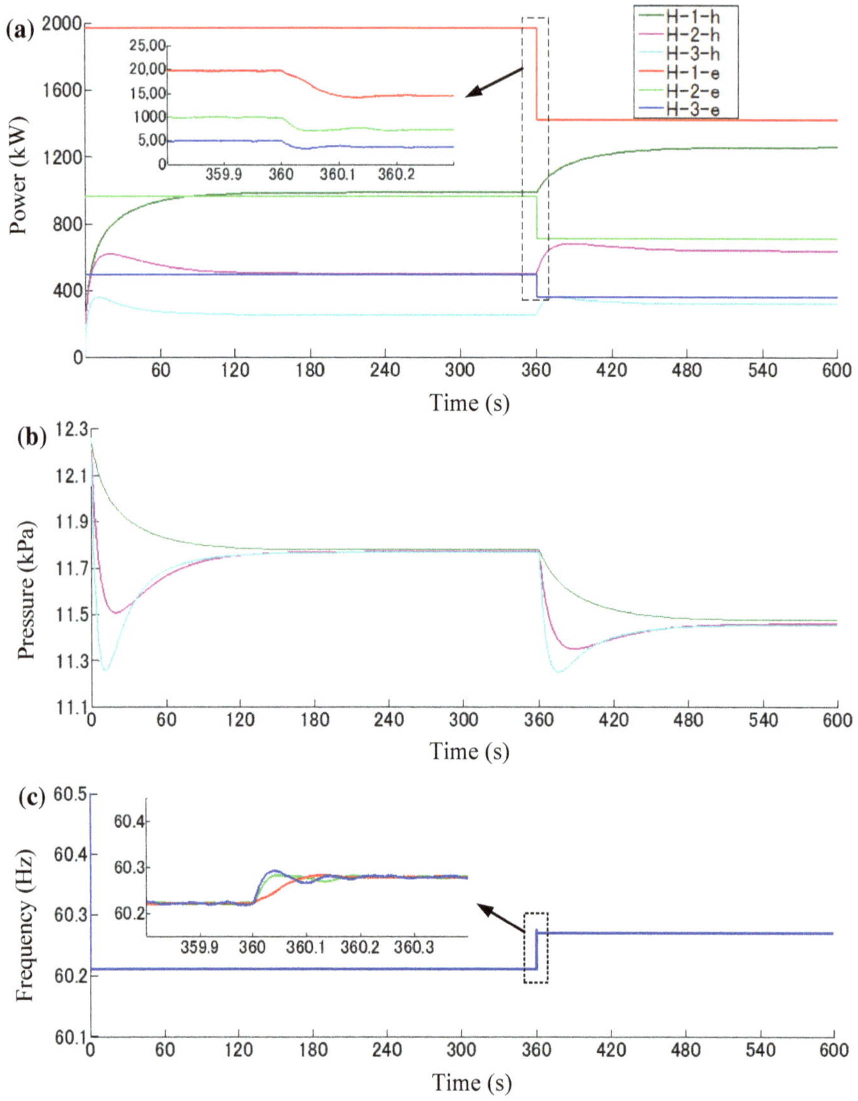

Fig. 7.12 The simulation result of case 4

$$\underbrace{\underbrace{gz_1}_{\text{Potential energy in}} + \underbrace{\frac{v_1^2}{2}}_{\text{Kinetic energy in}} + \underbrace{u_1}_{\text{Internal energy in}} + \underbrace{p_1 v_1}_{\text{Flow work in}} + \underbrace{w_{in}}_{\text{Work in}}}_{\text{Energy in}}$$

$$= \underbrace{\underbrace{gz_2}_{\text{Potential energy in}} + \underbrace{\frac{v_2^2}{2}}_{\text{Kinetic energy in}} + \underbrace{u_2}_{\text{Internal energy in}} + \underbrace{p_2 v_2}_{\text{Flow work in}} + \underbrace{w_{out}}_{\text{Work in}}}_{\text{Energy out}} \tag{7.31}$$

where g [m/s^2] stands for the gravitational acceleration, z [m] denotes height/altitude, v[m/s] is velocity, u [J/kg] represents specific internal energy, p [Pa] stands for pressure, v [m^3/kg] stands for specific volume, ω_{in}[J/kg] is the work per kg that goes into the system, ω_{out}[J/kg] stands for the work per kg that goes out of the system.

Exergy is a precise property for steady-state conditions. The value of exergy can be counted at any spot of the system. The exergy equation can be expressed as:

$$a = \underbrace{u - u_0}_{\text{Internal energy}} - \underbrace{T_0(s - s_0)}_{\text{Entropy}} + \underbrace{pv - p_0 v_0}_{\text{Work}}$$

$$+ \underbrace{\frac{v^2}{2}}_{\text{momentum}} + \underbrace{g(z - z_0)}_{\text{Gravity}} + \underbrace{\sum (\mu_{che} - \mu_0) n_{che}}_{\text{Chemical}} \tag{7.32}$$

where u[J/kg] is specific internal energy, T[K] is temperature, s[J/(K·kg)] is specific entropy, p[Pa] is pressure, v[m^3/kg] is specific volume, v[m/s] is velocity, g[m/s^2] is the gravitational acceleration, z[m] is height/altitude, μ_{che}[J/mole] is chemical potential and n_{che}[mole/kg] is quantity of moles per unit mass. Subscript '0' denotes the reference state.

The equation is usually used under situations where the chemical, gravitational and momentum can be ignored. In (7.32), the work contribution is involved in the enthalpy definition which programmed in (7.33). The specific exergy [J/kg] can be expressed as:

$$a = (h - h_0) - T_0(s - s_0) \tag{7.33}$$

where h is the specific enthalphy, $h = u + pv$. subscript '0' is the reference state. In order to count the valid potential work by taking the system from state 1 to state 2, the following equation is used:

$$a = (h_2 - h_1) - T_0(s_2 - s_1) \tag{7.34}$$

Comparison between energy and exergy are listed in the following chart: (Table 7.3)

Table 7.3 Comparison Between Energy And Exergy

Energy	Exergy
Related with physical properties, but not with the environment	Related with both physical properties and the environment
Unequal to zero while equilibrium with the environment (according to Einstein mass-energy relation: $E = mc^2$)	Equal to zero while equilibrium with the environment
Keep conservation in all processes	Conserve only in reversible process
Can neither produce nor disappear itself	In reversible processes can either be produced or lost, but can only be consumed in irreversible processes
Has many forms and all of them can be measured	Has many forms but can hardly be measured clearly
Only measured by quantity	Unity of quantity and quality

7.4.2 Exergy of Electricity, Gas and Heat

In this book, we focus on analysis of integrated electricity-gas-heat energy system of internet. Exergy of these three types of energy carrier are discussed next.

(1) Exergy of electricity

Electricity is usually treated as high-quality energy in the field of thermodynamics that its exergy is equal to energy and anergy is zero. From the perspective of electric system, electric power can be divided into active power and reactive power:

$$S_e = P_e + jQ_e \tag{7.35}$$

where S_e is the apparent power and P_e, Q_e are active and reactive power respectively. Active power is in charge of working while reactive power acts in excitation system. Based on this, in this paper, we treat active power as exergy of electric power:

$$Ex_e = P_e \tag{7.36}$$

(2) Exergy of chemical energy carrier

For chemical energy carrier such as biomass and gas, energy is usually calculated by the lower heating value (LHV):

$$P_{chem} = \dot{m} \cdot LHV \tag{7.37}$$

$\dot{m}$ is the mass flow. A factor ε is introduced to calculate the corresponding exergy:

$$Ex_{chem} = \varepsilon \cdot P_{chem} = \varepsilon \cdot \dot{m} \cdot LHV \tag{7.38}$$

ε varies according to the fuel type and is usually close to unity.

(3) Exergy of heat

Exergy of thermal load is directly related to the reference temperature and the temperature required. It can be calculated by multiplying the heat power Q by the Carnot factor η, as shown in (7.39)

$$Ex_h = \eta \cdot Q \tag{7.39}$$

The Carnot factor η is expressed as:

$$\eta = 1 - \frac{T_{ref}}{T} \tag{7.40}$$

where T_{ref} and T are the reference and required temperature respectively.

7.4.3 The Proposed Method

While optimizing integrated energy system, on the energy supply side, we hope more contribution of renewable energy and less of fossil energy as possible whether for the consideration of reducing energy costs or environment benefits. While on the energy consumption side, it is concerned that energy can be used effectively and efficiently which we measure by exergy mentioned above.

Based on these considerations, in this paper, the concept "quality character" of energy is proposed to analyze multiple energy systems, giving by:

$$\varphi = \frac{E_{X_{out}}}{E_{fos}} \tag{7.41}$$

φ is the defined system energy quality character. $E_{X_{out}}$ is the exergy demand of consumers i.e. the output exergy of system. E_{fos} is the input fossil energy.

While the output exergy is limited, we hope it is satisfied by renewable energy as much as possible. Then while the renewable power is limited, it is encouraged to contribute as much exergy as possible. Thus, the maximum value is expected for the proposed model.

The concept "quality character" reflects the level of system that converting renewable energy into available energy.

In the energy internet, load files are usually divided into two forms: electricity and heat. In this section, the heat load includes two types: domestic hot water (dhw) load and space heating (sh) load due to the different outlet temperature required.

$$\varphi = \frac{E_{X_{out}}}{E_{fos}} = \frac{E_{X_e} + E_{X_h}}{E_{fos}} = \frac{L_{P_e} + \left(1 - \frac{T_{ref}}{T_{dhw}}\right) \cdot L_{dhw} + \left(1 - \frac{T_{ref}}{T_{sh}}\right) \cdot L_{sh}}{\xi_{fos\_e} \cdot P_e + P_g + \xi_{fos\_h} \cdot P_h} \tag{7.42}$$

In order to take into account the parts of input electricity power P_e and heat power P_g that produced by fossil energy, the factor ξ is introduced. ξ_{fos_e} and ξ_{fos_h} represent for how much of the input power grid and heating supply system power is associated with fossil energy, respectively.

7.5 Conclusion

This chapter has proposed a new method to evaluate integrated energy systems which take full consideration of the utilization effect of renewable energy. Exergy analysis of multiple energy system has been introduced from the perspective of quality of energy. A new power flow expression of energy hub contain storage has been given in order to overcome the disadvantage of traditional power flow expression. This chapter has also proposed a droop control method of energy hub which contains thermal and electric droop control method. A general energy hub model has been formulated along with the multiple energy system and multiple energy hubs meet loads together in the system. The deducing process of droop control method of energy hub has been proposed to demonstrate the relationship between system parameters and energy hub outputs. The droop control method of energy hub makes energy hubs proportionally allocate their outputs alone and realize the decentralized control of the multiple energy system. The proposed method reduces the impact of communication and ensures the security operation of system. The three-dimensional diagram of droop control method of energy hub and two different kinds of operation strategies have been presented simultaneously to solve the energy hub operation problem that is difficult to dispose due to the intercoupling between hub inputs, outputs and system parameters. The proposed method has been tested on a multiple energy system with three energy hubs. Case studies have demonstrated that proposed method can proportionally allocate the thermal and electric outputs by the corresponding capacity of energy hubs and the results of case studies have showed that using different operation strategies get different operation consequences.

References

1. Martin Geidl, G. Andersson, Optimal power Flow of multiple energy carriers. IEEE Trans. Power Syst. **22**(1), 145–155 (2007)
2. Ralph Evins et al., New formulations of the 'energy hub' model to address operational constraints. Energy **73**(73), 387–398 (2014)
3. Mohammad Hossein Barmayoon et al., Energy storage in renewable-based residential energy hubs. IET Gener. Transm. Distrib. **10**(13), 3127–3134 (2016)
4. Iman Gerami Moghaddam, M. Saniei, E. Mashhour, A comprehensive model for self-scheduling an energy hub to supply cooling, heating and electrical demands of a building. Energy **94**, 157–170 (2016)
5. K. Orehounig, R. Evins, V. Dorer, Integration of decentralized energy systems in neighbourhoods using the energy hub approach. Appl. Energy **154**, 277–289 (2015)

6. Hongming Yang et al., Optimal operation of DES/CCHP based regional multi-energy prosumer with demand response. Appl. Energy **167**, 353–365 (2016)
7. Farhad Kamyab, S. Bahrami, Efficient operation of energy hubs in time-of-use and dynamic pricing electricity markets. Energy **106**(1), 343–355 (2016)
8. Aras Sheikhi et al., Integrated demand side management game in smart energy hubs. IEEE Trans. Smart Grid **6**(2), 675–683 (2015)
9. Shahab Bahrami, A. Sheikhi, From demand response in smart grid toward integrated demand response in smart energy hub. IEEE Trans. Smart Grid **7**(2), 650–658 (2016)
10. Aras Sheikhi, S. Bahrami, A.M. Ranjbar, An autonomous demand response program for electricity and natural gas networks in smart energy hubs. Energy **89**, 490–499 (2015)
11. Arsalan Najafi et al., Medium-term energy hub management subject to electricity price and wind uncertainty. Appl. Energy **168**, 418–433 (2016)
12. Samaneh Pazouki, M.R. Haghifam, Optimal planning and scheduling of energy hub in presence of wind, storage and demand response under uncertainty. Int. J. Electr. Power Energy Syst. **80**, 219–239 (2016)
13. Samaneh Pazouki, M.R. Haghifam, A. Moser, Uncertainty modeling in optimal operation of energy hub in presence of wind, storage and demand response. Int. J. Electr. Power Energy Syst. **61**, 335–345 (2014)
14. A. Vaccaro, C. Pisani, A.F. Zobaa, Affine arithmetic-based methodology for energy hub operation-scheduling in the presence of data uncertainty. Gener. Transm. Distrib. Iet **9**(13), 1544–1552 (2015)
15. A. Najafi, et al. A stochastic bilevel model for the energy hub manager problem. IEEE Trans. Smart Grid PP(99):1--1 (2016)
16. Azadeh Maroufmashat et al., Modeling and optimization of a network of energy hubs to improve economic and emission considerations. Energy **93**, 2546–2558 (2015)
17. Xiaping Zhang et al., Optimal expansion planning of energy hub with multiple energy infrastructures. IEEE Trans. Smart Grid **6**(5), 2302–2311 (2015)
18. M. Salimi et al., Optimal planning of energy hubs in interconnected energy systems: a case study for natural gas and electricity. Generation Transm. Distrib. Iet **9**(8), 695–707 (2015)
19. Amin Shabanpour-Haghighi, A.R. Seifi, Energy flow optimization in multicarrier systems. IEEE Trans. Ind. Inf. **11**(5), 1067–1077 (2015)
20. Boran Morvaj, R. Evins, J. Carmeliet, Optimising urban energy systems: Simultaneous system sizing, operation and district heating network layout. Energy **116**, 619–636 (2016)
21. A. Tuladhar, H. Jin, T. Unger, K. Mauch, Parallel operation of single phase inverter modules with no control interconnections, in *Proceedings of IEEE-APEC'97 Conference*, 23–27, vol. 1 (1997), pp. 94–100
22. M.N. Marwali, J.-W. Jung, A. Keyhani, Stability analysis of load sharing control for distributed generation systems. IEEE Trans. Energy Convers. **22**(3), 737–745 (2007)
23. J.M. Guerrero, J. Matas, L. García de Vicuña, M. Castilla, J. Miret, Wireless-control strategy for parallel operation of distributed-generation inverters. IEEE Trans. Ind. Electron. **53**(5), 1461–1470 (2006)
24. J. He, Y.W. Li, Analysis, design, and implementation of virtual impedance for power electronics interfaced distributed generation. IEEE Trans. Ind. Appl. **47**(6), 2525–2538 (2011)
25. Q.-C. Zhong, Robust droop controller for accurate proportional load sharing among inverters operated in parallel. IEEE Trans. Ind. Electron. **60**(4), 1281–1290 (2013)
26. Y. Shi et al., The parallel multi-inverter system based on the voltage-type droop control method. IEEE J. Emerg. Sel. Topics Power Electron. 4.4(2016):1332–1341

Chapter 8
Energy Flow Calculation of Energy Internet

Abstract In this chapter, an integrated energy network based on the Newton-Raphson method technique is developed to solve the energy flow problem. Referring to initial guess sensitivity issues of Newton method, a convergence theorem of Newton power flow is presented to improve the efficiency of calculation. Meanwhile, the proposed maximum iterations estimation theorem can ensure the rate of convergence. Proposed two theorems can be used to determine the convergence before calculation and directly select optimal initial guess from the feasible region. A case study is utilized to validate correctness and effectiveness of the proposed theorem. Furthermore, to solve the problem of low computing speed and high requirements of computing equipment in large-scale integrated energy networks, a distributed parallel computing method suitable for integrated energy networks is applied. By splitting the coupling nodes, the whole network is decomposed into many subnetworks. At the same time, multiple processors are used in parallel computing to improve the computation speed of energy flow and reduce the demand for a single processor.

8.1 Energy Flow in the Energy Internet

With the rapid development of economy, energy and environmental problems become increasingly prominent. How to achieve the clean and efficient use of energy has been a hot topic in recent years. Energy Internet (EI) as a new generation mode of energy use, is closely related to the development of human life and social economic due to its intensive, environmentally friendly, sustainable characters. EI is the product of deep integration among automation, electrical and other fields related to energy. It can be interpreted as a new energy supply form that combines physical entities (such as multiple energy networks) with advanced automatic control technology containing distributed cooperative control, intelligent optimization and information processing technology, etc. Compared to the traditional operation mode that electrical, nature gas and district heating systems are planned and operated separately, EI has realized the coordinated operation of multiple energy systems.

Q. Sun, *Energy Internet and We-Energy*, Renewable Energy Sources & Energy Storage, https://doi.org/10.1007/978-981-13-0523-8_8

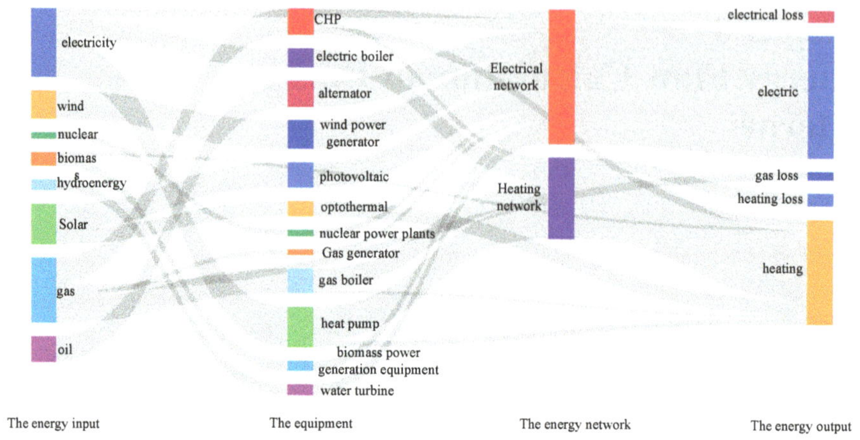

Fig. 8.1 Schematic of energy flows of energy internet

Using the concept of power flow in power grid, energy flow refers to the flow of various types of energy, which represent the mutual coupling, conversion transmission of electricity, heat, gas and other energy flows. Figure 8.1 shows the energy flow of the energy internet. As can be seen from the diagram, there are many kinds of energy sources. The production equipment of various kinds of energy (electric energy and thermal energy) is also very rich, and the coupling between networks is increasing, which makes the energy flow of the integrated energy system very complicated.

Energy flow is one of the important features for the Energy Internet. This not only brings many benefits to EI but also makes the system more complex and security issues more prominent. Energy flow analysis and calculation are the basis for researching various energy internet problems. Traditional power flow calculation method for power system as cannot satisfy the demand of coupled energy flow, so it is vital to find accurate and reasonable calculation method for energy flow problem.

The energy flow calculation of the energy internet is still in the primary stage. The research for a single network or two interconnected networks is abundant, but it is bare for three interconnected networks. To calculate energy flow quickly and accurately, the main research contents include: (1) Establishing the model of energy transmission for each network (2) Analyzing the coupling relationship among networks to find the relationship between the energy conversion (3) Improvement of the calculation method. Therefore, this chapter will carry on the concrete analysis of these contents.

8.2 Energy Internet Modeling

Energy Internet is a multi-energy transmission network consisting of power network, natural gas network, and heat network. The network structure has become more complex due to a variety of integrated network. To correctly analyze and calculate the energy flow of the energy internet, the first step is to have a complete understanding of the network and to model the mechanism of each network and coupling device.

8.2.1 *Electrical Network Model*

Power flow in the electrical network is formulated based on nodal power balance. The power flow model includes the Kirchhoff's current law (KCL), Kirchhoff's voltage law (KVL), and Ohm's law [1] :

$$\Delta P_i = P_i^{sp} - V_i \sum_{j=1}^{n_E} V_j \left(G_{ij} \cos \theta_{ij} + B_{ij} \sin \theta_{ij} \right) \tag{8.1}$$

$$\Delta Q_i = Q_i^{sp} - V_i \sum_{j=1}^{n_E} V_j \left(G_{ij} \sin \theta_{ij} - B_{ij} \cos \theta_{ij} \right) \tag{8.2}$$

where P_i^{sp} and Q_i^{sp} represent active power and reactive power, respectively. θ_{ij} is voltage angle difference of node i and j. V is the voltage. G_{ij} and B_{ij} are the conductance and susceptance on the line between node i and node j.

8.2.2 *Gas Network Model*

Natural gas network is comprised of gas producers, pipelines, compressors, and valves. The continuity of gas flow is expressed as

$$\mathbf{A_g f_g} = \mathbf{f_q} \tag{8.3}$$

where A_g is the gas network incidence matrix that relates the nodes to the branches. f_g is the gas flow in pipelines. f_q is the gas flow through each node injected from a source or discharged to a load.

The amount of gas flow in a pipeline depends on the pressures at both ends of the pipeline, expressed by the following equation [2]:

$$f_{ij} = k_{ij} sign(p_i, p_j) \sqrt{sign(p_i, p_j)(p_i^2 - p_j^2)} \tag{8.4}$$

The constant k_{ij} is related to properties of the pipeline. Where p_i is the gas pressure at node i. The sign function $sign(p_i, p_j)$ describes the direction of gas flow in pipelines, given by:

$$sign(p_i, p_j) = \begin{cases} +1, & if\ p_i \geq p_j \\ -1, & else \end{cases} \tag{8.5}$$

$\Delta \mathbf{p}$ is the pressure drop of a natural gas pipeline that can be expressed as

$$\Delta \mathbf{p} = -(\mathbf{A_g})^T \mathbf{p} \tag{8.6}$$

8.2.3 Heat Network Model

The heat network can be represented by two models which are hydraulic model and thermal model [3].

(1) *Hydraulic model*

The flow of water in the pipeline should meet the basic law of the network: Flow of each pipe at each node should meet the flow continuity equation, that is, the injection flow at the node is equal to the outflow. The loop pressure equation states that the sum of head losses around a closed loop must equal to zero.

$$\begin{cases} \mathbf{A_h m = m_q} \\ \mathrm{B_h}\mathbf{h_f} = 0 \end{cases} \tag{8.7}$$

where $\mathbf{A_h}$ is the heat network incidence matrix. $\mathbf{m}$ is the vector of the mass flow in each pipe. $\mathbf{m_q}$ is the vector of the mass flow through each node injected from a source or discharged to a load. $\mathrm{B_h}$ is the loop incidence matrix.$\mathbf{h_f}$ is the vector of the head losses. The head loss is the pressure change in the pipe which is due to friction. It can be:

$$\mathbf{h_f = Km|m|} \tag{8.8}$$

where $\mathbf{K}$ is the friction coefficient matrix of pipe. It is calculated from the diameter, the length and the roughness of a pipe. Hence, Eq. (8.7) is expressed as

$$\mathbf{B_h Km|m|} = \sum_{j=1}^{n_{pipe}} B_{ij} K_{ij} m_j |m_j| = 0 \tag{8.9}$$

where the subscript i is the index of loops and j is the index of pipes. n_{pipe} is the number of pipes. The heat pipes are divided into supply pipes and return pipes. In practical, the supply and return networks are identical.

(2) *Thermal mode*

The thermal model describes the relationship between water temperature and heat power in the pipe. It can be expressed as

$$\Phi = C_p \mathbf{m_q}(\mathbf{T_S} - \mathbf{T_O}) \tag{8.10}$$

where Φ is the vector of heat power. C_p is the specific heat of water. $\mathbf{T_S}$ is supply temperature. $\mathbf{T_O}$ is outlet temperature that is the temperature of the flow at the outlet of each node before mixing in the return network. The temperature drop in the pipe can be calculated using the following formula.

$$T_{end} = (T_{start} - T_a)e^{\frac{-\lambda L}{C_{pm}}} + T_a \tag{8.11}$$

where T_{end} and T_{start} are the temperatures at the end node and the start node of a pipe. T_a is the ambient temperature. λ is the heat transfer coefficient of pipes. L is the length of each pipe.

The water is mixed at the node and the temperature relationship before and after the mixing is as follows

$$\left(\sum \mathbf{m_{out}}\right)\mathbf{T_{out}} = \sum (\mathbf{m_{in}} \mathbf{T_{in}}) \tag{8.12}$$

where m_{out} is the mass flow rate in a pipe leaving the node. T_{out} is the temperature after mixing. m_{in} is the mass flow rate in a pipe injecting the node. T_{in} is the temperature of flow at the end of an incoming pipe. The existing calculation of the thermal model is assumed that the heat power and the outlet temperature of the load node are known, and the supply temperature and the heat power of source are known.

8.2.4 Coupling Device Model

The most important feature of the energy internet is the coupling and interaction among various energy networks. To analyze the energy internet accurately we must regard it as a unified whole, so the processing of the coupling device is very important.

(1) CHP is one of the most important devices in the energy internet which usually consumes natural gas to generate electricity and heat. There is a certain relationship between electricity and heat generated by CHPs, which considers the variable operating conditions of the cogeneration system:

$$P_i^{CHP} = \begin{cases} a_i^{CHP}\phi_i^{CHP} + b_i^{CHP}T_i^{s,CHP} + d_i^{CHP}, & u_i^{CHP}\psi_i^{CHP,max} \le \phi_i^{CHP} \le \phi_i^{CHP,\max} \\ a_i^{CHP}\phi_i^{CHP} + b_i^{CHP}T_i^{s,CHP} + d_i^{CHP} - \omega_{1,i}^{CHP}, & v_i^{CHP}\phi_i^{CHP,max} \le \phi_i^{CHP} \le u_i^{CHP}\phi_i^{CHP,\max} \\ a_i^{CHP}\phi_i^{CHP} + b_i^{CHP}T_i^{s,CHP} + d_i^{CHP} - \omega_{1,i}^{CHP} - \omega_{2,i}^{CHP}, & \phi_i^{CHP,min} \le \phi_i^{CHP} \le v_i^{CHP}\phi_i^{CHP,\max} \end{cases} \tag{8.13}$$

which:

$$\omega_{1,i}^{CHP} = \left(u_i^{CHP}\phi_i^{CHP,\max} - \phi_i^{CHP}\right)r_i^{CHP} \tag{8.14}$$

$$\omega_{2,i}^{CHP} = \left(v_i^{CHP}\phi_i^{CHP,\max} - \phi_i^{CHP}\right)s_i^{CHP} \tag{8.15}$$

where a_i^{CHP}, b_i^{CHP} and d_i^{CHP} are the constant coefficient of CHP. u_i^{CHP} and v_i^{CHP} are the limits for power generation changes. r_i^{CHP} and s_i^{CHP} are the positive coefficients of power generation when the equipment runs at variable operating conditions. The total efficiency of the CHP $\eta_i^{t,CHP}$ is constant, so the natural gas consumption of the CHP can be indicated as follows:

$$f_i^{CHP} = \frac{3.412}{GHV}\left(\frac{P_i^{CHP} + \phi_i^{CHP}}{\eta_i^{t,CHP}}\right) \tag{8.16}$$

GHV is a high calorific value that refers to the heat of fuel in the form of liquid water at the end of the combustion process. The common unit is BTU/m^3. 3.412 is the coefficient of the unit W converted to the unit BTU/m^3.

(2) Heat pump

Heat pump is a device that can obtain low heat energy from natural air or water to provide high heat energy by doing electrical work. The efficiency of the heat pump is higher than that of the gas boiler, but the cost is higher. The coefficient of performance of the heat pump satisfy the following formula:

$$c_{HP} = \frac{\phi_{HP}}{P_{HP}} \tag{8.17}$$

The coefficient of performance for a heat pump c_{HP} is the ratio of thermal power ϕ_{HP} to electrical power P_{HP}. The coefficient of performance is affected by the temperature of the heat source and the load.

(3) Electric boiler

The electric boiler consumes electricity to produce heat and its efficiency η_b is calculated by the following formula:

$$\eta_b = \frac{\phi_b}{P_b} \tag{8.18}$$

where ϕ_b is the heat production of electric boiler. P_b is the power consumption of electric boiler.

(4) Gas boiler

The gas boiler consumes gas to produce heat. The gas consumed by the gas boiler is shown as follows:

$$f_i^B = \frac{3.412}{GHV}\left(\frac{\phi_i^B + a_i^B \phi_i^{B,\max}}{b_i^B}\right) \tag{8.19}$$

where a_i^B and b_i^B are the constant parameters related to the performance of the gas boiler; $\phi_i^{B,\max}$ represents the output heat power of the gas boiler.

(5) Gas generator

The input power and output power relationships of the gas generator can be derived from the heat rate curve of the generator. The amount of fuel consumed by a generator can be determined by a formula that includes fuel GHV:

$$f_i^{GG} = \frac{1}{GHV}\left(\begin{array}{c} a_i^{GG}\left(P_i^{EG}\right)^2 + b^{GG} P_i^{EG} + c_i^{GG} + \\ \left|d_i^{GG} \sin\left[e_i^{GG}\left(P_i^{EG,min} - P_i^{EG}\right)\right]\right| \end{array}\right) \tag{8.20}$$

where a_i^{GG} , b_i^{GG} , c_i^{GG} , d_i^{GG} and e_i^{GG} are the coefficients of heat consumption of generator i; P_i^{EG} is the electrical power produced by generator i; $P_i^{EG,\min}$ is the minimum value of P_i^{EG}. The Eq. (8.20) considers the valve point effect of the generator. In the actual system, when the turbine inlet valve suddenly opened, which draws a ripple effect phenomenon that will be superimposed on the consumption characteristics curve of the unit. This is the valve point effect, and ignoring it can significantly affect the accuracy of the solution.

(6) Electric compressor

The electric compressor that consumes the electric is an important device in the natural gas network to maintain the pressure in the pipeline. The electric compressor is connected to the node i, and the electric required in a unit time can be calculated by the gas flow in the pipe. Reference [4] :

$$E_i^{GC} = \frac{151.47\,\pi_{0G}}{\eta_{GC}} \frac{\lambda_G}{T_{0G}\,\lambda_G - 1} z_G T_{GC} f_{ij}^{GL} \times \left(\left(H_i^{GC}\right)^{\frac{\lambda_G - 1}{\lambda_G}} - 1\right) \tag{8.21}$$

which η_{GC} is the efficiency of the compressor, π_{0G}, T_{0G} are the standard pressure and standard temperature, respectively, λ_G is the thermal efficiency of natural gas, T_{GC} is the temperature of the natural gas at the compressor, H_i^{GC} is the compression ratio can be calculated by the pressure of the node: $H_i^{GC} = \frac{\pi_i}{\pi_j}$. The unit of power calculated from the formula (8.21) is the horsepower. It needs to be converted into megawatts for the sake of calculation.

$$P_i^{GC} = \left(\frac{746 \times 10^{-6}}{3600}\right) E_i^{GC} \tag{8.22}$$

(7) Circulating pump

The circulating pump is used to generate impetus, so that the water in the heat pipe flows and circulates, and the energy consumed by the circulating pump can be calculated by [3]:

$$P_i^{HP} = \frac{\dot{m}_i^{HP} g H_P}{\eta_{HP}} \times 10^{-6} \tag{8.23}$$

Which $\dot{m}_i^{HP}$ is the mass flow of water through the pump. H_P is the lift of circulating pump. η_{HP} is the efficiency of the circulating pump.

The 7 coupling devices mentioned above are common coupling devices in the energy internet, but not every network contains all kinds of devices. And two kinds of coupling devices, namely, (8.6) and (8.7), are less energy efficient and often negligible.

8.3 Energy Flow Calculation Based on the Newton Method

8.3.1 Newton Raphson Method

Newton Raphson method [5] is an effective mathematical method for solving nonlinear algebraic equations. The main point is that the process of solving the nonlinear equation becomes a process of solving the corresponding linear equation repeatedly. The iterative process can be interpreted graphically, as shown in Fig. 8.2 below:

A nonlinear Eq. (8.24) is provided to simulate the computational process of Newton Raphson iteration:

$$f(x) = 0 \tag{8.24}$$

When solving this equation, the initial value of the solution $x^{(0)}$ is given. This value should be near the truth value and the error is $\Delta x^{(0)}$. The formula (8.24) can be written as follows:

$$x = x^{(0)} + \Delta x^{(0)} \tag{8.25}$$

$$f\left(x^{(0)} + \Delta x^{(0)}\right) = 0 \tag{8.26}$$

The formula (8.26) is expanded by Taylor series at $x^{(0)}$, and the following form is obtained:

$$f(x^{(0)} + \Delta x^{(0)}) = f(x^{(0)}) + f'(x^{(0)})\Delta x^{(0)} + \cdots + f^{(n)}(x^{(0)})\frac{(\Delta x(0))^n}{n!} = 0 \tag{8.27}$$

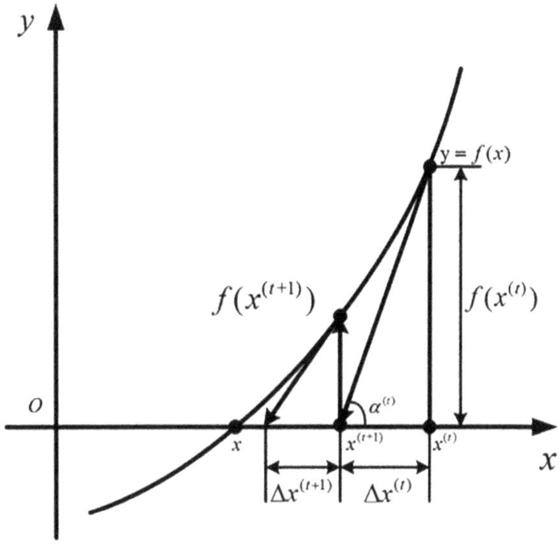

Fig. 8.2 The geometric interpretation of Newton method

which, $f^{(n)}(x^{(0)})$ is the n order derivative of the function at $x^{(0)}$. If the initial value is close to the truth, then the derivatives of the two orders above can be neglected and the upper formula can be simplified.

$$f(x^{(0)} + \Delta x^{(0)}) = f(x^{(0)}) + f'(x^{(0)})\Delta x^{(0)} = 0 \tag{8.28}$$

This is a linear equation for the modification of variables $\Delta x^{(0)}$. This is the correction equation in the energy flow, and it can be used to solve the variables $\Delta x^{(0)}$.

$$\Delta x^{(0)} = -\frac{f(x^{(0)})}{f'(x^{(0)})} \tag{8.29}$$

Thus, initial values $x^{(1)} = x^{(0)} - \Delta x^{(0)}$ can be obtained for the first time. There is still an error between the approximate solution and the truth. To approach the real solution further, $x^{(1)}$ is used as the initial value and the iterative calculation is continued until the criterion of convergence is satisfied:

$$\left|\Delta x^{(k)}\right| < \varepsilon \tag{8.30}$$

where ε is a given positive number. The curve in Fig. 8.2 is a function $y = f(x)$, whose solution is the intersection of the curve and the axis x. Making a tangent crossing the point $\left[x^{(k)}, y^{(k)} - f\left(x^{(k)}\right)\right]$ after the k iteration, and the intersection of the tangent and the axis x is the approximate solution $x^{(k+1)}$. It can be seen that Newton Raphson is a method of gradual linearization. It can not only solve the nonlinear equation of single variable, but also be applied to the nonlinear equation of many variables.

There are n simultaneous nonlinear equations:

$$\left.\begin{array}{l} f_1(x_1, x_2, \ldots, x_n) = 0 \\ f_2(x_1, x_2, \ldots, x_n) = 0 \\ \vdots \\ f_n(x_1, x_2, \ldots, x_n) = 0 \end{array}\right\} \tag{8.31}$$

Analog single variable processing methods can be obtained:

$$\begin{bmatrix} f_1\left(x_1^{(k)}, x_2^{(k)}, \ldots, x_n^{(k)}\right) \\ f_2\left(x_1^{(k)}, x_2^{(k)}, \ldots, x_n^{(k)}\right) \\ \vdots \\ f_n\left(x_1^{(k)}, x_2^{(k)}, \ldots, x_n^{(k)}\right) \end{bmatrix} = -\begin{bmatrix} \left.\frac{\partial f_1}{\partial x_1}\right|_k & \left.\frac{\partial f_1}{\partial x_2}\right|_k & \cdots & \left.\frac{\partial f_1}{\partial x_n}\right|_k \\ \left.\frac{\partial f_2}{\partial x_1}\right|_k & \left.\frac{\partial f_2}{\partial x_2}\right|_k & \cdots & \left.\frac{\partial f_2}{\partial x_n}\right|_k \\ & \vdots & & \\ \left.\frac{\partial f_n}{\partial x_1}\right|_k & \left.\frac{\partial f_n}{\partial x_2}\right|_k & \cdots & \left.\frac{\partial f_n}{\partial x_n}\right|_k \end{bmatrix} \begin{bmatrix} \Delta x_1^{(k)} \\ \Delta x_2^{(k)} \\ \vdots \\ \Delta x_n^{(k)} \end{bmatrix} \tag{8.32}$$

$$x_i^{(k+1)} = x_i^{(k)} + \Delta x_i^{(k)} \tag{8.33}$$

Formulas (8.31) and (8.32) can be abbreviated as:

$$F\left(X^{(k)}\right) = -J^{(k)} \Delta X^{(k)} \tag{8.34}$$

$$X^{(k+1)} = X^{(k)} + \Delta X^{(k)} \tag{8.35}$$

J is the Jacobian matrix. The element is the partial derivative of a function to a variable.

8.3.2 *Energy Flow Calculation Process*

In electrical system, each node contains four variables which are active power, reactive power, voltage amplitude and voltage phase angle. In the calculation, we must give the two quantities, and the other two quantities are unknown. First, select a node as the slack node. The voltage amplitude and phase angle of the node are given as reference values. The remaining nodes can be divided into nodes and nodes according to the given variables. The active power and reactive power of the node are given, and the voltage and phase angle of the node are the amount to be solved. Load nodes, photovoltaic, wind turbines and other new energy power generation devices

Table 8.1 Different node types

	Node type	Known	Unknown
Electricity	Slack	V,θ	P, Q
	PV	V, P	θ,Q
	PQ	P, Q	V,θ
Gas	Slack(Known pressure)	p	f
	Known-injection	f	p
Heat	Slack	T_s	Φ,T_r,m
	$\Phi\ T_S$	$\Phi\ T_S$	T_r,m
	$\Phi\ T_r$	$\Phi\ T_r$	T_s,m

can be classified into such nodes. Therefore, most nodes in power system belong to this kind of nodes. The nodes of the power plant and the adjustable reactive power supply can be used as nodes. Its active power and voltage amplitude are given. The reactive power and the voltage phase are the variables to be considered. The nodes need enough reactive power to maintain the given voltage amplitude, so this kind of node can also be called voltage control node. Before computing energy flow, the power loss in the network is unknown. In the network, one node's active power is unknown, and this node needs to balance the active power in the network, so it is called balanced node. In order to facilitate the calculation, the balance node and the slack node are selected in the same node.

In natural gas networks, the variables of each node are the gas flow f and the node pressure p. For the source node the pipeline pressure is known. It can be used as reference to other node pressure values of the whole network. This node is the balance node of the network that can provide gas balance network loss. For natural gas load nodes, they can be seen as nodes that are known to flow into natural gas, and the pipe pressure values of these nodes need to be calculated by energy flow.

In the heating network, the hydraulic model and thermal cycle model need to be considered so that the network parameters are more. Considering the actual situation of the heating network, it can be concluded that the water supply temperature of the equilibrium node is known, and can be used as the reference temperature of the whole network. ϕT_s nodes are usually heat source nodes which have known heat generating power and water supply temperature. It is necessary to calculate the backwater temperature and water flow of the nodes through the energy flow calculation. In this book, the heat power of all heat generating equipment is known (except for the balance node including the cogeneration unit and gas-fired boiler). Therefore, the total amount of gas consumed can be calculated, so the natural gas consumed by natural gas network is also fixed. In the heating network, ϕT_r the nodes are mainly load nodes and T_s and $\dot{m}$ need to be calculated by energy flow calculation.

The node types of electricity, gas and heat networks are shown in Table 8.1.

According to the model of energy internet in the previous section, the following formula can be obtained

$$\mathbf{F}(\mathbf{x}) = \begin{cases} \mathbf{P_i^{sp}} - V_i \sum_{j=1}^{n_E} V_j (G_{ij} \cos\theta_{ij} + B_{ij} \sin\theta_{ij}) = 0 \\ \mathbf{Q_i^{sp}} - V_i \sum_{j=1}^{n_E} V_j (G_{ij} \sin\theta_{ij} - B_{ij} \cos\theta_{ij}) = 0 \\ \mathbf{A_g f_g} - \mathbf{f_q^{sp}} = 0 \\ \mathbf{C_p A_h m_q (T_S - T_O)} - \mathbf{\Phi^{sp}} = 0 \\ \mathbf{B_h K m |m|} = 0 \\ \mathbf{C_s T_{s,load}} - \mathbf{b_s} = 0 \\ \mathbf{C_r T_{r,load}} - \mathbf{b_r} = 0 \end{cases} \tag{8.36}$$

where $\mathbf{P_i^{sp}}$, $\mathbf{Q_i^{sp}}$, $\mathbf{f_q^{sp}}$ and $\mathbf{\Phi^{sp}}$ are active power, reactive power, natural gas flow and heat power of the system. $\mathbf{C_s}$ and $\mathbf{C_r}$ are the correlation matrix of water supply network and return water network respectively. $\mathbf{b_s}$ and $\mathbf{b_r}$ are column vectors related to the heating temperature and the output temperature, respectively. The calculation method is described in [8].

The state variable of the system is:

$$x = \left[\theta \ V \ p \ \dot{m} \ T^{'}_{s,load} \ T^{'}_{r,load} \right]^T \tag{8.37}$$

J is the Jacobian matrix it can be expressed as

$$J = \begin{bmatrix} J_{ee} & J_{eg} & J_{eh} \\ J_{ge} & J_{gg} & J_{gh} \\ J_{he} & J_{hg} & J_{hh} \end{bmatrix} = \begin{bmatrix} \frac{\partial \Delta F_e}{\partial x_e^T} & \frac{\partial \Delta F_e}{\partial x_g^T} & \frac{\partial \Delta F_e}{\partial x_h^T} \\ \frac{\partial \Delta F_g}{\partial x_e^T} & \frac{\partial \Delta F_g}{\partial x_g^T} & \frac{\partial \Delta F_g}{\partial x_h^T} \\ \frac{\partial \Delta F_h}{\partial x_e^T} & \frac{\partial \Delta F_h}{\partial x_g^T} & \frac{\partial \Delta F_h}{\partial x_h^T} \end{bmatrix} \tag{8.38}$$

Subscripts e, g and h represent the electricity, gas and heat respectively. The diagonal elements in the Jacobian matrix represent the relationship between the flow of electricity, gas and heat systems themselves and their state quantities. The non-diagonal elements represent the coupling relationship among the three networks. Specific calculation process is shown in Fig. 8.3.

At the beginning of the calculation, the network parameters and the We-Energy parameters are input, and the incidence matrix of each network is formed on the basis of these data. Given the initial state parameter and then calculate ΔF. Determine whether the condition of the end iteration is satisfied, if it is not satisfied, calculate the extended Jacobian matrix and the modified equation $\Delta x^{(k+1)} = \left(J^{(k)}\right)^{-1} \Delta F^{(k)}$. Update x until the condition of the end calculation is satisfied and output the result.

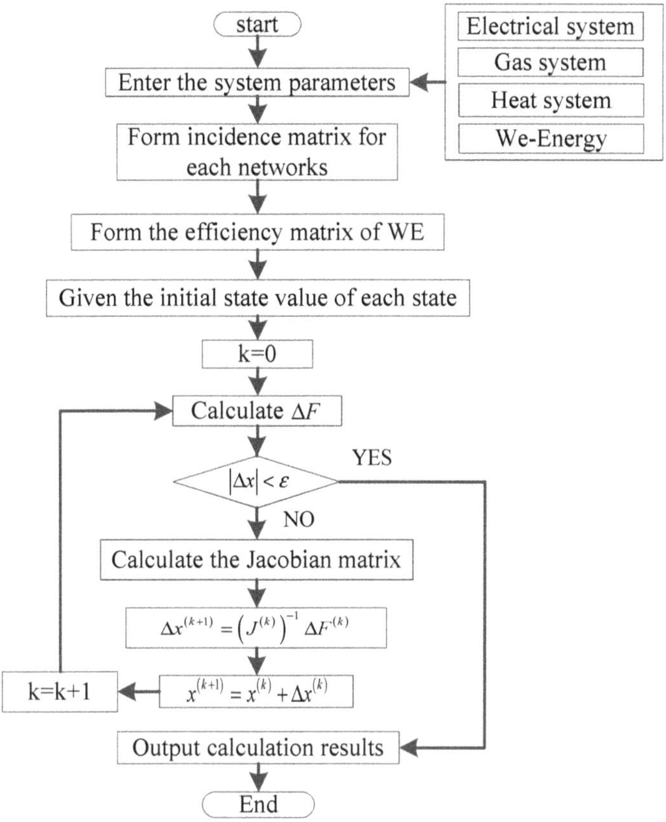

Fig. 8.3 Newton raphson algorithm flow chart

8.4 Convergence Analysis of Newton Method for Energy Flow Calculation

8.4.1 Convergence Theorem of Energy Flow

We need to find a way to determine whether the initial value selection is reasonable before iteration to make the energy flow calculation converge quickly.

Lemma 1 *([7]) Assume* $A, C \in L(R^n)$, *A is nonsingular,* $\left\|A^{-1}\right\| \leq \alpha$, $\|A - C\| \leq \beta$, *and* $\alpha\beta < 1$, *then*

$$\left\|C^{-1}\right\| \leq \frac{\alpha}{1 - \alpha\beta} \tag{8.39}$$

Theorem 1 *for energy flow equation* $F(x) = 0$, *both the initial values* x_0 *of the respective state variables and the Jacobian matrix* J_k *formed by the Newton method satisfy the following conditions:*

(1) the initial values of state variables is $x_0 \in D_0$, which makes the Jacobian matrix $J_0 = F'(x_0)$ nonsingular and $\left\|\mathbf{J}_0^{-1}\right\| \le \beta$, $\|J_m - J_n\| \le \gamma\|x_m - x_n\|$.

(2) $\begin{cases} \left\|J_0^{-1}[J_m - J_n]\right\| \le \omega\|x_m - x_n\| \\ \left\|J_0^{-1}F(x_0)\right\| \le \eta \end{cases}$, where ω and η are constant.

Then the energy flow calculation convergence operator ρ satisfies the following condition:

$$\rho = \omega\eta \le \frac{1}{2} \tag{8.40}$$

Let $t^* = \frac{1-\sqrt{1-2h}}{\omega}$, $t^{**} = \frac{1+\sqrt{1-2h}}{\omega}$, then Newton iterative sequence $x^{k+1} = x^k - \left[F'(x^k)\right]^{-1}F(x^k)$, $k = 0, 1, 2\cdots$, the resulting sequence $\{x^k\} \subset S(x^0, t^*) \subset D_0$ and converge to the only solution x^* in $S(x_0, t^{**}) \cap D_0$ satisfying $F(x) = 0$.

Proof Based on the following quadratic function (8.40)

$$j(u) = \frac{1}{2}\gamma u^2 - \frac{1}{\beta}u + \frac{\eta}{\beta} = 0 \tag{8.41}$$

the roots of equation are

$$u^* = \frac{1 - \sqrt{1 - 2\beta\gamma\eta}}{\beta\gamma}, u^{**} = \frac{1 + \sqrt{1 - 2\beta\gamma\eta}}{\beta\gamma} \tag{8.42}$$

Substitute (8.40) into (8.42)

$$u^* = \frac{1 - \sqrt{1 - 2\rho}}{\rho}\eta, u^{**} = \frac{1 + \sqrt{1 - 2\rho}}{\rho}\eta \tag{8.43}$$

The Newton iteration sequence is expressed by the t_k, that is,

$$u_{k+1} = u_k - \frac{j(u_k)}{j'(u_k)}, k = 0, 1, 2, \cdots, u_0 = 0 \tag{8.44}$$

It is easy to verify $u_k \to u^*$. Then we will prove $\|\Delta x_k - \Delta x^*\| \le u^* - u_k$. $\eta_k = u_{k+1} - u_k = -\frac{j(u_k)}{j'(u_k)}$, $\beta_k = -\frac{1}{j'(u_k)}$, $\rho_k = \beta_k\eta_k\gamma$, when $k = 0$, then $\eta_0 = \eta$, $\beta_0 = \beta$, $\rho_0 = \rho$. Take the derivative of Eq. (8.40) with respect to u:

$$j'(u) = \gamma u - 1/\beta \tag{8.45}$$

Substitute $j'(u_{k+1})$, $j'(u_k)$ into the (8.45):

$$j'(u_{k+1}) - j'(u_k) = \gamma(u_{k+1} - u_k) = \gamma\eta_k \tag{8.46}$$

Consider a function as follows:

$$\begin{aligned} & j(u_{k+1}) - [j(u_k) + j'(u_k)(u_{k+1} - u_k)] \\ & = \frac{1}{2}\gamma u_{k+1}^2 - \frac{1}{\beta}u_{k+1} - \frac{1}{2}\gamma u_k^2 + \frac{1}{\beta}u_k - (\gamma u_k - \frac{1}{\beta})(u_{k+1} - u_k) \\ & = \frac{1}{2}\gamma(u_{k+1} - u_k)^2 = \frac{1}{2}\gamma\eta_k^2 \end{aligned} \tag{8.47}$$

From (8.46), then:

$$j'(u_{k+1}) = \gamma\eta_k + j'(u_k) = \gamma\eta_k - \frac{1}{\beta_k} = \frac{\gamma\eta_k\beta_k - 1}{\beta_k} = \frac{\rho_k - 1}{\beta_k} \tag{8.48}$$

Substitute $\beta_k = -\frac{1}{j'(u_k)}$ and $\rho_k = \beta_k\eta_k\gamma$ into (8.47), (8.48), then:

$$\begin{cases} \beta_{k+1} = -\frac{1}{j'(u_{k+1})} \\ \eta_{k+1} = -\frac{j(u_{k+1})}{j'(u_{k+1})} = \frac{1}{2}\frac{\gamma\beta_k\eta_k^2}{1-\rho_k} = \frac{1}{2}\frac{\rho_k}{1-\rho_k}\eta_k \\ \rho_{k+1} = \gamma\beta_{k+1}\eta_{k+1} = \frac{\gamma\beta_k}{1-\rho_k}\cdot\frac{\rho_k\eta_k}{2(1-\rho_k)} = \frac{\rho_k^2}{2(1-\rho_k)^2} \end{cases} \tag{8.49}$$

Notice that $0 \le \rho \le \frac{1}{2}$ and $1 \le \frac{1-\sqrt{1-2\rho}}{\rho} \le 2$, then:

$$\|\Delta x_1 - \Delta x_0\| = \left\|[J_0]^{-1}\right\| \le \eta_0 = \eta = |u_1 - u_0| \tag{8.50}$$

The first iteration is $x^{(1)} \in S\left(x^{(0)}, \delta\right)$, then we use (8.47) and (8.50) to get the following:

$$\begin{aligned} & \|F(x^{(1)})\| = \|F(x^{(1)}) - F(x^{(0)}) - F'(v)(x^{(1)} - x^{(0)})\| \\ & \le \frac{1}{2}\gamma\left\|x^{(1)} - x^{(0)}\right\|^2 \le \frac{1}{2}\gamma\eta_0^2 \\ & v \in \left(x^{(0)}, x^{(1)}\right) \end{aligned} \tag{8.51}$$

Since $\left\|J_0^{-1}\right\| \le \beta$ and $\|J_1 - J_0\| \le \gamma\left\|x^{(1)} - x^{(0)}\right\| \le \gamma\eta_0$, J_1^{-1} exists and $\left\|J_1^{-1}\right\| \le \frac{\beta_0}{1-\rho_0} = \beta_1$ can be obtained from Lemma 1.

From the (8.42) and (8.50), it can be obtained as follows:

$$\begin{aligned} & \left\|x^{(2)} - x^{(0)}\right\| \le \left\|x^{(2)} - x^{(1)}\right\| + \left\|x^{(1)} - x^{(0)}\right\| \\ & \le \left(\frac{1 - \sqrt{1 - 2\rho}}{\rho} - 1\right)\eta + \eta = \frac{1 - \sqrt{1 - 2\rho}}{\rho}\eta \le \delta \end{aligned} \tag{8.52}$$

where $x^{(2)} \in S(x_0, \delta)$ is state variables.

Similar to Eq. (8.52), $x_k \in S(x_0, \delta)$, k=0, 1, 2... can be generalized by mathematical derivation as follows:

$$\begin{cases} \|F(x_k)\| \le \frac{1}{2}\gamma\eta_{k-1}^2 \\ \left\|J_k^{-1}\right\| \le \beta_k \\ \left\|J_k^{-1}F(x_k)\right\| \le \eta_k \\ \|\Delta x_{k+1} - \Delta x_k\| \le t_{k+1} - t_k \end{cases} \quad (8.53)$$

For any positive integer m, n, k, which satisfies $m = n + k$. According to the formula (8.53), thus:

$$\begin{aligned} \left\|x^{(m)} - x^{(n)}\right\| &\le \left\|x^{(n+k)} - x^{(n+k-1)}\right\| + \cdots + \left\|x^{(k+1)} - x^{(k)}\right\| \\ &\le |u_{n+k} - u_{n+k-1}| + \cdots + |u_{n+1} - u_n| \\ &\le |u_m - u_n| \end{aligned} \quad (8.54)$$

$\left\{x^{(k)}\right\}$ is Cauchy convergence sequence from the convergence of t_k , so $\left\{x^{(k)}\right\}$ exists limit, which is x^*, the following can be obtained when $k \to \infty$:

$$\left\|x^* - x^{(k)}\right\| \le \left|u^* - u_k\right| \quad (8.55)$$

Using the formula in the first row in Eq. (8.55) and paying attention to $\eta_k \to 0$ and the continuity of $F(x^*)$, thus $F(x^*)$=0, that is to say, x^* is the solution of the energy flow equation.

Theorem 1 is the convergence theorem of the energy flow calculation. Before calculating the energy flow, the power flow operator is calculated to determine whether the initial value can reach a convergence solution for the energy flow calculation. $\rho < 0.5$ is a sufficient condition for the energy flow to be solvable, and the operator is calculated after the initial selection of the initial value. If the convergence theorem is not satisfied, the initial value is re-selected, and the process is repeated until the initial value satisfies this condition. This reduces the difficulty of initial selection to calculate the energy flow and avoids unnecessary calculation.

On the basis of this convergence theorem, it is also important to note that the magnitude of the voltage can be set 1 $p.u.$ for a PQ node where the voltage is unknown for the power network. The voltage amplitude of the PV node is controllable for voltage keep constant. The phase angle of the voltage can be zero. For a thermal network, the mass flow rate for all pipelines can be initialized to 1, and the supply and return temperature of all nodes can be initialized at random. When the initial value is chosen, the convergence operator ρ is calculated. If the convergence condition is not satisfied, the initial value is re-trained until the initial value satisfies the convergence condition.

However, for natural gas networks, specific analysis is required, and the initial selection of state variables for natural gas networks requires special attention. The gas flow through the gas infrastructure is a function of the pressure difference at the end of the pipe. In the extreme case, if the initial value of the pressure at both ends of the pipe is chosen to be the same, a sick Jacobian matrix will be produced. In this case, the linearized energy flow in the Jacobian matrix will produce an empty diagonal element. In order to clarify this problem more clearly, we calculate the derivative of f_{ij} for the pressure p as follows:

$$\frac{\partial f_{ij}}{\partial p_i} = \frac{K_{ij} p_i}{\left(sign(p_i, p_j) \times \left(p_i^2 - p_j^2\right)\right)^{-0.5}} \tag{8.56}$$

From the formula (8.56), it can be obtained if the initial value of a natural gas network pipe pressure is too flat, the value of this derivative is too large or zero, which will cause the Jacobian matrix to be singular. The strategy to solve this problem is to ensure the maintenance of the pipeline pressure difference of 5% ~ 10% at delivery end and collect end during the preliminary selection of pipe pressure.

Based on the Newton method, the convergence theorem of the energy flow is calculated. The iterative number estimation of the energy flow of Newton- Raphson method is proposed for the Newton-Raphson method.

Theorem 2 *If the convergence operator* $\rho < \frac{1}{2}$ *for the energy flow and*

$$max\left\{\Delta x_l^{(k)}\right\} < \varepsilon \tag{8.57}$$

where $\Delta x_l^{(k)}$ is the variation in the ith node for the two iterations, ε is iterative accuracy. The maximum iterations of the energy flow equation can be expressed as:

$$k = Int\left[\log_2\left(\log_\theta \frac{\varepsilon\theta}{\eta + \varepsilon\theta}\right)\right] + 1, \theta = \frac{1 - \sqrt{1 - 2\rho}}{1 + \sqrt{1 - 2\rho}} \tag{8.58}$$

Proof Taylor series expansion of the energy flow equation can be expressed as:

$$\begin{aligned} f\left(u^*\right) &= f(u_k) + f'(u_k)\left(u^* - u_k\right) + \frac{\gamma}{2}\left(u^* - u_k\right)^2 \\ &= f(u_k) + f'(u_k)(u_{k+1} - u_k) + f'(u_k)\left(u^* - u_{k+1}\right) + \frac{\gamma}{2}\left(u^* - u_k\right)^2 \end{aligned} \tag{8.59}$$

where $\gamma = f''(\xi), \xi \in (u_k, u^*)$. The following can be derived from (8.59):

$$\begin{aligned} u^* - u_{k+1} &= -\frac{\gamma}{2f'(u_k)}\left(u^* - u_k\right)^2 \\ &= \frac{\gamma\beta_k}{2}\left(u^* - u_k\right)^2 = \frac{\rho_k}{2\eta_k}\left(u^* - u_k\right)^2 \end{aligned} \tag{8.60}$$

Taking k times iteration to (8.60), for $k = 0, 1, 2, \cdots,$

$$u^* - u_{k+1} = \prod_{j=0}^{k} \left(\frac{\rho_{k-j}}{2\eta_{k-j}} \right)^{2j} (u^* - u_0)^{2^{k+1}} u_0 = 0 \tag{8.61}$$

It follows that:

$$u^* - u_{k+1} = \prod_{j=0}^{k} \left(\frac{\rho_{k-j}}{2\eta_{k-j}} \right)^{2j} (u^*)^{2^{k+1}} \tag{8.62}$$

According to formula (8.49):

$$\frac{\rho_j}{2\eta_j} = \frac{\eta_j}{\eta_{j-1}^2}, j = 1, 2, \cdots, \tag{8.63}$$

Substitute it into the above formula, we have:

$$\begin{aligned} u^* - u_{k+1} &= \prod_{j-0}^{k-1} \left(\frac{\eta_{k-j}}{\eta_{k-j-1}^2} \right)^{2j} \left(\frac{\rho}{2\eta} \right)^{2k} (u^*)^{2^{k+1}} \\ &= \eta_k \left(\frac{\rho}{2} \right)^{2^k} \left(\frac{1 - \sqrt{1 - 2\rho}}{\rho} \right)^{2^{k+1}} \end{aligned} \tag{8.64}$$

If

$$\theta_k = \frac{1 - \sqrt{1 - 2\rho_k}}{1 + \sqrt{1 - 2\rho_k}} \tag{8.65}$$

It is easy to get $\theta_k = \theta_{k-1}^2 = \cdots = \theta^{2^k}$, $\rho_k = \frac{2\theta_k}{(1+\theta_k)^2}$,
From the formula (8.49) we obtain:

$$\eta_k = \frac{\theta^{2^{k-1}}}{1 + \theta^{2^k}} \eta_{k-1} = \prod_{j=0}^{k-1} \frac{\theta^{2^j}}{1 + \theta^{2^{j+1}}} \eta = \frac{\theta^{2^k - 1}}{\sum_{j=0}^{2^{k-1}-1} \theta^{2j}} \eta \tag{8.66}$$

Substitute (8.63) into the (8.61), and $\frac{1-\sqrt{1-2\rho}}{\rho} = 1 + \theta$, we obtain

$$\left\| x_k - x^* \right\| \le \frac{\theta^{2^k - 1}}{\sum_{j=0}^{2^{k-1}-1} \theta^{2j}} \eta \tag{8.67}$$

Substitute above into $\|x^* - x_k\| \le u^* - u_k$, it can be concluded as following:

$$\left\| x_k - x^* \right\| \le \sqrt{\varepsilon} = \frac{\theta^{2^k - 1}}{\sum_{j=0}^{2^{k-1}-1} \theta^{2j}} \eta \tag{8.68}$$

Namely:

$$k = Int\left[\log_2\left(\log_\theta \frac{\varepsilon\theta}{\eta + \varepsilon\theta}\right)\right] + 1 \tag{8.69}$$

The Theorem 2 is proved completely.

The maximum number of iterations is calculated by Theorem 2, which may be larger than the actual number of iterations. The maximum error is 1 by simulation. Thus, we can determine that the iterations estimated by this theorem must obtain a convergence solution. Theorem 2 can be used to estimate the number of convergence after using Theorem 1 to verify the convergence of the trend, and then we can know the initial value of the energy flow equation for the convergence rate is good or bad, and then calculate the power flow, which avoid the redundant calculation of the energy flow.

8.4.2 The Newton Method Energy Flow Calculation Process with Initial Value Selection

In this section, according to the Newton-Raphson energy flow calculation method, the steps of the Newton method initial estimation are added. The initial value of the selected state variable is estimated before the energy flow is calculated and the maximum number of iterations is estimated to improve the computational efficiency of the energy flow. The proposed energy flow calculation process with initial value estimation is shown in Fig. 8.4.

The specific process of calculating the energy flow of Newton method with initial value is as follows:

(1) First, input system data include natural gas network, power network, heat network and we-energy. Determine the node types of mixed energy network, taking into account the access mode of distributed power at the same time, putting all the node information into the system.
(2) Select the initial values for each state variable.
(3) Calculate the initial Jacobian matrix based on the selected initial value.
(4) Judge whether the convergence operator determined by the initial value selection satisfies the conditions according to the convergence theorem. If $\rho \le 0.5$, turn to the next step, otherwise, return to step (2) to re-select the initial value.
(5) The maximum number of iterations is estimated according to the iteration number estimation theorem.
(6) Calculate the correction equation of the network ΔF.

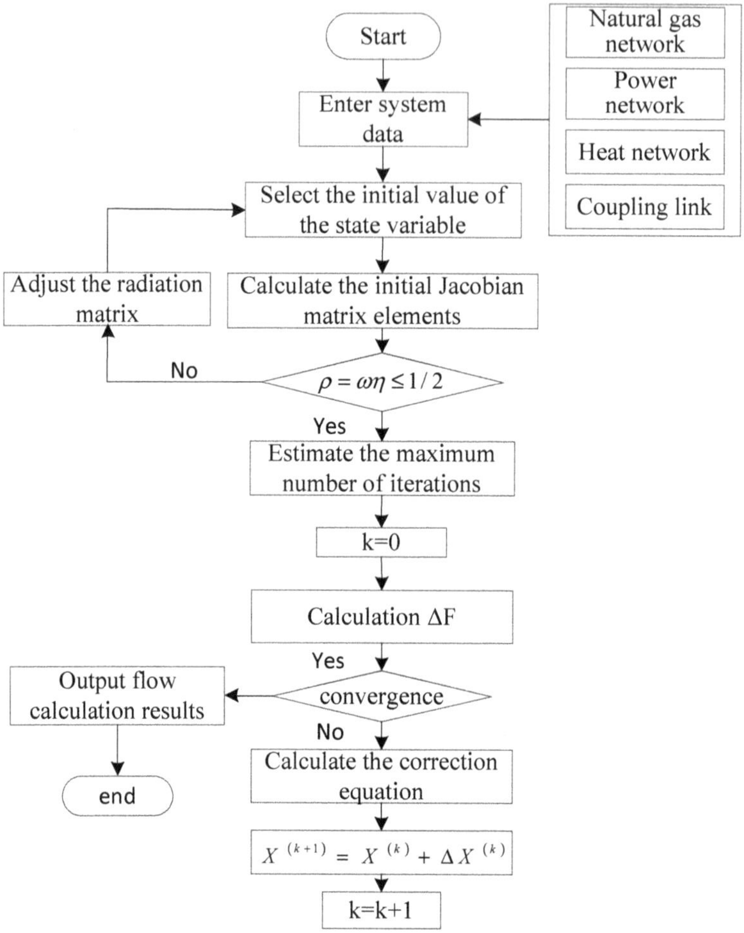

Fig. 8.4 The energy flow calculation process with initial value estimation

(7) Fix each node by $x^{k+1} = x^k + \Delta x^k$.
(8) Verify whether the result satisfies the convergence precision. If it satisfies, output the results of the calculation of each branch flow, otherwise it will turn to the sixth step for iteration of the next round.

8.4.3 *Case Study*

The hybrid energy system is solved by the Newton-Raphson method, and a 9-node hybrid energy system is used as an example in simulation. The hybrid energy system includes natural gas systems, power systems, and centralized thermal systems. Each

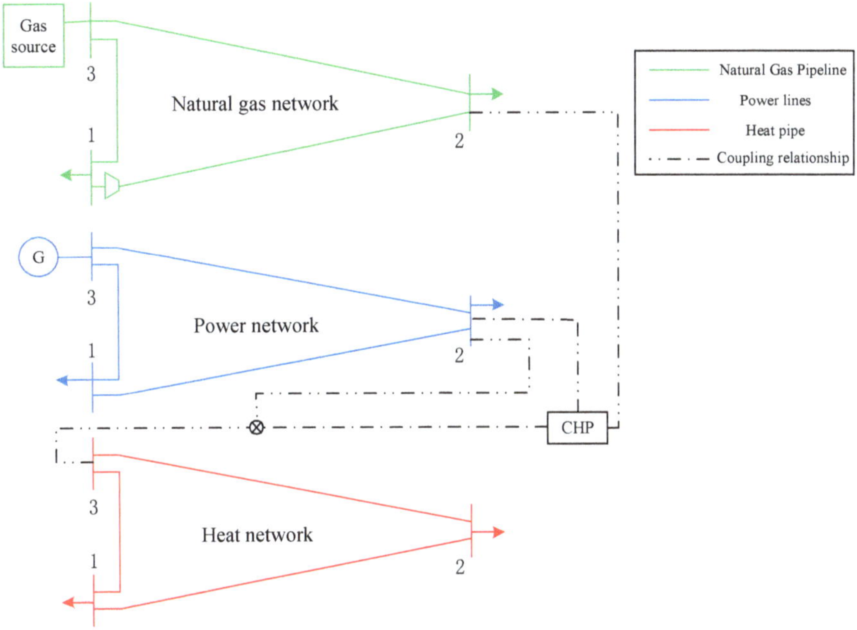

Fig. 8.5 The 9 nodes electricity/gas/heat hybrid energy system topology

subsystem is connected by three nodes and three buses. In a natural gas system, there is a turbo compressor at node 1 whose compression ratio has been given: A CHP unit and a circulating pump are connected between node 2 in the power system and node 3 in the centralized thermal system. As shown in the following figure (Fig. 8.5).

In the 3-node power system, the impedance of each transmission line is $Z = 0.09 + j0.157$.Node 1 is PQ node, node 2 is PV node, and node 3 is balanced node.$P_1 = 0.15\,\mathbf{MW}$, $Q_1 = 0.0493$ MW, $P_2 = 0.15\,\mathbf{MW}$, $V_2 = 1.05\,\mathbf{V}$, $V_3 = 1.02\,\mathbf{V}$. The above values are per unitary values.

In the natural gas system, the length and diameter of the natural gas pipeline are 300 and 150 mm, respectively; the pressure of the node is $\pi_3 = 50\,\mathbf{bar}$; the power consumed by the node load is $f_2^{GD} = 10MSCM/h$, $f_3^{GD} = 20MSCM/h$; compression ratio is $H_1 = 1.3$; the temperature of natural gas is $T_a = 281.15\,°\mathbf{K}$, the proportion of natural gas is $G = 1.309$ and gas compression coefficient is $Z_a = 0.8$. Node 3 is balance node.

In the centralized heat system, the diameter of heat pipe is 150 mm, the length of 3-1 pipe is 400 m, the length of 1-2 pipe is 400 m and the length of 2-3 pipe is 600 m; load thermal power is $\phi_1 = \phi_2 = 0.3\mathbf{MW}$, the temperature of the equilibrium node is $T_3^s = 100\,°\mathbf{C}$, the temperature of the load is $T_2^r = T_3^r = 50\,°\mathbf{C}$; specific heat capacity of water is $c_p = 4182\,\mathbf{J/(kg \bullet K)} = 4.182 \times 10^{-3}\,\mathbf{MJ/(kg \bullet K)}$, the temperature of pipe surface is $T_g = 10\,°\mathbf{C}$, the density of water is $\rho = 958.4\,\mathbf{kg/m^3}$, the heat transfer coefficient is $\lambda = 0.2$,the roughness of pipe is $\varepsilon = 0.00125$, the kinematic

Table 8.2 The calculation results

Node	Measurement value	Contrast value	Measurement error
Power systems			
Voltage amplitude V	1.0195	1.0232	− 0.0037
Voltage phase angle θ_1	2.9014	2.7652	0.1362
Voltage phase angle θ_2	3.1032	3.0852	0.0018
Branch head power S_{31}	−24.8902+j12.9100	−24.5409+j15.3363	0.3493 − j2.4263
Branch end power S_{13}	25.8114 − j11.8822	25.2653 − j14.0669	0.5461+j2.1847
Branch head power S_{12}	−10.0528 − j10.9337	−10.2654 − j14.0163	–0.2126+j3.0826
Branch end power S_{21}	10.2436+j12.0764	10.5269+j14.4744	–0.2833 − j2.3980
Branch head power S_{23}	36.9341+j0.0412	36.6029+j0.0701	0.3312 − j0.0289
Branch end power S_{32}	−35.7226+j1.6406	−35.5092+j1.8463	−0.2134 − j0.2057
Natural gas system			
Pressure π_1	41.5160	40.8247	0.6913
Pressure π_2	48.0827	46.8612	1.2215
pipe flow f_{31}	14.2387	14.7514	−0.5127
Pipe flow f_{12}	12.5263	14.7473	−2.221
Pipe flow f_{32}	7.0075	8.9099	−1.9024
Pressure π_3	41.3451	40.8247	0.0395

viscosity of water is $\mu = 0.294 \times 10^{-6}$, gravity acceleration is $g = 9.81$ **kg • m/s**^{2}. Node 3 is the balance node.

At the same time, the thermoelectric conversion ratio of the CHP unit is $c_m = 1.3$, the circulation pump efficiency is $\eta_p = 0.65$, $\eta_e = 1.2$.

According to the above parameters, the energy flow calculation is carried out by using the Newton method with the initial value estimated in this chapter, and the results of the numerical simulation of the 9-node small-scale mixed energy system in the literature [8] are used as the contrast, set the iteration accuracy to 10^{-3}, the simulation result of the energy flow can be obtained (Tables 8.2 and 8.3).

This section verifies the validity of the proposed convergence theorem on the basis of the 9-node hybrid energy system. Initial value selection is $V_i = 1 + j0$. Then calculate convergent operator ρ, the convergence operator ρ is calculated from the table less than 0.5, the Newton's energy flow calculation converges at this time, which is in accordance with the method proposed in this paper.

On this basis, the computational convergent operator ρ is calculated, the initial voltage amplitude and phase angle are set to 1 and 0 respectively, and the initial Jacobian matrix is as follows:

Table 8.3 The calculation results

Heat system			
Mass flow rate m_{31}	1.6420	1.5894	0.0526
Mass flow rate m_{12}	0.2067	0.2175	0.2067
Mass flow rate m_{32}	41.5160	46.8612	0.0343
Supply temperature T_1^s	98.9576	1.3056	−0.2200
Supply temperature T_2^s	97.1401	98.9233	−0.0014
Return to temperature T_1^r	49.5583	97.1000	0.0069
Return to temperature T_3^r	49.1251	48.3511	0.0119
Thermal power ϕ_1	0.3011	49.3451	−0.2200
Thermal power ϕ_2	0.3284	0.3025	−0.0014
Thermal power ϕ_3	0.6355	0.3215	0.0069

$$
J_0 = \begin{bmatrix}
10.21903 & -5.10926 & 5.821771 & 0 & 0 & 0 & 0 & 0 & 0 & 0 & 0 \\
-5.12878 & 10.4016 & -2.90435 & 0.162953 & 0 & 0.162953 & 0 & 0 & 0 & 0 & 0 \\
-5.52177 & 2.938539 & 9.657371 & 0 & 0 & 0 & 0 & 0 & 0 & 0 & 0 \\
0 & 0 & 0 & -0.21868 & 0.218681 & 0 & -0.00574 & 0 & 0 & 0 & 0 \\
0 & 0 & 0 & 0 & -0.19697 & -0.19697 & 0 & -0.00637 & 0 & 0 & 0 \\
0 & 0 & 0 & -0.05712 & -0.00967 & 0.071155 & 0 & 0 & 0 & 0 & 0 \\
0 & 0 & 0 & 0 & 0 & 0 & -1 & 0 & 0 & 0 & 0 \\
0 & 0 & 0 & 0 & 0 & 0 & 0.199193 & -1.52306 & 0 & 0 & 0 \\
0 & 0 & 0 & 0 & 0 & 0 & 0 & 0 & -1.58937 & 0.199193 & 0 \\
0 & 0 & 0 & 0 & 0 & 0 & 0 & 0 & 0 & -1 & 0 \\
0 & 0 & 0 & 0.135794 & 0 & 0.135794 & 0 & 0 & 0 & 0 & 37.12887
\end{bmatrix}
$$

The norm of J_0^{-1} is $\left\| J_0^{-1} \right\| = 12.5754$.
Due to:

$$
F(x^{(0)}) = \begin{matrix} [0.1500\ 0.0493\ 1.0500\ \ 1.0200\ \ 5.0000\ 2.0000\ 0.3000 \\ 1.0000\ \ 5.0000\ 3.0000\ \ 5.0000]^T \end{matrix}
$$

$$
\left\| J_0^{-1} F(x^{(0)}) \right\| = 39.9894
$$

Then, when $m = 2, n = 1, \eta \geq 39.9894$.

$$J_1 = \begin{bmatrix} 10.11803 & -5.01026 & 5.731771 & 0 & 0 & 0 & 0 & 0 & 0 & 0 & 0 \\ -5.01978 & 10.3116 & -2.82435 & 0.059953 & 0 & 0.062853 & 0 & 0 & 0 & 0 & 0 \\ -5.41677 & 2.82853 & 9.567371 & 0 & 0 & 0 & 0 & 0 & 0 & 0 & 0 \\ 0 & 0 & 0 & -0.12868 & 0.125681 & 0 & -0.00464 & 0 & 0 & 0 & 0 \\ 0 & 0 & 0 & 0 & -0.09597 & -0.08697 & 0 & -0.00527 & 0 & 0 & 0 \\ 0 & 0 & 0 & -0.04612 & -0.00767 & 0.065155 & 0 & 0 & 0 & 0 & 0 \\ 0 & 0 & 0 & 0 & 0 & 0 & -1 & 0 & 0 & 0 & 0 \\ 0 & 0 & 0 & 0 & 0 & 0 & 0.097193 & -1.41706 & 0 & 0 & 0 \\ 0 & 0 & 0 & 0 & 0 & 0 & 0 & 0 & -1.47837 & 0.098193 & 0 \\ 0 & 0 & 0 & 0 & 0 & 0 & 0 & 0 & 0 & -1 & 0 \\ 0 & 0 & 0 & 0.036794 & 0 & 0.036794 & 0 & 0 & 0 & 0 & 36.13887 \end{bmatrix}$$

$$J_2 = \begin{bmatrix} 10.20903 & -5.10926 & 5.811771 & 0 & 0 & 0 & 0 & 0 & 0 & 0 & 0 \\ -5.11878 & 10.3016 & -2.89435 & 0.152953 & 0 & 0.152953 & 0 & 0 & 0 & 0 & 0 \\ -5.51177 & 2.908539 & 9.647371 & 0 & 0 & 0 & 0 & 0 & 0 & 0 & 0 \\ 0 & 0 & 0 & -0.20868 & 0.208681 & 0 & -0.00474 & 0 & 0 & 0 & 0 \\ 0 & 0 & 0 & 0 & -0.18697 & -0.18697 & 0 & -0.00537 & 0 & 0 & 0 \\ 0 & 0 & 0 & -0.04712 & -0.00867 & 0.070155 & 0 & 0 & 0 & 0 & 0 \\ 0 & 0 & 0 & 0 & 0 & 0 & -1 & 0 & 0 & 0 & 0 \\ 0 & 0 & 0 & 0 & 0 & 0 & 0.189193 & -1.51306 & 0 & 0 & 0 \\ 0 & 0 & 0 & 0 & 0 & 0 & 0 & 0 & -1.57937 & 0.189193 & 0 \\ 0 & 0 & 0 & 0 & 0 & 0 & 0 & 0 & 0 & -1 & 0 \\ 0 & 0 & 0 & 0.125794 & 0 & 0.125794 & 0 & 0 & 0 & 0 & 37.11887 \end{bmatrix}$$

We can get the following results:

$$\|J_2 - J_1\| = 1.2621$$
$$\left\|x^{(2)} - x^{(1)}\right\| = 0.0910$$

When

$$\eta = 39.9894, \omega = 0.0122$$

Then

$$\rho = \omega\beta = 0.4878 < 0.5$$

8.5 Parallel Distributed Energy Flow Calculation Method

With the growing size of the energy internet, in the process of calculating its energy flow, a large number of network data and the size of the Jacobi matrix will cause the calculation slow, while the performance requirements of the computer are also very high. Therefore, this section proposes a parallel distributed energy flow calculation method(PD) based on Newton Raphson method (NR). The parallel computation is realized by network decomposition, which accelerates the calculation speed of energy flow.

8.5.1 Network Decomposition Principle

When using the Newton method to calculate the energy flow of the large-scale energy internet, the Jacobian matrix is large. This leads to decrease in the computational speed. Decomposing the network into multiple subnetworks can reduce the network size and increase the computational speed.

When calculating the energy flow, using the Newton method to linearize the nonlinear equation we can get the following formula:

$$Ax = b \tag{8.70}$$

For a network with n nodes, the formula (8.70) can be developed:

$$a_{i1}x_1 + a_{i2}x_2 + \cdots a_{ii}x_i + \cdots + a_{in}x_n = b_i \tag{8.71}$$

If the network is decomposed into two networks at the node, then the formula (8.71) can also be decomposed into the following form:

$$a_{i1}x_1 + a_{i2}x_2 + \cdots a_{ii}^{'}x_i = b_i^{'} + y_i \tag{8.72}$$

$$a_{ii}^{''}x_i + \cdots + a_{in}x_n = b_i^{''} - y_i \tag{8.73}$$

where $a_{ii}^{\prime} + a_{ii}^{\prime\prime} = a_{ii}$, $b_i^{'} + b_i^{''} = b_i$, y_i is an unknown quantity. From the above formula, it can be concluded that when the whole network is divided into sub-network α and sub-network β, the nodes i appear in the two sub-networks at the same time. Equation (8.72) represents the sub-network α, and (8.73) represents the sub-network β. All nodes in the entire network are divided into two parts contained in the two formulas. Thus, the formula (8.70) can be changed as follows:

$$Bx^{'} = b^{'} + Ky \tag{8.74}$$

The matrix B is a block diagonalized matrix, where x' and b' are the vectors after the division of the nodes. Each split node will introduce a new unknown quantity y and a column of related coefficient matrices K. There are only three cases of elements in each column of the matrix, 0; -1;$+1$.

When there are m split nodes in the network to divide the entire network into s subnetworks:

B is a $[(m+n) \times (m+n)]$ block of diagonal matrix; K is a $[(m+n) \times m]$ matrix; x' is a $[(m+n) \times 1]$ vector; b' is a $[(m+n) \times 1]$ vector.

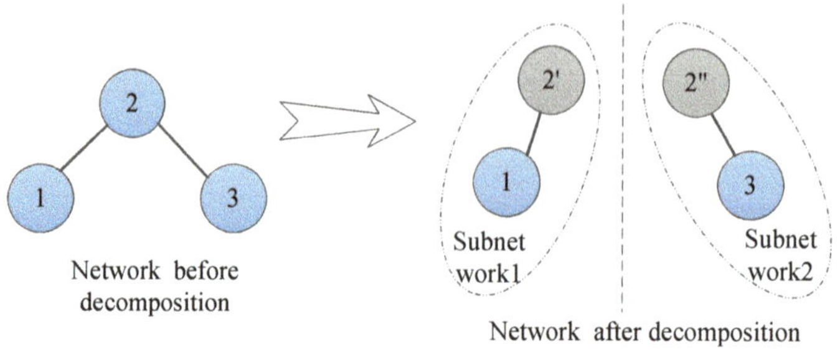

Fig. 8.6 Schematic of three nodes network decomposition

Next, the network decomposition principle is illustrated by an example of a simple three-node network as shown in Fig. 8.6.

As shown in Fig. 8.6, the node 2 is connected to node 1 and node 3, and node 1 and node 3 are not connected. Therefore, it can be easily seen that the network can be divided into two sub-networks through the node 2, and the node 2 is divided into two virtual nodes, namely node 2' and node 2'', respectively.

According to the network topology of Fig. 8.6, the equation can be expanded as follows:

$$\begin{bmatrix} a_{11} & a_{12} & a_{13} \\ a_{21} & a_{22} & a_{23} \\ a_{31} & a_{32} & a_{33} \end{bmatrix} \begin{bmatrix} x_1 \\ x_2 \\ x_3 \end{bmatrix} = \begin{bmatrix} b_1 \\ b_2 \\ b_3 \end{bmatrix} \tag{8.75}$$

From the figure we can see the coefficient $a_{13} = a_{31} = 0$. Therefore, the formula (8.74) becomes:

$$\begin{bmatrix} a_{11} & a_{12} & 0 \\ a_{21} & a_{22} & a_{23} \\ 0 & a_{32} & a_{33} \end{bmatrix} \begin{bmatrix} x_1 \\ x_2 \\ x_3 \end{bmatrix} = \begin{bmatrix} b_1 \\ b_2 \\ b_3 \end{bmatrix} \tag{8.76}$$

Select node 2 as the split node, divide the network into two subnetworks:

$$\begin{bmatrix} a_{11} & a_{12} & 0 & 0 \\ a_{21} & a_{22}^{'} & 0 & 0 \\ 0 & 0 & a_{22}^{''} & a_{23} \\ 0 & 0 & a_{32} & a_{33} \end{bmatrix} \begin{bmatrix} x_1 \\ x_2 \\ x_2 \\ x_3 \end{bmatrix} = \begin{bmatrix} b_1 \\ b_2^{'} \\ b_2^{''} \\ b_3 \end{bmatrix} + \begin{bmatrix} 0 \\ +1 \\ -1 \\ 0 \end{bmatrix} y \tag{8.77}$$

The choice of split nodes depends on the topology of the network. Firstly, the network coupling point is chosen as the splitting node to decompose the coupling network into three approximate networks, namely approximate power grid, approximate gas network and approximate heat network. These approximate networks can be further decomposed according to their own network topology.

8.5.2 *Distributed Parallel Algorithm Based on Newton Method*

1. The mathematical principles of distributed parallel algorithms

In the process of energy flow calculation using Newton method, the nonlinear equation is linearized, and the following form are obtained:

$$-J\Delta x = \Delta F \tag{8.78}$$

Using the network decomposition principle in 8.5.1, the hybrid network can be decomposed into the following form:

$$B(\Delta x)' - Ky = (\Delta F)' \tag{8.79}$$

Transform the formula (8.78) into the following form:

$$(\Delta x)' = B^{-1}(\Delta F)' - B^{-1}K\left(K^T B^{-1} K\right)^{-1} K^T B^{-1}(\Delta F)' \tag{8.80}$$

Since the matrix in the formula (8.79) is a partitioned diagonal matrix, the corresponding matrices K and vectors $(\Delta x)'$ and vectors $(\Delta F)'$ are decomposed, so the formula (8.79) can be implemented in parallel computation. The main steps of calculation $(\Delta x)'$ are as follows:

(1) Calculate $B^{-1}\left[(\Delta F)',\ K\right]$: In the calculation of energy flow, the network is decomposed into many subnetworks, that is to say, the matrix B becomes a block diagonal matrix, and the matrices K and vectors $(\Delta F)'$ are decomposed. The calculation process can be performed simultaneously by many sub-processors.
(2) Calculate $(-y) = \left(K^T B^{-1} K\right)^{-1} K^T B^{-1}(\Delta F)'$: Based on the calculation of the values of $B^{-1}K$ and $B^{-1}(\Delta F)'$ in the first step, the values of $(-y)$ can be obtained.
(3) Calculate the final solution vector $(\Delta x)'$: on the basis of the first two steps, each processor calculates the subnet $\Delta x_i = B_{ii}^{-1}\Delta F_i - B_{ii}^{-1}Ky$.

When the coupling node K is split and decomposed into the nodes i in the subnet 1 and the node j in the subnet 2, the network ΔF_K^{sp} can be arbitrarily decomposed into ΔF_i^{sp} and ΔF_j^{sp}

$$\begin{aligned}\Delta F_K = F_K^{sp} - F_K^{cal} &= \left(F_i^{sp} + F_i^{sp}\right)\\ &\quad - \left(F_i^{cal} + F_i^{cal}\right) = \Delta F_i + \Delta F_j\end{aligned} \tag{8.81}$$

2. The implementation process of distributed parallel algorithm

The implementation of parallel algorithm is illustrated by splitting the network from the coupling node as an example. The integrated energy network is divided into three subnetworks through the splitting of coupled nodes, and four processors need to be calculated simultaneously in parallel computation. One processor acts as the primary processor, and the other three processors act as subprocessors. A part of the algorithm can be implemented in parallel with the subprocessor. The three subprocessors calculate the power network, the thermal network and the natural gas network respectively, and a part of the algorithm flow is implemented in the main processor. In the calculation process, attention should be paid on the consistency and synchronization of the data algorithm.

The program and data for calculating energy flow are stored in each processor. The calculated data is returned to the main processor. The main processor calculates and checks whether the calculation results are globally convergent, and then sends the results to the subprocessor. Each subprocessor calculates the energy flow of each approximate network to complete a cycle.

Repeat this process until the calculation results satisfy the conditions.

The main processor steps are as follows:

(1) Broadcast start information.
(2) Wait for the results from the subprocessor.
(3) After receiving the results from the processor:

(a) Whether the results are convergent, and calculate $(-y)$.
(b) Broadcast results $(-y)$ to the subprocessor and judge whether the results are globally convergent.
(c) If it is not convergent then execute step (2), if it is convergent then stop.

The steps of the subprocessor are as follows:

(1) Performing the initial work of the energy flow calculation after receipt of the start command. For example, input network data and form association matrices.
(2) Calculate ΔF and form Jacobi matrix.
(3) Send the local convergence results to the main processor.
(4) Compute $B^{-1}\left[(\Delta F)',\ K\right]$ and wait for instructions from the main processor.
(5) Calculate $\Delta x_i = B_{ii}^{-1}\Delta F_i - B_{ii}^{-1}Ky$.
(6) If the result is convergent, then stop the calculation, or return to step (2).

The calculation flowchart of the distributed parallel computing is shown in Fig. 8.7:

The judgment process of global convergence is that judging whether $|\Delta F| < \varepsilon$ or not by the subprocessor for non-split nodes. For split nodes, the calculation results are merged by the main processor,ΔF_i and ΔF_j are subnodes after division, and the condition of global convergence is $\left|\Delta F_i + \Delta F_j\right| < \varepsilon$.

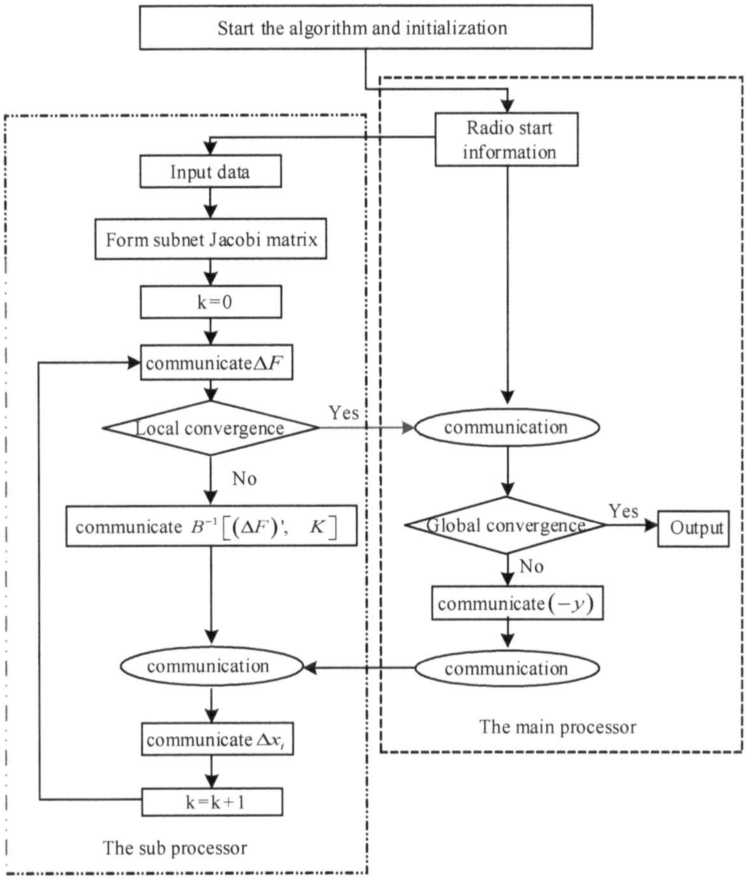

Fig. 8.7 The flow chart of parallel computing

8.5.3 *Case Study*

To verify the distributed parallel energy flow calculation method proposed in this chapter, we use the integrated energy network in [8] to compare the traditional serial Newton method with the method proposed in this paper.

The electrical network uses a standard IEEE - 14 bus grid. The generators at node 1 and node 2 are all gas generators, and node 1 is balance node. A typical gas network is the Belgian natural gas network. The network data of the electrical network and the natural gas network can be obtained in the literature [9] and [10].The heat network model is shown in Fig. 8.8. The length of each pipeline is shown in the figure, and the pipe diameter is 150 mm.

The type and location of the coupling devices among the three networks are given in Table 8.4

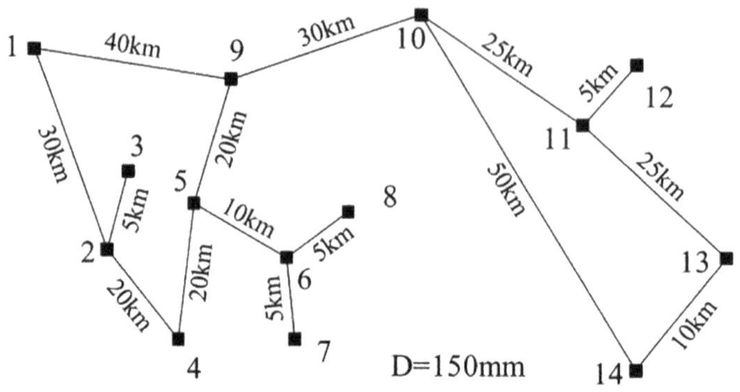

Fig. 8.8 Schematic diagram of the heating network

Table 8.4 Structure of network equipment

Device	Node of Electrical network	Node of gas network	Node of heat network
Gas generators	1	12	–
Gas generators	2	19	–
Compressor	4	9	–
Compressor	5	18	–
Gas boiler, pump	12	3	1
CHP, gas boiler, pump	6	15	4
CHP, gas boiler, pump	13	7	9
CHP, gas boiler, pump	14	6	10
CHP, gas boiler, pump	8	10	13

Use four computers with the same configuration. The processor model is Intel (R) Core (TM) i5-6500 CPU@3.20 GHz; 4.00 GB (RAM) is installed. Four computers are used to build the MATLAB distributed parallel computing platform. One is the main processor. The others are the subprocessor. MATLAB Distributed Computing Server and Parallel Computing toolbox are installed.

The results of the distributed parallel computation proposed in this paper are compared with the results of case2. a in the literature [9]. The specific results of the node related parameters are shown in Table 8.5

From the calculation results of the above table, it can be seen that the simulation results are basically consistent with the traditional method. Therefore, we can conclude that the distributed parallel computing method proposed in this paper is accurate and effective for the energy flow of the energy internet.

Table 8.6 lists the required computational time for energy flow of the traditional serial Newton method and the proposed parallel method, respectively. The time of the parallel algorithm is divided into computation time and communication time. From

Table 8.5 The calculation results

Node	Electrical network				Gas network		Heat network			
	V (p.u.)		θ (°)		p (bar)		T_s(°C)		T_r (°C)	
	Basis	Contrast	Basis	Contrast	Basis	Contrast	Basis	Contrast	Basis	Contrast
1	1.0600	1.0600	0	0	56.000	56.000	120.000	120.000	47.075	47.077
2	1.0400	1.0398	−3.1321	−3.1311	55.963	55.985	115.960	115.962	48.969	48.971
3	0.9692	0.9691	−10.374	−10.380	55.805	55.825	115.271	115.307	50.000	50.000
4	0.9917	0.9904	−7.7379	−7.7383	53.992	54.002	123.857	123.852	49.271	49.305
5	1.0002	1.0012	−6.4250	−6.4301	52.809	52.909	120.734	120.734	49.557	49.562
6	1.0300	1.0321	−10.241	−10.252	52.055	52.105	120.119	120.120	49.778	49.766
7	0.9988	0.9987	−9.3856	−9.3866	52.165	52.215	119.609	119.615	50.000	50.000
8	1.0000	1.0000	−8.3719	−8.3729	49.426	49.438	119.358	119.335	50.000	50.000
9	0.9907	0.9904	−10.911	−10.935	48.887	48.896	122.409	122.408	48.445	48.459
10	0.9897	0.9875	−11.098	−11.072	56.827	56.902	124.921	124.922	48.396	48.396
11	1.0058	1.0056	−10.797	−10.801	55.650	55.666	121.626	121.627	48.666	48.683
12	1.0190	1.0200	−10.944	−10.963	53.838	53.847	120.838	120.916	50.000	50.000
13	1.0200	1.0195	−10.902	−10.900	52.732	52.703	124.281	124.281	49..814	49..824
14	1.0000	1.0000	−11.326	−11.326	52.558	52.509	118.572	118.612	50.000	50.000
15					51.208	51.234				
16					49.540	49.526				
17					54.580	54.581				
18					45.099	45.128				
19					31.843	31.887				
20					29.692	29.707				

Table 8.6 Calculating time

	Communication time (s)	Calculating time (s)	Total time(s)
NR	–	0.146	0.146
PD	0.051	0.048	0.099

Table 8.6, it is obtained that we can see the speed of the distributed parallel method is faster than the traditional method for the energy flow and the calculating time is obtained by multiple calculating then average the results. The computational speed of the two methods is affected by the configuration of the computer. The speed of parallel computing is greatly influenced by the speed of communication. Therefore, not the more processors are used, the less time it takes. The larger the scale of the network, the more obvious the advantages of the algorithm.

8.6 Conclusion

A general framework of modeling steady energy flow problems of IES is formulated in this chapter. Based on the accurate calculation of the energy flow using Newton method, the Newton method with initial value estimation is used to calculate the energy flow. In addition, for the large-scale integrated energy system, a distributed parallel computing method and a split node selection method are proposed, which can further increase the calculation speed.

References

1. J.D. Glover, M.S. Sarma, T. Overbye, *Power System Analysis & Design* (SI version, Cengage Learn, 2012)
2. Q. Li, S. An, T. W. Gedra, Solving natural gas loadflow problems using electric loadflow techniques, in *Proceedings of North American Power Symposium* (2003)
3. M. Pirouti, Modelling and analysis of a district heating network, Ph.D. dissertation, School of Engineering, Cardiff University, Cardiff, U.K. (2013)
4. B. Bakhouya, D. De Wolf, *Solving gas transmission problems by taking compressors into account* (Univ, Littoral Opal Coast, Dunkerque, France, 2008)
5. W.H. Press, S.A. Teukolsky, W.T. Vetterling, B.P. Flannery, Numerical Recipes 3rd Edition: The Art of Scientific Computing (Cambridge University Press, Cambridge, 2007)
6. M. Qadrdan, M. Abeysekera, M. Chaudry et al., Role of power-to-gas in an integrated gas and electricity system in Great Britain[J]. Int. J. Hydrog. Energy **40**(17), 5763–5775 (2015)
7. Y.F. Su, H.Y. Zhou, A geometric result for approximating fixed points of nonlinear mappings by iteration sequence[J]. Acta Mathematica Sinica. **49**(6), 1321–1326 (2006)

8. A. Shabanpour-Haghighi, A.R. Seifi, An integrated steady-state operation assessment of electrical, natural gas, and district heating networks[J]. IEEE Trans. Power Syst. **31**(5), 3636–3647 (2016)
9. D. De Wolf, Y. Smeers, The gas transmission problem solved by an extension of the simplex algorithm. Manag. Sci. **46**(11), 1454–1465 (2000)
10. The standard IEEE 14-bus test system. http://www.ee.washington.edu/research/pstca/

Chapter 9
Distributed Optimal Energy Management for Energy Internet

Abstract In this chapter, a novel energy management framework for Energy Internet with many energy bodies is presented, which features multi-coupling of different energy forms, diversified energy roles and peer-to-peer energy supply/demand, etc. The energy body as an integrated energy unit, which may have various functionalities and play multiple roles at the same time, is formulated for the system model development. Forecasting errors, confidence intervals and penalty factor are also taken into account to model renewable energy resources to provide trade-off between optimality and possibility. Furthermore, a novel distributed-consensus alternating direction method of multipliers (ADMM) algorithm, which contains a dynamic average consensus algorithm and distributed ADMM algorithm, is presented to solve the optimal energy management problem of energy internet. The proposed algorithm can effectively handle the problems of power-heat-gas-coupling, global constraint limits and non-linear objective function. With this effort, not only the optimal energy market clearing price but also the optimal energy outputs/demands can be obtained through only local communication and computation. Simulation results are presented to illustrate the effectiveness of the proposed distributed algorithm.

9.1 Introduction

The concept of "Energy Internet" is recently presented to meet the challenges of developing sustainable and environmentally friendly energy resources, hybrid energy utilization model, flexible energy management and secure system control. Several literatures about energy internet (EI), mainly focusing on system architecture [1], energy router [2] and voltage control [3], etc., have been proposed in recent years. Likewise, energy management problem (EMP), as one of fundamental issues in power system, is encouraged to be revisited to achieve the envisioned EI conception.

EMP is typically formulated as an optimization problems. A number of approaches have been proposed to solve the above problem and can be classified into two categories, i.e., centralized approaches and distributed approaches. With regard to centralized approaches [4–7], they have the advantage in the aspect of obtaining the optimal solution. However, as the physical power system tends to become distributed,

Q. Sun, *Energy Internet and We-Energy*, Renewable Energy Sources & Energy Storage, https://doi.org/10.1007/978-981-13-0523-8_9

centralized approaches have to confront many challenges. Specifically, the centralized approaches, relying on a central controller, require high bandwidth communication capabilities to act on system-wide gathered information, resulting in high susceptibility to single-point failures and modeling errors. Moreover, either the physical or communication network of the future power grid is likely to have a variable topology, which may undermine the efficacy of the centralized approach. In addition, the distributed energy resources owners are generally unwilling to disclose their own information to the external centralized controller to protect their privacy [8].

On the contrary, the distributed approaches, with better robustness [9], faster computation and less communication [10], are promising options for the future power system. A number of literatures, concerned about solving the EMP in a distributed manner, have been proposed for smart grid or microgrid recently, which can be classified into two main categories. The one, which does not consider the demand response, is also called economic dispatch problem (EDP), whose object is minimizing the total operating cost while meeting some equality and inequality constraints. From this prospective, the λ-consensus algorithm has been firstly proposed in [11], which can solve the traditional EDP in a fully distributed manner. The work in [12] has presented a new distributed algorithm which does not rely on the projection scheme and also can achieve the underlying power flow control. In order to address non-quadratic cost function, a distributed bisection method and a distributed algorithm based on projected gradient method have been presented in [13, 14], respectively. Moreover, the ramping rate limits [15], transmission losses [16], finite-time [17], non-convex conditions [18] and frequency recovery [19] have been further taken into consideration and studied in the EDP. The other one aims at maximizing the social welfare through the coordination of the suppliers' generations and customers' demands [20–22]. Since the demand response is taken into consideration, the power system operation becomes more flexible and economic.

Different from the studies above for the traditional system or smart grid, the EMP for EI is a relatively new but difficult problem, which will bring many challenges. On one hand, EI is a hybrid energy network with many new characteristics which can be summarized in three aspects. Firstly, EI integrates various kinds of energy networks. As the main energy medium, it is urgent for power, heat and gas flows to be co-planned. Secondly, each unit in EI can be not only an energy consumer but also an energy supplier. The forms of energy generation and consumption are tending to be diversified. Thirdly, distributed energy suppliers or consumers within EI have peer-to-peer relationship, which is different from the traditional, vertical supply-demand relationship. The detailed features and requirements of EI will be discussed in section II-A. Therefore, how to establish a system model, which can describe aforementioned features of EI in a better way, is one of the major challenges. On another hand, EI should meet the plug-and-play nature and topology variabilities, etc. Thus, it is desirable to develop a distributed method to solve the EMP of EI. More importantly, different from smart grid, there exists strong coupling among power, heat and gas in the process of energy-generation, energy-conversion and energy-consumption for EI. Meanwhile, there also exist many global constraint limits and non-linear objective function. However, the existing distributed methods discussed above can not deal with this kind of distributed non-linear coupling optimization problem. Therefore, how to establish a fully distributed method to solve the EMP of EI is another major challenge.

To address above challenges, this chapter focuses on the static EMP of the future EI considering multi-energy-networks and intermittency of renewable energy resources, where a novel distributed-consensus-ADMM algorithm is proposed to solve this problem in a fully distributed fashion. The major contributions of this chapter are as follows:

(1) The energy management framework is developed for the future EI featuring multiple couplings of different energy forms, diversified energy roles and peer-to-peer energy supply/ demand, etc. Subsequently, the energy body, seen as both energy supplier and customer, is presented for system model development.

(2) Forecasting errors, confidence intervals and penalty factor are also considered to model renewable energy resource. With this effort, the optimality and generation possibility of the renewable energy can be traded off by designing the penalty on curtailment of renewable generation.

(3) By dividing the global computation process into the distributed participants, our implementation fashion can make each participant locally calculate its optimal operation. It can result in enhanced reliability, flexibility and scalability, etc., and is more suitable for the future EI to integrate distributed multi-energy resources.

(4) The proposed algorithm can effectively overcome the weakness of conventional distributed ADMM algorithm which assumes part of global information is known via designing a dynamic average consensus algorithm for estimating the global information. Moreover, the optimality and convergence analysis are strictly proved.

(5) We get the explicit iterative form of coupled variables by transforming and decoupling the max-min problem, and make the coupled inequality constraints be solved by locally calculating sub-min problem. With this effort, the strongly power-heat-gas-coupling problem can be effectively solved by implementing the proposed algorithm.

9.2 Energy Management Framework of EI

9.2.1 *Structure and Features Analysis*

The anticipated structure of EI, shown in Fig. 9.1, consists of various kinds of distributed energy suppliers and/or customers which are referred as energy bodies (EBs). Fed by multi-energy-carriers, the energy resources of each EB can be divided into four classes, i.e., power-only devices which contain an equivalent distributed renewable generator (DRG), distributed fuel generator (DFG) and distributed power storage device (DPSD), heat-only devices which contain an equivalent distributed renewable heating device (DRHD), distributed fuel heating device (DFHD) and distributed heat

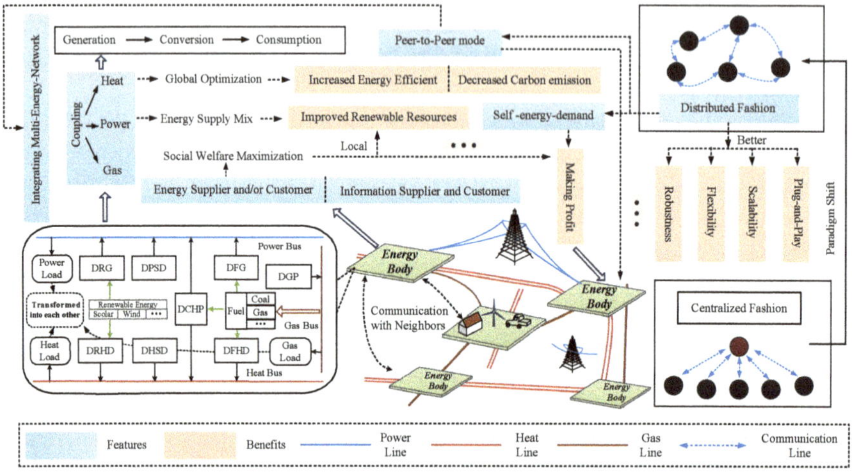

Fig. 9.1 Structure, features and benefits of energy internet

storage device (DHSD), an equivalent distributed combined heat and power (DCHP) device and an equivalent distributed gas producer (DGP). Meanwhile, the energy loads (ELs) of each EB can be divided into three classes, i.e., power load, heat load and gas load, in which every type of the load contains an equivalent controllable load and a must-run load. Note that each EB contains at least one type of energy resource or load which are connected to the corresponding energy bus. More importantly, the EB can be seen as small as a house or as large as a town based on the requirements of different scenarios. In order to achieve the anticipated EI, the major features and requirements of the proposed energy management framework of EI are summarized as follows:

(1) As for EI, one of the features is to integrate multi-energy-networks, such as electricity, transportation, nature gas, oil and heat networks, etc. Since power, gas and heat are the major energy that can be flexibly transferred and directly consumed, it is urgent that they can be co-managed and be the primary energy medium of energy-generation, energy-transmission and energy-consumption, which brings many advantages. For example, since there are many kinds of alternative energy resources in EI, consumers can diversify their energy supply mix and determine the amount of energy utilization in an overall optimal manner. This will create significant incentives to increase the consumer participation and put full use of the function of "virtual storage" caused by demand dispatch to flatten out the variability of energy generation. Moreover, The co-planning benefits improvement of the energy efficiency and economy and decreasing carbon emission, etc.

(2) In the anticipated EI, each EB may be equipped with its own energy generation, conversion and storage devices and schedulable loads, so it plays a role of not only energy (i.e. power, heat and gas) supplier but also energy consumer. For example, a EB may be a power supplier and a heat consumer at the same time.

Meanwhile, motivated by the Internet, each EB is also both the information supplier and information consumer, and they only exchange information with its neighbors through local network. Moreover, all EBs are cooperative to achieve the maximization of social welfare while meeting a set of local and global constraints. Then, each EB can determine its corresponding energy-production or/and energy-consumption. It is worth noting that each EB may compensate its self-energy-demand, and even make additional profit by selling the part of overabundance energy to other EB. In this case, it can be seen as an energy-supplier. Otherwise, it is seen as an energy consumer. In such an energy management framework, EB can somehow weaken dependency on utility grid and promote network flattening; meanwhile, there will be more interactions and flexibility among EBs. More importantly, driven by profit, it is able to create sufficient incentives for each EB to make self-installation of renewable energy resources, resulting in increased penetration of renewable energy resources.

(3) To integrate and coordinate the large number EBs with the plug-and-play feature, both the physical and communication topology of EI are subject to variability and uncertainty. Therefore, it's imperative to develop a distributed control strategy, with decision made by the EBs themselves rather than a central controller. This also means that each EB should have the function of local information processing and computing. The distributed implementation fashion benefits for enhancement of system reliability, flexibility for possible expansions of distributed energy resources, and scalability in terms of the computation and communications burden, etc. Moreover, in the proposed energy management framework, it achieves the vertical energy supply mode to peer-to-peer mode. To be specific, EBs have peer-to-peer relationship. The renewable and conventional fuel devices are in peer-to-peer mode to accomplish their own function. In addition, the diversified energy resources, including electric power, gas and heat, are also in the peer-to-peer mode to collectively achieve their optimization, management and control.

From the discussion above, the major differences of EI and the smart grid on EMP are summarized as follows. Firstly, the smart grid mainly focuses on the optimization and management of electrical power on power system aspect, in which the electrical power is the major energy medium in the process of energy-generation, energy-conversion and energy-consumption. However, EI focuses on integrating various kinds of energy resources to fulfil the co-planning of different energy networks, which contains multi-type coupled energy mediums such as power, heat and gas. In addition, each energy network in EI is in peer-to-peer position to fulfil its functionalities. Secondly, each unit of smart grid can be only seen as an electrical power supplier or consumer in one scheduling horizon. On the contrast, the EB in EI, regarded as both energy supplier and consumer, can play multiple roles at the same time. Lastly but not least, both smart grid and EI are eager to develop distributed methods for the EMP. However, the strongly power-heat-gas coupling situation of EI, which exists in the objective function and constraint limits, vastly increases the solving difficulty and is more complicated than electricity.

9.2.2 System Model

(1) DRG and DRHD models: For DRGs, the solar radiation and wind are the major energy sources with zero fuel cost. Thus, the direct cost of renewable generators is often the operating cost which is proportional to the produced power. It is worth noting that renewable generators cannot be regarded as dispatchable units due to the intermittency and randomness features. To take them into consideration in our EMP, the forecast technologies are used in this chapter. The mean forecasting value, seen as the reference during each scheduling horizon, can be calculated from day-ahead forecast curve given by

$$p_{ij,t}^{re,\dagger} = \left(\int_t^{t+T} p_{ij,\bar{t}}^{re,\dagger} d\bar{t} \right) / (T), \tag{9.1}$$

where $p_{ij,\bar{t}}^{re,\dagger}$ is the forecasting power generation of jth DRG of ith EB at time $\bar{t}$; T is the scheduling horizon. Note that real value may not be the forecasting value because of the forecast error. In this chapter, it is assumed that the forecast error, denoted by $\Delta p_{ij,t}^{re,\dagger}$, obeys Gaussian distribution whose feasibility analysis has been discussed in [23]. Then, the corresponding probability density function can be modeled as

$$f\left(\Delta p_{ij,t}^{re,\dagger}\right) = \frac{1}{\sqrt{2\pi}\delta_{ij,t}^{re}} \exp\left(-\left(\Delta p_{ij,t}^{re,\dagger}\right)^2 / (2\delta_{ij,t}^{re})^2 \right). \tag{9.2}$$

Furthermore, let the confidence level be $100(1 - \wp_{ij})\%$, then we can get the confidence intervals $[p_{ij,t}^{re,\dagger,down}, p_{ij,t}^{re,\dagger,up}]$ by using the method proposed in [24]. Therein, $\wp_{ij}$ denotes the corresponding significance level. In addition, we assume that the power generation of each DRG can be fully consumed by loads, then we can let

$$p_{ij,t}^{re} = p_{ij,t}^{re,\dagger} + \Delta p_{ij,t}^{re,\dagger} \rightarrow \underline{p}_{ij,t}^{re} \leq p_{ij,t}^{re} \leq \overline{p}_{ij,t}^{re}, \tag{9.3}$$

where $\underline{p}_{ij,t}^{re} = p_{ij,t}^{re,\dagger,down} + p_{ij,t}^{re,\dagger}$, $\overline{p}_{ij,t}^{re} = p_{ij,t}^{re,\dagger} + p_{ij,t}^{re,\dagger,up}$, and $p_{ij,t}^{re}$ is the power output of jth DRG of ith EB.

Based on the above discussion, the cost function of DRG is modeled as

$$C\left(p_{ij,t}^{\text{re}}\right) = b_{ij} p_{ij,t}^{\text{re}} + \varepsilon_{ij} \exp\left(\gamma_{ij} \frac{\overline{p}_{ij,t}^{\text{re}} - p_{ij,t}^{\text{re}}}{\overline{p}_{ij,t}^{\text{re}} - \underline{p}_{ij,t}^{\text{re}}} \right), \tag{9.4}$$

where $b_{ij} > 0$ and $\varepsilon_{ij} > 0$ are the cost coefficients, $\gamma_{ij} < 0$ is the penalty factor. The first item of the above equation denotes the direct operating cost while the second one denotes the penalty on curtailment of renewable generation. The purpose and significance of (9.4) are discussed in Remark 1. In addition, the modeling process and cost function of DRHD are designed accordingly as for DRG. And let $h_{ij,t}^{re}$ and $C(h_{ij,t}^{re})$ denote the corresponding heat generation and cost function, respectively.

Remark 1 Compared with DFGs, the operating cost of each DRG is very small. As a result, from the viewpoint of optimization, it may tend to run at its corresponding upper bound, i.e., $\overline{p}^{re}_{ij,t}$. However, the higher value the $p^{re}_{ij,t}$ is calculated, the lower possibility the practical power generation has to meet this requirement. On the contrary, if the renewable generator runs at point $\underline{p}^{re}_{ij,t}$, there is $100(1-\wp_{ij})\%$ possibility to believe that its practical capability can meet this power output, but that may decrease the optimality. Thus, the exponential penalty on curtailment of renewable generation is designed to provide trade-off between the optimality and possibility. In this way, the DFG and DRG can be in the same form.

Based on the aforementioned discussions, the concept of the cost function model of DRGs is described in Fig. 9.2. Further, a simple example is supplied to clearly illustrate the idea of (9.4). We consider a system composed of a DRG with direct operating cost $C_D(p^{\text{re}}) = 0.1095p^{\text{re}}$ [25], a DFG with cost function $C(p^{\text{fe}}) = 0.04(p^{\text{fe}})^2 + 25p^{\text{fe}} + 99 + 50\exp(0.01p^{\text{fe}})$ $(30\,(\text{p.u.}) \le p^{\text{fe}} \le 150\,(\text{p.u.}))$ [13] and a must-run load $L = 220$ (p.u.). The confidence level and the forecasting value are respectively set as 90% and $93.90(p.u.)$, s.t., 84.3 (p.u.) $\le p^{\text{re}} \le 103.5$ (p.u.). Before considering the penalty on curtailment of renewable generation, the goal is to minimize $Goal = C_D(p^{\text{re}}) + C(p^{\text{fe}})$ while meeting the supply-demand balance constraint, i.e., $p^{\text{re}} + p^{\text{fe}} = L$, and the inequality constraints mentioned above. Since the direct operating cost is proportional to p^{re}, the incremental cost of the DRG is a constant, i.e., 0.1095 which is very small compared to the one of DFG. The calculation results show that $p^{\text{re}} = 103.5$ (p.u.) and $p^{\text{fe}} = 116.5$ (p.u.). In this version, although it can obtain better optimality or less cost, its corresponding possibility is very small. In other words, only when the real value is over 103.5 (p.u.), the practical capability can meet this requirement. Note that the lower value of p^{re} is, the higher possibility becomes. Thus, to guarantee possibility, we can let $p^{\text{re}} = 84.3$ (p.u.), resulting in decreased optimization. Different from the two cases, the exponential

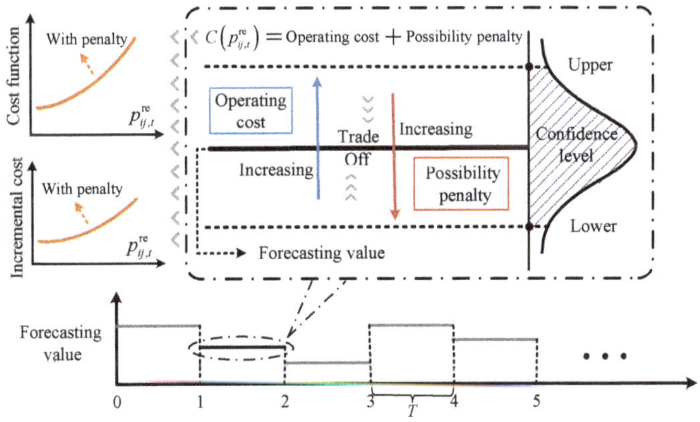

Fig. 9.2 Cost function model of DRG

penalty on curtailment of renewable generation, set as $748\exp(-1.1\frac{103.5-p^{\mathrm{re}}}{103.5-84.3})$, is further taken into consideration. Then, the scheduling problem is re-calculated with results of $p^{\mathrm{re}}=125.19$ (p.u.) and $p^{\mathrm{fe}}=94.81$ (p.u.). From Fig. 9.2, it can be seen that, after considering the penalty on curtailment, the incremental cost of DRG exponentially increases as the increasing of p^{re}. It also means that the higher value of p^{re} is, the greater penalty becomes. As a consequence, the optimality and possibility can be traded off in a suitable position, but not the upper or lower bound to unilaterally ensure the optimization or possibility only.

(2) DFG, DFHD and DGP models: To deal with and investigate the ramping rate limits of DFGs, its discrete form is always formulated into a knapsack problem [15]. For DFG, its cost function is modeled as the following non-linear function [4, 13],

$$C\left(p_{ij,t}^{\mathrm{fe}}\right)=a_{ij}\left(p_{ij,t}^{fe}\right)^2+b_{ij}p_{ij,t}^{fe}+c_{ij}+\varepsilon_{ij}\exp\left(\eta_{ij}p_{ij,t}^{fe}\right) \tag{9.5}$$

$$p_{ij,t}^{fe,\min}\le p_{ij,t}^{fe}\le p_{ij,t}^{fe,\max} \tag{9.6}$$

$$-p_{ij,t}^{fe,ramp}\le p_{ij,t}^{fe}-p_{ij,t-1}^{fe}\le p_{ij,t}^{fe,ramp}, \tag{9.7}$$

where a_{ij}, b_{ij}, c_{ij}, ε_{ij} and η_{ij} represent nonnegative cost coefficients which are derived from the energy emission of the thermal unit [4]; $p_{ij,t}^{fe}$ is the power output of jth DFG of ith EB; $p_{ij,t}^{fe,\min}$ and $p_{ij,t}^{fe,\max}$ are the lower and upper bounds; and $p_{ij,t}^{fe,ramp}$ is the ramp rate limit. In addition, if the above DFG is a gas turbine, the amount of gas consumption, denoted by $gas_{ij,t}^{p}$, cannot exceed the gas transmission maximum capability denoted by $gas_{ij,t}^{p,max}$. For a gas turbine, the relationship between power generation and the amount of gas consumption can be approximatively evaluated by the following equation

$$gas_{ij,t}^{p}=\Theta\left(\mu_{ij}\left(p_{ij,t}^{fe}\right)^2+\nu_{ij}p_{ij,t}^{fe}+\theta_{ij}\right), \tag{9.8}$$

where μ_{ij}, ν_{ij}, θ_{ij}, ξ_{ij} and ω_{ij} are heat rate coefficients. Θ is the conversion ratio from MW to SCM/h, and the value of Θ is about 84.

The model of DFG can be applied for DFHD. However, in different condition, the ramp rate limits of DFHDs are not taken into consideration [4]. And let $h_{ij,t}^{fe}$ and $C(h_{ij,t}^{fe})$ denote the corresponding heat generation and cost function, respectively. If the heating device is a gas boiler, the amount of gas consumption can be approximatively evaluated by

$$gas_{ij,t}^{h}=\Theta\left(h_{ij,t}^{fe}/\eta_{ij,t}^{fe}\right), \tag{9.9}$$

where $\eta_{i,t}^{fe}$ is the efficiency. And we let $gas_{ij,t}^{h,max}$ represent the gas transmission maximum capability.

Finally, the cost function of DGP is modeled as

$$C\left(g_{ij,t}^{g}\right) = a_{ij}\left(g_{ij,t}^{g}\right)^{3} + b_{ij}\left(g_{ij,t}^{g}\right)^{2} + d_{ij}g_{ij,t}^{g} + c_{ij} \tag{9.10}$$

$$0 \leq g_{ij,t}^{g,\min} \leq g_{ij,t}^{g} \leq g_{ij,t}^{g,\max}, \tag{9.11}$$

where a_{ij}, b_{ij}, c_{ij} and d_{ij} are nonnegative cost coefficients; $g_{ij,t}^{g}$ is the gas output of jth DGP of ith EB; and $g_{ij,t}^{g,\min}$ and $g_{ij,t}^{g,\max}$ are the corresponding lower and upper bounds. It is not difficult to verify that function (9.10) is convex in the region determined by (9.11).

(3) DCHP model: The cost function of DCHP is often modeled as the following convex function [5]

$$\begin{aligned} C\left(p_{ij,t}^{chp}, h_{ij,t}^{chp}\right) =& a_{ij}\left(p_{ij,t}^{chp}\right)^{2} + b_{ij}p_{ij,t}^{chp} + \alpha_{ij}\left(h_{ij,t}^{chp}\right)^{2} \\ &+ \beta_{ij}\left(h_{ij,t}^{chp}\right) + \sigma_{ij}p_{ij,t}^{chp}h_{ij,t}^{chp} + c_{ij} \end{aligned} \tag{9.12}$$

$$e_{ij,m}p_{ij,t}^{chp} + f_{ij,m}h_{ij,t}^{chp} + z_{ij,m} \geq 0, (m = 1, 2, 3, 4) \tag{9.13}$$

$$-p_{ij,t}^{chp,ramp} \leq p_{ij,t}^{chp} - p_{ij,t-1}^{chp} \leq p_{ij,t}^{chp,ramp}, \tag{9.14}$$

where a_{ij}, b_{ij}, α_{ij}, β_{ij}, σ_{ij} and c_{ij} are nonnegative cost coefficients; $e_{ij,m}$, $f_{ij,m}$ and $z_{ij,m}$ are the coefficients of mth linear inequality constraint of DCHP caused by feasible operating region; $p_{ij,t}^{chp}$ and $h_{ij,t}^{chp}$ are the power and heat outputs of jth DCHP of ith EB; and $p_{ij,t}^{chp,ramp}$ is the ramp rate limit. If the DCHP unit is fed by nature gas, the total gas consumption can be approximatively evaluated by

$$gas_{ij,t}^{chp} = \Theta\left(\left(p_{ij,t}^{chp} + h_{ij,t}^{chp}\right)/\eta_{ij,t}^{chp}\right), \tag{9.15}$$

where $\eta_{ij,t}^{chp}$ is the total efficient of the DCHP. $gas_{ij,t}^{chp,max}$ represents its gas transmission maximum capability.

(4) DPSD and DHSD models: We denote $p_{ij,t}^{\text{be}}$ and $SOC_{ij,t}^{\text{be}}$ as the exchanged power and the stored energy of jth DPSD of ith EB at time t, respectively. Therein, we let $p_{ij,t}^{\text{be}}$ be positive for discharging and negative for charging actions. Each DPSD, which cannot be charged and discharged simultaneously, should meet the following dynamic constraints [26, 27].

$$-p_{ij,t}^{\text{ch},\max} \leq p_{ij,t}^{\text{be}} \leq p_{ij,t}^{\text{ds},\max} \tag{9.16}$$

$$SOC_{ij,t}^{\min} \le SOC_{ij,t}^{\mathrm{be}} \le SOC_{ij,t}^{\max} \tag{9.17}$$

$$SOC_{ij,t}^{\mathrm{be}} = SOC_{ij,t-1}^{\mathrm{be}} - (\varsigma_{ij}^{\mathrm{ch}}\delta_{ij,t-1}^{\mathrm{ch}} + \frac{1}{\varsigma_{ij}^{\mathrm{ds}}}\delta_{ij,t-1}^{\mathrm{ds}})p_{ij,t-1}^{\mathrm{be}}T \tag{9.18}$$

$$\delta_{ij,t-1}^{\mathrm{ch}} + \delta_{ij,t-1}^{\mathrm{ds}} \le 1, \tag{9.19}$$

where $p_{ij,t}^{\mathrm{ch,max}}$ and $p_{ij,t}^{\mathrm{ds,max}}$ are the maximum charging and discharging rates; $SOC_{ij,t}^{\min}$ and $SOC_{ij,t}^{\max}$ are the lower and upper bounds for allowed energy stored in the corresponding DPSD; $\varsigma_{ij}^{\mathrm{ch}}$ and $\varsigma_{ij}^{\mathrm{ds}}$ are charging and discharging coefficients; $\delta_{ij,t-1}^{\mathrm{ch}}$, $\delta_{ij,t-1}^{\mathrm{ds}} \in \{0, 1\}$ denote the operation state of the corresponding DPSD determined by the previous scheduling horizon. Therein, $\delta_{ij,t-1}^{\mathrm{ch}} = 1$ or $\delta_{ij,t-1}^{\mathrm{ds}} = 1$ represents charging or discharging state.

During each scheduling horizon, each DPSD shall be charged when the buying price is cheaper and vice versa. According to [21], the following cost function is used to capture the operations as follows

$$C\left(p_{ij,t}^{\mathrm{be}}\right) = a_{ij}(p_{ij,t}^{\mathrm{be}} + b_{ij})^2, \tag{9.20}$$

where a_{ij} and b_{ij} are the cost coefficients. In addition, the model above can also be applied for DHSD. And let $h_{ij,t}^{be}$ and $C(h_{ij,t}^{be})$ denote the corresponding heat output and cost function, respectively.

(5) EL models: During each scheduling horizon, it is assumed that the amount of equivalent must-run power, heat and gas loads, denoted by $p_{ij,t}^{mrl}$, $h_{ij,t}^{mrl}$ and $g_{ij,t}^{mrl}$, are fixed. The utility function of EL demand is the sum of three quadratic functions of power, heat and gas demands as follows

$$U_{i,t} = U\left(p_{ij,t}^{cl}\right) + U\left(h_{ij,t}^{cl}\right) + U\left(g_{ij,t}^{cl}\right) \tag{9.21}$$

$$\begin{cases} 0 \le p_{ij,t}^{cl} \le p_{ij,t}^{l,\max} - p_{ij,t}^{mrl} \\ 0 \le h_{ij,t}^{cl} \le h_{ij,t}^{l,\max} - h_{ij,t}^{mrl} \\ 0 \le g_{ij,t}^{cl} \le g_{ij,t}^{l,\max} - g_{ij,t}^{mrl} \end{cases}, \tag{9.22}$$

where

$$\begin{cases} U\left(p_{ij,t}^{cl}\right) = a_{ij}^{p}\left(p_{ij,t}^{cl} + p_{ij,t}^{mrl}\right)^2 + b_{ij}^{p}\left(p_{ij,t}^{cl} + p_{ij,t}^{mrl}\right) \\ U\left(h_{ij,t}^{cl}\right) = a_{ij}^{h}\left(h_{ij,t}^{cl} + h_{ij,t}^{mrl}\right)^2 + b_{ij}^{h}\left(h_{ij,t}^{cl} + h_{ij,t}^{mrl}\right) \\ U\left(g_{ij,t}^{cl}\right) = a_{ij}^{g}\left(g_{ij,t}^{cl} + g_{ij,t}^{mrl}\right)^2 + b_{ij}^{g}\left(g_{ij,t}^{cl} + g_{ij,t}^{mrl}\right) \end{cases},$$

$p_{ij,t}^{cl}$, $h_{ij,t}^{cl}$ and $g_{ij,t}^{cl}$ represent controllable power, heat and gas load demands, respectively. a_{ij}^{p}, a_{ij}^{h} and a_{ij}^{g} are negative utility coefficients while b_{ij}^{p}, b_{ij}^{h} and b_{ij}^{g} are positive utility coefficients. $p_{ij,t}^{l,\max}$, $h_{ij,t}^{l,\max}$ and $g_{ij,t}^{l,\max}$ are upper bounds of power, heat and gas loads, respectively.

It is worth noting that some energy load demands can be satisfied by various forms of energy supplies. In other words, power, heat and gas load demands can be transformed into each other in some circumstances. Inspired by [28], the concept above can be mathematically expressed as

$$\begin{cases} \phi_{ij,g\ to\ p}^{\min} \le p_{ij,t}^{cl} / \left(p_{ij,t}^{cl} + g_{ij,t}^{cl} / \Theta \right) \le \phi_{ij,g\ to\ p}^{\max} \\ \phi_{ij,h\ to\ p}^{\min} \le p_{ij,t}^{cl} / \left(p_{ij,t}^{cl} + h_{ij,t}^{cl} \right) \le \phi_{ij,h\ to\ p}^{\max} \\ \phi_{ij,g\ to\ h}^{\min} \le h_{ij,t}^{cl} / \left(h_{ij,t}^{cl} + g_{ij,t}^{cl} / \Theta \right) \le \phi_{ij,g\ to\ h}^{\max} \end{cases}, \tag{9.23}$$

where $\phi_{ij,g\ to\ p}^{\min}$, $\phi_{ij,h\ to\ p}^{\min}$ and $\phi_{ij,g\ to\ h}^{\min}$, and $\phi_{ij,g\ to\ p}^{\max}$, $\phi_{ij,h\ to\ p}^{\max}$ and $\phi_{ij,g\ to\ h}^{\max}$ express the corresponding lower and upper translating percentages.

9.2.3 Energy Management of EI

In this chapter, we mainly focus on the hourly EMP of EI to achieve the co-planning of power, heat and gas. The objective function is to cooperatively maximize the social welfare shown in equation (9.24), while meeting all of the inequality constraints mentioned above and three global equality constraints shown in (9.25).

$$\max \quad F = \sum_{i=1}^{n} (W_{i,t}) \tag{9.24}$$

$$\sum_{i=1}^{n} \Delta p_{i,t} = 0; \quad \sum_{i=1}^{n} \Delta h_{i,t} = 0; \quad \sum_{i=1}^{n} \Delta g_{i,t} = 0, \tag{9.25}$$

where

$$\begin{cases} W_{i,t} = U_{i,t} - C_{i,t}(\bullet) + pr_t \Delta p_{i,t} + hr_t \Delta h_{i,t} + gr_t \Delta g_{i,t} \\ C_{i,t}(\bullet) = C\left(p_{ij,t}^{re}\right) + C\left(p_{ij,t}^{fe}\right) + C\left(h_{ij,t}^{re}\right) + C\left(h_{ij,t}^{fe}\right) \\ +C\left(p_{ij,t}^{be}\right) + C\left(h_{ij,t}^{be}\right) + C\left(p_{ij,t}^{chp}, h_{ij,t}^{chp}\right) + C\left(g_{ij,t}^{g}\right) \\ \Delta p_{i,t} = p_{ij,t}^{re} + p_{ij,t}^{fe} + p_{ij,t}^{be} + p_{ij,t}^{chp} - p_{ij,t}^{mrl} - p_{ij,t}^{cl} \\ \Delta h_{i,t} = h_{ij,t}^{re} + h_{ij,t}^{fe} + h_{ij,t}^{be} + h_{ij,t}^{chp} - h_{ij,t}^{mrl} - h_{ij,t}^{cl} \\ \Delta g_{i,t} = g_{ij,t}^{g} - g_{ij,t}^{mrl} - g_{ij,t}^{cl} \end{cases},$$

$W_{i,t}$ is the benefit function of ith EB. $\Delta p_{i,t}$, $\Delta h_{i,t}$ and $\Delta g_{i,t}$ are the power, heat and gas supply-demand mismatches of ith EB, respectively. pr_t, hr_t and gr_t are the market power, heat and gas clearing price at time t, respectively. Note that the ith

EB serves as a power, heat or gas supplier when $\Delta p_{i,t}$, $\Delta h_{i,t}$ or $\Delta g_{i,t}$ is positive and vice versa.

Remark 2 The coupling relationships among power, heat and gas of EI are embodied in three aspects: (1) For some energy conversion devices, the natural gas is the clean fuel for power and heat generation; (2) For DCHPs, the power and heat are cooperatively generated; meanwhile, they also should run in the corresponding feasible operating region; (3) Power, heat and gas demands can be transformed into each other in some situations.

For convenience, let $p_{ij,t}$ $(h_{ij,t})$ represent the power (heat) generation of DRG (DRHD), DFG (DFHD), DPSD (DHSD), DCHP or controllable power (heat) load demand of ith EB, and let $g_{ij,t}$ represent the gas supply of DGP or controllable gas load demand of ith EB. The number of decision variables of different participant may be different, e.g., DFG, DCHP and EL have one, two and three decision variables, respectively. Without loss of generality, we let $\boldsymbol{e}_{ij,t} = \left[p_{ij,t}, h_{ij,t}, g_{ij,t}\right]^T$ be a three-dimensional decision variable. Then, each participant of each EB can extend its variable dimensions to the form of $\boldsymbol{e}_{ij,t}$. Furthermore, let $\boldsymbol{q}_{i,t} = \left[p_{ij,t}^{mrl}, h_{ij,t}^{mrl}, g_{ij,t}^{mrl}\right]^T$ and $\boldsymbol{Er}_t = [pr_t, hr_t, gr_t]^T$. We rewrite $V\left(\boldsymbol{e}_{ij,t}\right)$ as cost function or negative utility function of each participant, i.e., $V\left(\boldsymbol{e}_{ij,t}\right) = C_{i,t}\left(\bullet\right)$ or $V\left(\boldsymbol{e}_{ij,t}\right) = U_{i,t}$. Then, by some operations, the EM of EI can be re-written as

$$
\min\ F = \sum_{i=1}^{n}\sum_{j=1}^{m_i} V\left(\boldsymbol{e}_{ij,t}\right)
$$

$$
s.t.\ \begin{cases} \sum\limits_{i=1}^{n}\sum\limits_{j=1}^{m_i} \boldsymbol{Y}_{ij}\boldsymbol{e}_{ij,t} = \boldsymbol{Q} \\ \Phi\left(\boldsymbol{e}_{ij,t}\right) \le 0,\ \boldsymbol{e}_{ij,t} \in R^3 \Leftrightarrow \boldsymbol{e}_{ij,t} \in \boldsymbol{X}_{ij} \end{cases}, \tag{9.26}
$$

where $\boldsymbol{Q} = \sum_{i=1}^{n} \boldsymbol{q}_{i,t}$ represents the sum of must-run energy demands; m_i is the number of participates of ith EB; $\boldsymbol{X}_{ij}$, determined by $\Phi\left(\boldsymbol{e}_{ij,t}\right)$, is the local closed convex set for $\boldsymbol{e}_{ij,t}$. $\boldsymbol{Y}_{ij} = -\boldsymbol{I}$ if $\boldsymbol{e}_{ij,t}$ represents the controllable energy load demand; otherwise, $\boldsymbol{Y}_{ij} = \boldsymbol{I}$. $\Phi\left(\boldsymbol{e}_{ij,t}\right)$ represents the local inequality constraints related to $\boldsymbol{e}_{ij,t}$. Besides $\boldsymbol{e}_{ij,t}$, $\boldsymbol{Er}_t$ is also an important variable which is expected to be calculated in a distributed fashion. Although, it is not included in problem (9.26) caused by the three global equality constraints (9.25), we can also find its optimal solution by implementing the proposed algorithm, which will be discussed in Remark 4.

9.2.4 Short-Term Power Scheduling Adjustment

Note that the variations of power loads and the fluctuations of DRGs are always seen as 5-min or 15-min, while the heat and gas loads variations are always seen as being hourly [28–30]. Thus, on the basis of the hourly EMP of EI, an adjustment strategy for power-only devices and controllable power loads is further developed for adapting to the shorter power scheduling problem, i.e., 5-min considered in this chapter. Then, the 5-min power loads variations and fluctuations of DRGs can be accommodated by dispatching the power generations of DFGs, DPSDs and controllable power loads. In this chapter, the goal is to make the 5-min power scheduled result of each power participant follows its corresponding hourly scheduled result as close as possible. Meanwhile, in general, the larger the capacity upper is, the heavier task the participant has to smooth out the variations and fluctuations. This concept above can be mathematically expressed as

$$\min Obj = \sum_{i=1}^{n} \Big(\frac{1}{p_{ij,t}^{fe,\max}} (p_{ij,t-\hbar}^{fe} - p_{ij,t}^{fe})^2 + \varpi_{ij} (p_{ij,t-\hbar}^{be} - p_{ij,t}^{be})^2 + \frac{1}{p_{ij,t}^{l,\max}} (p_{ij,t-\hbar}^{cl} - p_{ij,t}^{cl})^2 \Big), \quad (9.27)$$

where $\varpi_{ij} = \min\{\frac{1}{p_{ij,t}^{ch,max}}, \frac{1}{p_{ij,t}^{ds,max}}\}$; the subscript "$_{t-\hbar}$" is utilized to represent the $\hbar$th 5-min interval of tth hourly interval. For example, $p_{ij,t-\hbar}^{fe}$ is the $\hbar$th 5-min power output of jth DFG of ith EB during tth hour, and the similar definitions for $p_{ij,t-\hbar}^{be}$, $p_{ij,t-\hbar}^{re}$, $p_{ij,t-\hbar}^{chp}$, $p_{ij,t-\hbar}^{mrl}$ and $p_{ij,tm}^{cl}$. In addition, problem (9.27) should satisfy the 5-min power supply-demand balance constraint given by

$$\sum_{i=1}^{n} (p_{ij,t-\hbar}^{re} + p_{ij,t-\hbar}^{fe} + p_{ij,t-\hbar}^{be} + p_{ij,t-\hbar}^{chp} - p_{ij,t-\hbar}^{mrl} - p_{ij,t-\hbar}^{cl}) = 0. \quad (9.28)$$

Since DCHPs generate not only power but also heat, they are not employed to participate in the 5-min power scheduling. Thus, the value of $p_{ij,t-\hbar}^{chp}$ is the same as $p_{ij,t}^{chp}$ determined by its corresponding hourly scheduled result. Moreover, all the DFGs, DPSDs and controllable power loads also should meet their 5-min timescale operation constraints which are similar to the corresponding hourly ones.

9.3 Distributed Algorithm

9.3.1 Graph Theory

Consider an EI with n EBs where each EB has m_i participants. A graph $G = (\mathscr{V}, \mathscr{E})$ is used to model the EI, where $\mathscr{V} = \{v_{ij} \mid i = 1, \ldots, n; j = 1, \ldots, m_i\}$ is a set of nodes representing participants and $\mathscr{E} \subset \mathscr{V} \times \mathscr{V}$ is a set of edges. Therein, the edge

$(v_{ij}, v_{\bar{i}j})$ denotes that ijth node and $\bar{i}j$th node can exchange information between each other as needed. The set of neighbors of ijth node is denoted by N_{ij} with cardinality $|N_{ij}|$. In this chapter, we assume that graph G is connected.

9.3.2 Main Algorithm

In this chapter, we consider that problem (9.26) is solvable, i.e., there exist a set of values $\{\boldsymbol{e}_{ij,t} \mid i = 1, \ldots, n;\ j = 1, \ldots, m_i\}$ satisfying all the equality and inequality constraints. Moreover, it is worth noting that (9.26) is a convex optimization problem and $\Phi\left(\boldsymbol{e}_{ij,t}\right)$ in (9.26) is local and affine function. Thus, the refined Salter condition is satisfied [31] (pp. 226–227). That also means (9.26) has a zero duality gap. As a consequence, the primal problem and its dual problem hold the same optimal solution which is the saddle point of the Lagrangian [32]. Further, in order to solve problem (9.26), we can transform it into its dual problem given by

$$\min \sum_{i=1}^{n}\sum_{j=1}^{m_i}\left[\Gamma\left(\boldsymbol{\lambda}\right) + \boldsymbol{\lambda}^T \boldsymbol{\kappa}^*\right], \tag{9.29}$$

where $\Gamma\left(\boldsymbol{\lambda}\right) = \max\limits_{\boldsymbol{e}_{ij,t} \in \boldsymbol{X}_{ij}} \left\{-V\left(\boldsymbol{e}_{ij,t}\right) - \boldsymbol{\lambda}^T \boldsymbol{Y}_{ij} \boldsymbol{e}_{ij,t}\right\}$, $\boldsymbol{\kappa}^* = \boldsymbol{Q} / \sum_{i=1}^{n} m_i$, $\boldsymbol{\lambda} = [\lambda_1, \lambda_2, \lambda_3]^T$ is the Lagrange dual variable, and λ_1, λ_2 and λ_3 represent the power, heat and gas dual variables, respectively. Note that $\boldsymbol{\lambda}$ is global variable, which is not suitable for distributed implementation. To solve (9.29) in a distributed fashion, a set of auxiliary variables are added into (9.29) which is rewritten as

$$\begin{aligned} &\min \sum_{i=1}^{n}\sum_{j=1}^{m_i}\left[\Gamma\left(\boldsymbol{\lambda}_{ij}\right) + \boldsymbol{\lambda}_{ij}^T \boldsymbol{\kappa}^*\right] \\ &s.t.\ \boldsymbol{\lambda}_{ij} = \varphi_{ij,\bar{i}j};\ \boldsymbol{\lambda}_{\bar{i}j} = \varphi_{ij,\bar{i}j} \quad \forall \bar{i}j \in N_{ij}; \forall ij \in \mathscr{V}, \end{aligned} \tag{9.30}$$

where $\Gamma\left(\boldsymbol{\lambda}_{ij}\right) = \max\limits_{\boldsymbol{e}_{ij,t} \in \boldsymbol{X}_{ij}} \left\{-V\left(\boldsymbol{e}_{ij,t}\right) - \boldsymbol{\lambda}_{ij}^T \boldsymbol{Y}_{ij} \boldsymbol{e}_{ij,t}\right\}$; $\boldsymbol{\lambda}_{ij}$ is local estimated dual variable. Equations (9.29) and (9.30) are equivalent optimization problems in essence. The reason is that, for each pair of neighbors, we add an auxiliary variable and let $\boldsymbol{\lambda}_{ij} = \varphi_{ij,\bar{i}j} = \boldsymbol{\lambda}_{\bar{i}j}$, such that, over connected graph, all $\boldsymbol{\lambda}_{ij}$ must be equal to the same value which is the $\boldsymbol{\lambda}$ in (9.29). In this way, (9.30) holds a dissociable structure. Then, the quadratically augmented Lagrangian function of (9.30) is given by

$$L\left[\{\boldsymbol{\lambda}_{ij}\}, \{\varphi_{ij,\bar{i}j}\}, \boldsymbol{\mu}\right] = \sum_{i=1}^{n}\sum_{j=1}^{m_i}\Big\{\left(\Gamma\left(\boldsymbol{\lambda}_{ij}\right) + \boldsymbol{\lambda}_{ij}^T \boldsymbol{\kappa}^*\right)$$

$$+ \sum_{\bar{i}\bar{j}\in N_{ij}} \left[\boldsymbol{\mu}_{ij,\bar{i}\bar{j}}^{T}\left(\boldsymbol{\lambda}_{ij}-\boldsymbol{\varphi}_{ij,\bar{i}\bar{j}}\right)+\boldsymbol{\mu}_{\bar{i}\bar{j},ij}^{T}\left(\boldsymbol{\lambda}_{\bar{i}\bar{j}}-\boldsymbol{\varphi}_{ij,\bar{i}\bar{j}}\right)\right.$$
$$\left.\left. + \frac{c}{2}\left(\left\|\boldsymbol{\lambda}_{ij}-\boldsymbol{\varphi}_{ij,\bar{i}\bar{j}}\right\|_2^2+\left\|\boldsymbol{\lambda}_{\bar{i}\bar{j}}-\boldsymbol{\varphi}_{ij,\bar{i}\bar{j}}\right\|_2^2\right)\right]\right\}, \tag{9.31}$$

where $\boldsymbol{\mu} := \left\{\left\{\boldsymbol{\mu}_{ij,\bar{i}\bar{j}}\right\},\left\{\boldsymbol{\mu}_{\bar{i}\bar{j},ij}\right\}\right\}$, $\boldsymbol{\mu}_{ij,\bar{i}\bar{j}}$ and $\boldsymbol{\mu}_{\bar{i}\bar{j},ij}$ are corresponding Lagrange multipliers.

We first assume that $\Gamma\left(\boldsymbol{\lambda}_{ij}\right)$ can be calculated, then the following distributed algorithm is proposed to obtain the optimal $\boldsymbol{\lambda}_{ij}$ of problem (9.30), i.e.,

$$\boldsymbol{v}_{ij}(k) = \boldsymbol{v}_{ij}(k-1) + c\sum_{\bar{i}\bar{j}\in N_{ij}}\left[\boldsymbol{\lambda}_{ij}(k)-\boldsymbol{\lambda}_{\bar{i}\bar{j}}(k)\right] \tag{9.32}$$

$$\boldsymbol{\kappa}_{ij}(k+1) = \sum_{i=1}^{n}\sum_{j=1}^{m_i} w_{ij,\bar{i}\bar{j}}\boldsymbol{\kappa}_{\bar{i}\bar{j}}(k) \tag{9.33}$$

$$\boldsymbol{\lambda}_{ij}(k+1) = \arg\min_{\boldsymbol{\lambda}_{ij}}\left\{\Gamma\left(\boldsymbol{\lambda}_{ij}\right)+\boldsymbol{\lambda}_{ij}^{T}\boldsymbol{\kappa}_{ij}(k+1)+\boldsymbol{v}_{ij}^{T}(k)\boldsymbol{\lambda}_{ij}\right.$$
$$\left. +c\sum_{\bar{i}\bar{j}\in N_{ij}}\left\|\boldsymbol{\lambda}_{ij}-\left[\boldsymbol{\lambda}_{ij}(k)+\boldsymbol{\lambda}_{\bar{i}\bar{j}}(k)\right]/2\right\|_2^2\right\} \tag{9.34}$$

where c is a constant for the algorithm; $\boldsymbol{v}_{ij}(k) = 2\sum_{\bar{i}\bar{j}\in N_{ij}}\boldsymbol{\mu}_{ij,\bar{i}\bar{j}}(k)$; $w_{ij,\bar{i}\bar{j}} = 2/\left(\left|N_{ij}\right|+\left|N_{\bar{i}\bar{j}}\right|+\Lambda\right)$ if $\bar{i}\bar{j}\in N_{ij}$, $w_{ij,\bar{i}\bar{j}} = 1-\sum_{\bar{i}\bar{j}\in N_{ij}}\left(2/\left(\left|N_{ij}\right|+\left|N_{\bar{i}\bar{j}}\right|+\Lambda\right)\right)$ if $\bar{i}\bar{j} = ij$, otherwise, $w_{ij,\bar{i}\bar{j}} = 0$ [21]. Therein, Λ is a small number. The initial values are set as $\boldsymbol{\mu}_{ij,\bar{i}\bar{j}}(-1) = \boldsymbol{\mu}_{\bar{i}\bar{j},ij}(-1) = \boldsymbol{v}_{ij}(-1) = \boldsymbol{\lambda}_{ij}(0) = \mathbf{0}$ and $\boldsymbol{\kappa}_{ij}(0) = \boldsymbol{q}_{i,t}/m_i$. In this way, $\boldsymbol{\mu}_{ij,\bar{i}\bar{j}}(k) = -\boldsymbol{\mu}_{\bar{i}\bar{j},ij}(k)$ is always satisfied, and then we can get the following fact which will be used for the convergence analysis,

$$\sum_{i=1}^{n}\sum_{j=1}^{m_i}\boldsymbol{v}_{ij}(k) = \sum_{i=1}^{n}\sum_{j=1}^{m_i}\sum_{\bar{i}\bar{j}\in N_{ij}}\left(\boldsymbol{\mu}_{ij,\bar{i}\bar{j}}(k)-\boldsymbol{\mu}_{\bar{i}\bar{j},ij}(k)\right) = 0.$$

Theorem 1 *We assume the communication graph G is connected and $\Gamma\left(\boldsymbol{\lambda}_{ij}\right)$ can be calculated. Then for all $ij\in\mathscr{V}$, staring form the corresponding initial conditions mentioned above, the algorithm composed of* (9.32), (9.33) *and* (9.34) *can make $\boldsymbol{\lambda}_{ij}(k)\to\boldsymbol{\lambda}^{*}$ as $k\to\infty$.*

Proof According to the definition of $w_{ij,\bar{i}\bar{j}}$, (9.33) makes up an average-consensus protocol with $\boldsymbol{\kappa}_{ij}(0) = \boldsymbol{q}_{i,t}/m_i$. Then, after a sufficient long time K, we can get

$$\begin{aligned}\boldsymbol{\kappa}_{ij}(K) &= \sum_{i}^{n}\sum_{j}^{m_i}\boldsymbol{\kappa}_{ij}(0)/\sum_{i}^{n} m_i\\ &= \sum_{i}^{n}\left(m_i\boldsymbol{q}_{i,t}/m_i\right)/\sum_{i}^{n} m_i = \boldsymbol{\kappa}^*,\end{aligned} \tag{9.35}$$

which implies that if we construct a variable $\boldsymbol{\kappa}_{ij}$ for each participant, then, starting from $\boldsymbol{\kappa}_{ij}(0) = \boldsymbol{q}_{i,t}/m_i$, $\boldsymbol{\kappa}_{ij}(k)$ can converge to $\boldsymbol{\kappa}^*$ by implementing (9.33). Note that each participant can easily access $\boldsymbol{q}_{i,t}/m_i$.

Now, for (9.31), ADMM results in the following updating iterative process

(1) Step 1: Updating of local estimate

$$\begin{aligned}\boldsymbol{\lambda}_{ij}(k+1) &= \arg\min_{\boldsymbol{\lambda}_{ij}} L\left[\{\boldsymbol{\lambda}_{ij}\}, \{\boldsymbol{\varphi}_{ij,\bar{i}\bar{j}}(k)\}, \boldsymbol{\mu}(k)\right]\\ &= \arg\min_{\boldsymbol{\lambda}_{ij}}\Bigg\{\Gamma\left(\boldsymbol{\lambda}_{ij}\right) + \boldsymbol{\lambda}_{ij}^T\boldsymbol{\kappa}_{ij}(k+1)\\ &\quad + 2\sum_{\bar{i}\bar{j}\in N_{ij}}\boldsymbol{\mu}_{ij,\bar{i}\bar{j}}^T(k)\,\boldsymbol{\lambda}_{ij}\\ &\quad + c\sum_{\bar{i}\bar{j}\in N_{ij}}\left\|\boldsymbol{\lambda}_{ij} - \boldsymbol{\varphi}_{ij,\bar{i}\bar{j}}(k)\right\|_2^2\Bigg\}.\end{aligned} \tag{9.36}$$

The reason for obtaining the above equation is that participant $\bar{i}\bar{j}$ is the neighbor of participant ij; meanwhile, participant ij is also the neighbor of $\bar{i}\bar{j}$. Moreover, we use $\boldsymbol{\kappa}_{ij}(k+1)$ computed by (9.33) to replace $\boldsymbol{\kappa}^*$, since each participant can not directly access $\boldsymbol{\kappa}^*$.

(2) Step 2: Updating of auxiliary

$$\begin{aligned}\boldsymbol{\varphi}_{ij,\bar{i}\bar{j}}(k+1) &= \arg\min_{\boldsymbol{\varphi}_{ij,\bar{i}\bar{j}}} L\left[\{\boldsymbol{\lambda}_{ij}(k+1)\}, \{\boldsymbol{\varphi}_{ij,\bar{i}\bar{j}}\}, \boldsymbol{\mu}(k)\right]\\ &= \arg\min_{\boldsymbol{\varphi}_{ij,\bar{i}\bar{j}}}\Bigg\{-(\boldsymbol{\mu}_{ij,\bar{i}\bar{j}}^T(k) + \boldsymbol{\mu}_{\bar{i}\bar{j},ij}^T(k))\boldsymbol{\varphi}_{ij,\bar{i}\bar{j}}\\ &\quad + \frac{c}{2}\Big(\left\|\boldsymbol{\lambda}_{ij}(k+1) - \boldsymbol{\varphi}_{ij,\bar{i}\bar{j}}\right\|_2^2\\ &\quad + \left\|\boldsymbol{\lambda}_{\bar{i}\bar{j}}(k+1) - \boldsymbol{\varphi}_{ij,\bar{i}\bar{j}}\right\|_2^2\Big)\Bigg\}\\ &= \frac{1}{2c}(\boldsymbol{\mu}_{ij,\bar{i}\bar{j}}(k) + \boldsymbol{u}_{\bar{i}\bar{j},ij}(k))\\ &\quad + \frac{1}{2}(\boldsymbol{\lambda}_{ij}(k+1) + \boldsymbol{\lambda}_{\bar{i}\bar{j}}(k+1)).\end{aligned} \tag{9.37}$$

(3) Step 3: Updating of Lagrange multiplier

$$\begin{aligned}\boldsymbol{\mu}_{ij,\bar{ij}}(k+1) =& \boldsymbol{\mu}_{ij,\bar{ij}}(k) \\ &+ c(\boldsymbol{\lambda}_{ij}(k+1) - \boldsymbol{\varphi}_{ij,\bar{ij}}(k+1)) \\ \boldsymbol{\mu}_{\bar{ij},ij}(k+1) =& \boldsymbol{\mu}_{\bar{ij},ij}(k) \\ &+ c(\boldsymbol{\lambda}_{\bar{ij}}(k+1) - \boldsymbol{\varphi}_{ij,\bar{ij}}(k+1)).\end{aligned} \tag{9.38}$$

The next is to verify that (9.33) and (9.36)–(9.38) can be simplified to (9.32)–(9.34). Note that, for (9.38), if the initial values of $\boldsymbol{\mu}_{ij,\bar{ij}}$ and $\boldsymbol{\mu}_{\bar{ij},ij}$ are set to zero, then $\boldsymbol{\mu}_{ij,\bar{ij}}(k+1) = -\boldsymbol{\mu}_{\bar{ij},ij}(k+1)$ is always satisfied for all $k \geq 0$. Thus, (9.37) can be further simplified to

$$\boldsymbol{\varphi}_{ij,\bar{ij}}(k+1) = \frac{1}{2}(\boldsymbol{\lambda}_{ij}(k+1) + \boldsymbol{\lambda}_{\bar{ij}}(k+1)). \tag{9.39}$$

By substituting (9.39) in (9.38), $\boldsymbol{\mu}_{ij,\bar{ij}}(k+1)$ can be re-written as the following form

$$\begin{aligned}\boldsymbol{\mu}_{ij,\bar{ij}}(k+1) =& \boldsymbol{\mu}_{ij,\bar{ij}}(k) \\ &+ \frac{c}{2}[\boldsymbol{\lambda}_{ij}(k+1) - \boldsymbol{\lambda}_{\bar{ij}}(k+1)].\end{aligned} \tag{9.40}$$

Furthermore, let $\boldsymbol{v}_{ij}(k) = 2\sum_{\bar{ij}\in N_{ij}} \boldsymbol{\mu}_{ij,\bar{ij}}(k)$. Then, according to (9.40), we can get

$$\begin{aligned}\boldsymbol{v}_{ij}(k) &= 2\sum_{\bar{ij}\in N_{ij}} \boldsymbol{\mu}_{ij,\bar{ij}}(k-1) + c\sum_{\bar{ij}\in N_{ij}} \left[\boldsymbol{\lambda}_{ij}(k) - \boldsymbol{\lambda}_{\bar{ij}}(k)\right] \\ &= \boldsymbol{v}_{ij}(k-1) + c\sum_{\bar{ij}\in N_{ij}} \left[\boldsymbol{\lambda}_{ij}(k) - \boldsymbol{\lambda}_{\bar{ij}}(k)\right].\end{aligned} \tag{9.41}$$

Finally, by plugging (9.39) and $\boldsymbol{v}_{ij}(k) = 2\sum_{\bar{ij}\in N_{ij}} \boldsymbol{\mu}_{ij,\bar{ij}}(k)$ in (9.36), $\boldsymbol{\lambda}_{ij}(k+1)$ is simplified to

$$\begin{aligned}\boldsymbol{\lambda}_{ij}(k+1) = \arg\min_{\boldsymbol{\lambda}_{ij}} \Bigg\{ & \Gamma\left(\boldsymbol{\lambda}_{ij}\right) + \boldsymbol{\lambda}_{ij}^T \boldsymbol{\kappa}_{ij}(k+1) + \boldsymbol{v}_{ij}^T(k)\,\boldsymbol{\lambda}_{ij} \\ &+ c\sum_{\bar{ij}\in N_{ij}} \left\|\boldsymbol{\lambda}_{ij} - \left[\boldsymbol{\lambda}_{ij}(k) + \boldsymbol{\lambda}_{\bar{ij}}(k)\right]/2\right\|_2^2 \Bigg\}.\end{aligned} \tag{9.42}$$

It can be seen that (9.41) and (9.42) are the same as (9.32) and (9.34), respectively. The proof is completed.

However, in Theorem 1, it is very difficult to directly calculate $\boldsymbol{\lambda}_{ij}(k+1)$ of (9.34) in fact, because $\Gamma\left(\boldsymbol{\lambda}_{ij}\right)$ is a maximum subproblem determined by $\boldsymbol{e}_{ij,t}$. Taking the explicit form of $\boldsymbol{\lambda}_{ij}(k+1)$ into account, (9.34) is rewritten as

$$\begin{aligned}
\boldsymbol{\lambda}_{ij}(k+1) &= \arg\min_{\boldsymbol{\lambda}_{ij}} \max_{\boldsymbol{e}_{ij,t}\in\boldsymbol{X}_{ij}} \Big\{-V\left(\boldsymbol{e}_{ij,t}\right) \\
&- \boldsymbol{\lambda}_{ij}^{T}\boldsymbol{Y}_{ij}\boldsymbol{e}_{ij,t} + \boldsymbol{\lambda}_{ij}^{T}\boldsymbol{\kappa}_{ij}(k+1) + \boldsymbol{v}_{ij}^{T}(k)\boldsymbol{\lambda}_{ij} \\
&+ c\sum_{\bar{i}\bar{j}\in N_{ij}} \left\|\boldsymbol{\lambda}_{ij} - \left[\boldsymbol{\lambda}_{ij}(k) + \boldsymbol{\lambda}_{\bar{i}\bar{j}}(k)\right]/2\right\|_2^2\Big\} \\
&= \arg\max_{\boldsymbol{e}_{ij,t}\in\boldsymbol{X}_{ij}} \min_{\boldsymbol{\lambda}_{ij}} \Big\{-V\left(\boldsymbol{e}_{ij,t}\right) \\
&+ c\left|N_{ij}\right| \left\|\boldsymbol{\lambda}_{ij} - \Theta/\left(2\left|N_{ij}\right|\right)\right\|_2^2 - c\left\|\Theta\right\|_2^2/\left(4\left|N_{ij}\right|\right) \\
&+ c\sum_{\bar{i}\bar{j}\in N_{ij}} \left\|\left[\boldsymbol{\lambda}_{ij}(k) + \boldsymbol{\lambda}_{\bar{i}\bar{j}}(k)\right]/2\right\|_2^2\Big\},
\end{aligned} \tag{9.43}$$

where $\Theta = \sum_{\bar{i}\bar{j}\in N_{ij}} \left(\boldsymbol{\lambda}_{ij}(k) + \boldsymbol{\lambda}_{\bar{i}\bar{j}}(k)\right) - \left(\boldsymbol{v}_{ij}(k) - \boldsymbol{Y}_{ij}\boldsymbol{e}_{ij,t} + \boldsymbol{\kappa}_{ij}(k+1)/c\right)$.

The reason for obtaining the above equation is that the objective function is concave for $\boldsymbol{e}_{ij,t}$ when $\boldsymbol{\lambda}_{ij}$ is given; meanwhile, the one is convex for $\boldsymbol{\lambda}_{ij}$ when $\boldsymbol{e}_{ij,t}$ is given. Thus, the min-max problem and its corresponding max-min problem can be transformed into each other. Then, based on the final form of (9.43), the calculation of $\boldsymbol{\lambda}_{ij}$ and $\boldsymbol{e}_{ij,t}$ can be decoupled as follows:

$$\begin{aligned}
\boldsymbol{\lambda}_{ij}(k+1) = \Big[&\sum_{\bar{i}\bar{j}\in N_{ij}} \left(\boldsymbol{\lambda}_{ij}(k) + \boldsymbol{\lambda}_{\bar{i}\bar{j}}(k)\right) - \big(\boldsymbol{v}_{ij}(k) \\
&- \boldsymbol{Y}_{ij}\boldsymbol{e}_{ij,t}(k+1) + \boldsymbol{\kappa}_{ij}(k+1)\big)/c\Big]/(2\left|N_{ij}\right|),
\end{aligned} \tag{9.44}$$

$$\begin{aligned}
\boldsymbol{e}_{ij,t}(k+1) = \arg&\min_{\boldsymbol{e}_{ij,t}\in\boldsymbol{X}_{ij}} \Big\{V\left(\boldsymbol{e}_{ij,t}\right) + c\left\|\Theta\right\|_2^2 \\
&/\left(4\left|N_{ij}\right|\right) - c\sum_{\bar{i}\bar{j}\in N_{ij}} \left\|\left[\boldsymbol{\lambda}_{ij}(k) + \boldsymbol{\lambda}_{\bar{i}\bar{j}}(k)\right]/2\right\|_2^2\Big\}.
\end{aligned} \tag{9.45}$$

Thanks to (9.44) and (9.45), our main algorithm, referred as distributed-consensus-ADMM algorithm, is further proposed and summarized as in Algorithm 1.

Algorithm 1: Distributed-Consensus-ADMM Algorithm

Initialize:
Set $\boldsymbol{v}_{ij}(-1) = \mathbf{0}$, $\boldsymbol{\lambda}(0) = 0$, $\boldsymbol{e}_{ij,t}$ any admissible values and $\boldsymbol{\kappa}_{ij}(0) = \boldsymbol{q}_{i,t}/m_i$ for each participant.
Iteration: $(k \geq 0)$
1) Transmit $\boldsymbol{\lambda}_{ij}(k)$ and $\boldsymbol{\kappa}_{ij}(k)$ to its neighbors
2) Update $\boldsymbol{v}_{ij}(k)$ based on (9.32)
3) Update $\boldsymbol{\kappa}_{ij}(k+1)$ based on (9.33)
4) Update $\boldsymbol{e}_{ij,t}(k+1)$ by solving problem (9.45)
4) Update $\boldsymbol{\lambda}_{ij}(k+1)$ based on (9.44)
5) Let $k = k + 1$, turn to step 1.

Remark 3 The distributed-consensus-ADMM algorithm is a fully distributed algorithm, which only requires the local communication and calculation among nodes without a central controller or a leader. To be specific, each node only needs to exchange the information of $\boldsymbol{\lambda}_{ij}(k)$ and $\boldsymbol{\kappa}_{ij}(k)$ with its neighbors. And the updating of variables $\boldsymbol{v}_{ij}(k)$, $\boldsymbol{\kappa}_{ij}(k+1)$, $\boldsymbol{e}_{ij,t}(k+1)$ and $\boldsymbol{\lambda}_{ij}(k+1)$ are implemented via local calculation based on their own and neighbors information.

Theorem 2 *Let G be a connected graph. If the problem of* (9.26) *is feasible, then the Algorithm 1 is stable, and each* $\boldsymbol{e}_{ij,t}$ *can converge to its optimal point, i.e.,*

$$\boldsymbol{e}_{ij,t}(k) \to \boldsymbol{e}^*_{ij,t}, \quad k \to \infty.$$

Proof In Theorem 1, it has been verified that $\boldsymbol{\lambda}_{ij}$ can converge a common value, i.e., $\boldsymbol{\lambda}^*$. The rest is to verify that $\boldsymbol{e}_{ij,t}(k+1)$ can converge to its corresponding optimal point. To begin with, according to the KKT (Karush-Kuhn-Tucker) condition of (9.45), we can get

$$\begin{aligned} &\lim_{k\to\infty} \Bigg\{\partial V(\boldsymbol{e}_{ij,t}(k+1)) + \boldsymbol{Y}_{ij}^T\Bigg[\sum_{\bar{i}\bar{j}\in N_{ij}} \big(\boldsymbol{\lambda}_{ij}(k) + \boldsymbol{\lambda}_{\bar{i}\bar{j}}(k)\big) \\ &\quad - \big(\boldsymbol{v}_{ij}(k) - \boldsymbol{Y}_{ij}\boldsymbol{e}_{ij,t}(k+1) + \boldsymbol{\kappa}_{ij}(k+1)\big)/c\Bigg] \\ &\quad /(2\left|N_{ij}\right|)\Bigg\}/\partial \boldsymbol{e}_{ij,t}(k+1) = 0, \quad \boldsymbol{e}_{ij,t} \in \boldsymbol{X}_{ij}, \end{aligned} \tag{9.46}$$

$$\begin{aligned} \Rightarrow &\lim_{k\to\infty} \left\{\left[\partial V\left(\boldsymbol{e}_{ij,t}(k+1)\right) + \boldsymbol{Y}_{ij}^T\boldsymbol{\lambda}^*\right]/\partial \boldsymbol{e}_{ij,t}(k+1)\right\} \\ &= \mathbf{0}, \quad \boldsymbol{e}_{ij,t} \in \boldsymbol{X}_{ij}. \end{aligned} \tag{9.47}$$

Moreover, upon summing (9.44), we can get

$$\sum_{i=1}^{n}\sum_{j=1}^{m_i} \boldsymbol{Y}_{ij}\boldsymbol{e}_{ij,t}(k+1) - \sum_{i=1}^{n}\sum_{j=1}^{m_i} \boldsymbol{\kappa}_{ij}(k+1)$$
$$= c\sum_{i=1}^{n}\sum_{j=1}^{m_i}\sum_{\bar{i}\bar{j}\in N_{ij}} \left(2\boldsymbol{\lambda}_{ij}(k+1) - \boldsymbol{\lambda}_{ij}(k) - \boldsymbol{\lambda}_{\bar{i}\bar{j}}(k)\right)$$
$$+ \sum_{i=1}^{n}\sum_{j=1}^{m_i} \boldsymbol{v}_{ij}(k), \quad \boldsymbol{e}_{ij,t} \in \boldsymbol{X}_{ij}.$$

It is worth noting that, as $k \to \infty$, $\boldsymbol{\lambda}_{ij}(k+1) = \boldsymbol{\lambda}_{ij}(k) = \boldsymbol{\lambda}_{\bar{i}\bar{j}}(k) = \boldsymbol{\lambda}^*$, $\boldsymbol{\kappa}_{ij}(k+1) = \boldsymbol{\kappa}^*$ and the sum of $\boldsymbol{v}_{ij}(k)$ is equal to zero mentioned above, then we can obtain

$$\lim_{k\to\infty} \sum_{i=1}^{n}\sum_{j=1}^{m_i} \boldsymbol{Y}_{ij}\boldsymbol{e}_{ij,t}(k+1) - \boldsymbol{Q} = \boldsymbol{0}, \quad \boldsymbol{e}_{ij,t} \in \boldsymbol{X}_{ij}. \tag{9.48}$$

Note that (9.47) and (9.48) are the optimality conditions of (9.26). Thus, we can conclude that $\boldsymbol{e}_{ij,t}(k+1) \to \boldsymbol{e}^*_{ij,t}$ as $k \to \infty$.

The proof is completed.

The proposed distributed-consensus-ADMM algorithm, which can guarantee improved reliability, robustness, flexibility and scalability, and provide better plug-and-play functionality, is a promising option for applying in the future EI. Firstly, the work of each node is based only on the local sharing of information among neighbors to find the optimal operating point, resulting in alleviated single-node congestion. Moreover, the EI can work well even under single and even several communication link failure(s), as long as the communication network maintains connection. Thus, the EI becomes more robust against single-point failure by implementing the proposed algorithm. Secondly, a node can join in the EI by establishing communication links with its local neighbors only. Consequently, it is more flexible for possible integration and expansions of different energy resources. Thirdly, regardless of the scale of the EI, each node only performs a local optimization. It also means that the communication and computation burden can be divided among the distributed processors, thus being able to encourage better scalability. Lastly but not least, distributed implementation fashion can provide desired plug-and-play capability to face the topology variability of the future EI.

It is worth noting that problem (9.27), with all of its equality and inequality constraints, doesn't have complex coupling relationship among variables, which can be seen as a simplified version of problem (9.26). Therefore, the proposed distributed-consensus-ADMM algorithm can also be employed to solve the 5-min power scheduling problem.

Remark 4 In our algorithm, $\boldsymbol{\lambda}$ has important physical significance, that is $\boldsymbol{\lambda}$ is the same as negative energy market clearing price, i.e.,

$$\boldsymbol{\lambda} = -\boldsymbol{Er}_t, \tag{9.49}$$

which implies that each participant can estimate the global energy market clearing price in a distributed fashion because they can distributively calculate system $\boldsymbol{\lambda}$. The proof of (9.49) is as follows: In the interest of clarity, we analyze the case where the local inequality constraints are not taken into account. For a given energy selling (or bidding) price denoted by $\boldsymbol{Er}_{ij,t}$. The estimated profit function of EB i is

$$W_{i,t} = \sum_{j=1}^{m_i} W_{ij,t} = \sum_{j=1}^{m_i} (-V\left(\boldsymbol{e}_{ij,t}\right) + \boldsymbol{Er}_{ij,t}^T (\boldsymbol{Y}_{ij}\boldsymbol{e}_{ij,t} - \boldsymbol{q}_{i,t})).$$

The corresponding maximum profit can be calculated by

$$\frac{\partial W_{ij,t}}{\partial \boldsymbol{e}_{ij,t}} = -\frac{\partial V\left(\boldsymbol{e}_{ij,t}\right)}{\partial \boldsymbol{e}_{ij,t}} + \boldsymbol{Er}_{ij,t}^T \boldsymbol{Y}_{ij} = \boldsymbol{0}. \tag{9.50}$$

To obtain maximum profit, each participant tends to adjust its energy outputs (or energy demands) until the marginal cost (or utility) is equal to $\boldsymbol{Er}_{ij,t}$. Stimulated by energy supplier-demand mismatches, each participant will adjust its selling (or bidding) to meet the supplier-demand balance. For example, a supplier is willing to increase its selling price and produce more energy for deficit market energy demand. We consider that the market is comparatively competitive. Thus, the market is cleared when all the selling and bidding price are equal to the same value, i.e. $\boldsymbol{Er}_{ij,t} = \boldsymbol{Er}_t$ for all participates. And $\boldsymbol{Er}_t$, i.e., so-called energy market clearing price, is determined by the conditions

$$\begin{cases} \dfrac{\partial V\left(\boldsymbol{e}_{ij,t}\right)}{\partial \boldsymbol{e}_{ij,t}} - \boldsymbol{Er}^T \boldsymbol{Y}_{ij} = \boldsymbol{0} \\ \sum_{i=1}^{n} \sum_{j=1}^{m_i} \boldsymbol{Y}_{ij}\boldsymbol{e}_{ij,t} = \boldsymbol{Q} \end{cases}. \tag{9.51}$$

Now, we consider the Lagrangian function of (9.26), i.e.,

$$L(F) = \sum_{i=1}^{n} \sum_{j=1}^{m_i} V\left(\boldsymbol{e}_{ij,t}\right) + \boldsymbol{\lambda}^T (\sum_{i=1}^{n} \sum_{j=1}^{m_i} \boldsymbol{Y}_{ij}\boldsymbol{e}_{ij,t} - \boldsymbol{Q}). \tag{9.52}$$

The optimal operating point is determined by the following KKT conditions

$$\begin{cases} \dfrac{\partial V\left(\boldsymbol{e}_{ij,t}\right)}{\partial \boldsymbol{e}_{ij,t}} + \boldsymbol{\lambda}^T \boldsymbol{Y}_{ij} = \boldsymbol{0} \\ \sum_{i=1}^{n} \sum_{j=1}^{m_i} \boldsymbol{Y}_{ij}\boldsymbol{e}_{ij,t} = \boldsymbol{Q} \end{cases}. \tag{9.53}$$

Compared (9.51) with (9.53), it can be observed that $\boldsymbol{\lambda} = -\boldsymbol{Er}_t$. That indicates the calculation of $\boldsymbol{\lambda}$ and $\boldsymbol{Er}_t$ is equivalent.

Remark 5 In our algorithm, the updating of $\boldsymbol{\lambda}(k+1)$ in (9.34) and (9.44) doesn't include $\boldsymbol{\kappa}^* = \boldsymbol{Q}/\sum_{i=1}^{n} m_i$, which implies this chapter doesn't require the assumption that each participates can access $\boldsymbol{\kappa}^*$. However, the existing distributed ADMM algorithms, e.g., [33, 34], etc., need this requirement unless $\boldsymbol{\kappa}^* = 0$. Note that, in the EI, $\boldsymbol{\kappa}^*$ can not directly be known by each participant because it is a global constant during each scheduling horizon. In this chapter, a dynamic average consensus algorithm, i.e., (9.33), is used to estimate the $\boldsymbol{\kappa}^*$ in a fully distributed fashion. Then,

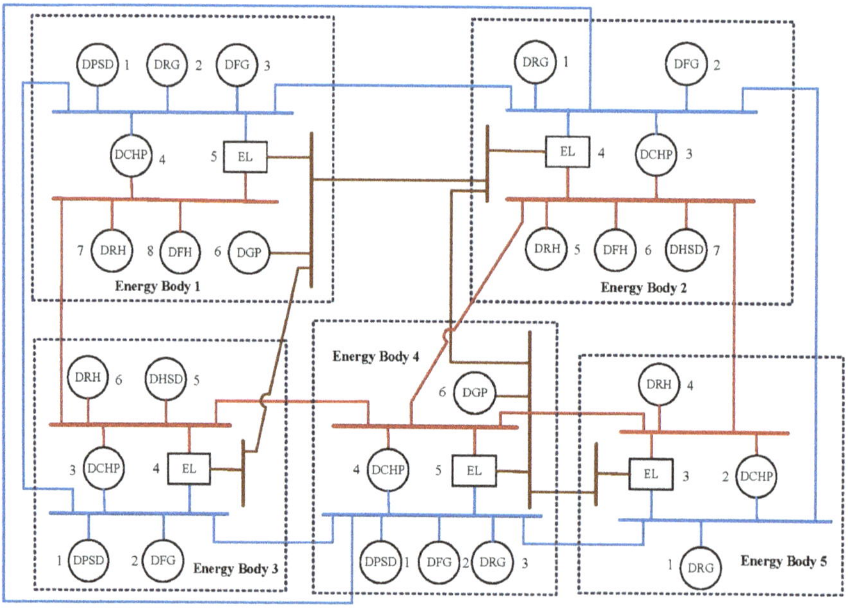

Fig. 9.3 A test system with five EBs

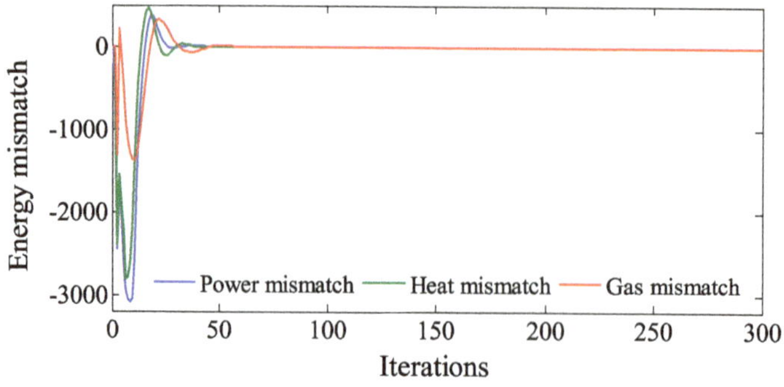

Fig. 9.4 Energy mismatch in case 1

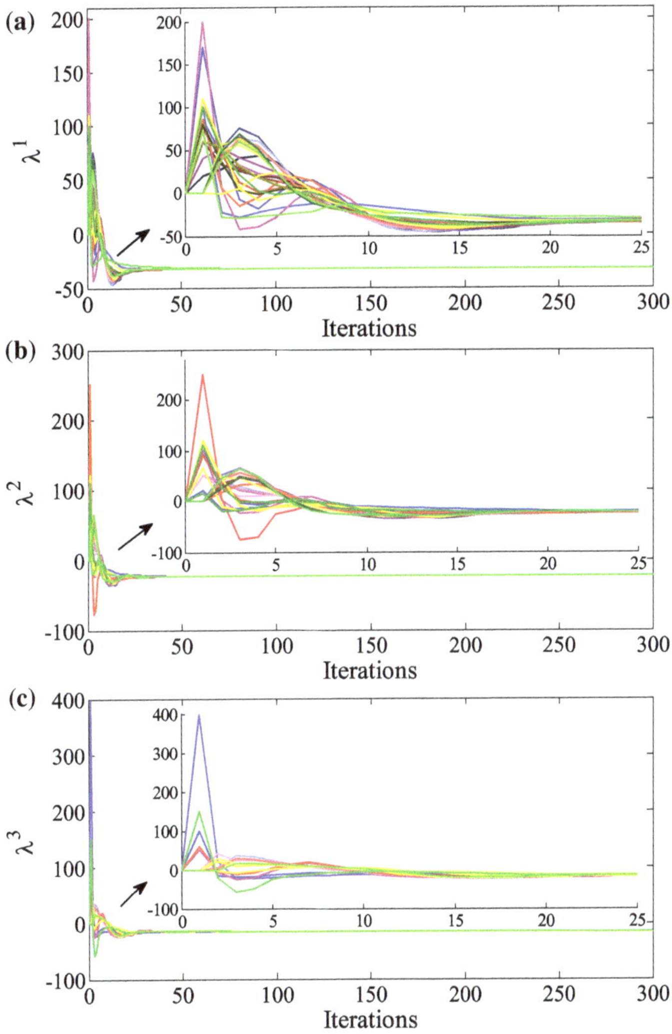

Fig. 9.5 Lagrange dual variable in case 1: **a** power dual variable, **b** heat dual variable, **c** gas dual variable

$\kappa_{ij}(k)$ is employed to design our distributed ADMM algorithm. In the end, not only the optimal energy outputs/demands but also the optimal energy market clearing price can be calculated by decoupling the max-min problem.

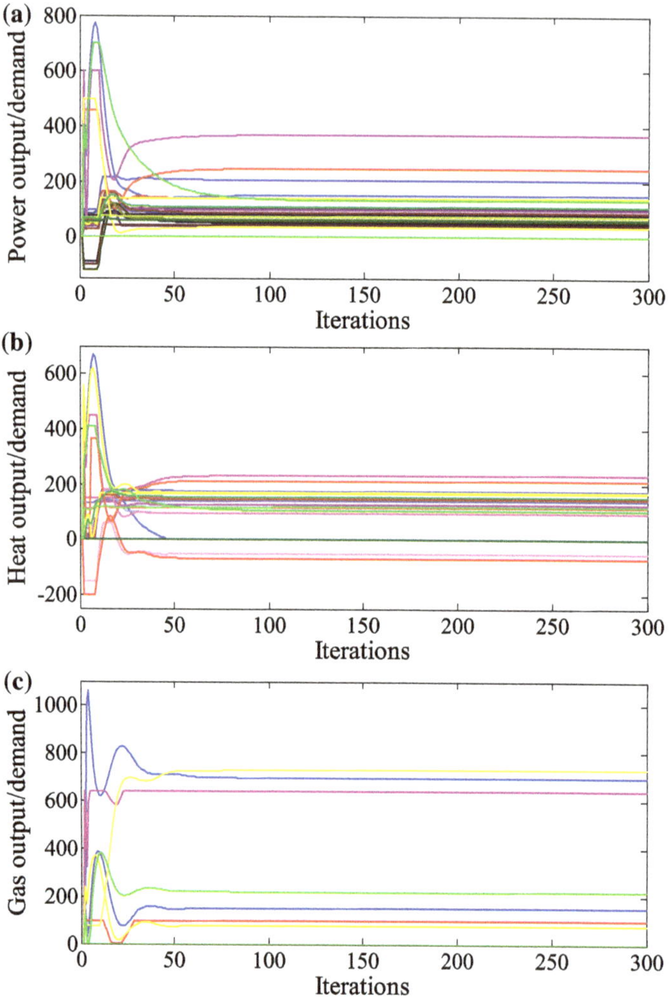

Fig. 9.6 Energy outputs/demands in case 1: **a** power outputs/demands, **b** heat outputs/demands, **c** gas outputs/demands

9.4 Simulation Results

The configuration of the test system with five EBs is shown in Fig. 9.3. Energy translating percentages and the heat rate coefficients can be found in [28, 35], respectively. We unify the energy scale, i.e., 1 p.u. = 1 MW for power (or heat), 1 p.u.= 84SCM/h for gas and 1 p.u.= 1 $/MWh for price.

Table 9.1 Results of energy outputs and demands

	Num	p_{ij,t_0}	h_{ij,t_0}	g_{ij,t_0}	Profits
EB1	N.11	48.7267	0.0000	0.0000	18323
	N.12	99.9955	0.0000	0.0000	
	N.13	81.9282	0.0000	0.0000	
	N.14	205.9113	175.0625	0.0000	
	N.15	150.9603	0	151.1224	
	N.16	0.0000	0.0000	697.1795	
	N.17	0.0000	140.4004	0.0000	
	N.18	0.0000	93.8493	0.0000	
$\Delta p_{1,t_0} = 135.6014$		$\Delta h_{1,t_0} = 285.3122$		$\Delta g_{1,t_0} = 496.0571$	
EB2	N.21	79.8126	0.0000	0.0000	16007
	N.22	40.0000	0.0000	0.0000	
	N.23	97.5582	124.7394	0.0000	
	N.24	368.7102	234.9917	640.0000	
	N.25	0.0000	149.9000	0.0000	
	N.26	0.0000	94.9338	0.0000	
	N.27	0.0000	−50.9579	0.0000	
$\Delta p_{2,t_0} = -256.3394$		$\Delta h_{2,t_0} = -66.3764$		$\Delta g_{2,t_0} = -700.0000$	
EB3	N.31	43.0000	0.0000	0.0000	7032
	N.32	86.3168	0.0000	0.0000	
	N.33	139.1637	145.6758	0.0000	
	N.34	247.9711	214.2604	100.0000	
	N.35	0.0000	−66.9453	0.0000	
	N.36	0.0000	115.3434	0.0000	
$\Delta p_{3,t_0} = -64.4906$		$\Delta h_{3,t_0} = -155.1865$		$\Delta g_{3,t_0} = -180.0000$	
EB4	N.41	57.3022	0.0000	0.0000	15329
	N.42	63.8648	0.0000	0.0000	
	N.43	74.8884	0.0000	0.0000	
	N.44	137.9981	168.0150	0.0000	
	N.45	34.6088	169.6302	77.3927	
	N.46	0.0000	0.0000	731.8679	
$\Delta p_{4,t_0} = 199.4447$		$\Delta h_{4,t_0} = -91.6152$		$\Delta g_{4,t_0} = 604.4752$	
EB5	N.51	59.7181	0.0000	0.0000	965
	N.52	107.4697	155.1540	0.0000	
	N.53	131.4040	104.8087	220.5323	
	N.54	0.0000	117.5203	0.0000	
$\Delta p_{5,t_0} = -14.2161$		$\Delta h_{5,t_0} = 27.8656$		$\Delta g_{5,t_0} = -220.5323$	

9.4.1 Convergence and Profit Analysis

In this case study, the focus is on showing the convergence of the proposed algorithm and analyzing the profits for EBs. The must-run energy loads for EB1 to EB5 are initialized with [150 (p.u.), 124 (p.u.), 50 (p.u.)], [105 (p.u.), 150 (p.u.), 60 (p.u.)], [85 (p.u.), 135 (p.u.), 80 (p.u.)], [100 (p.u.), 90 (p.u.), 50 (p.u.)] and [50 (p.u.), 140 (p.u.), 0 (p.u.)], respectively. Running the proposed distributed-consensus-ADMM algorithm at t_0, the power, heat and gas mismatches, lagrange dual variables and outputs/demands are obtained as shown in Figs. 9.4, 9.5 and 9.6, respectively. From Fig. 9.4, It can be observed that the estimated power, heat and gas mismatches, i.e., $\sum_{i=1}^{n} \Delta p_{i,t}$, $\sum_{i=1}^{n} \Delta h_{i,t}$ and $\sum_{i=1}^{n} \Delta g_{i,t}$ in (9.25), converge to zero after about 75 iterations. This results also mean that the three global equality constraints are satisfied to further verify the power heat and gas supply-demand balance subplot. More importantly, the estimated power, heat and gas dual variables of each participant can converge to three different values, i.e., $\lambda_1 = -32.6887$ (p.u.), $\lambda_2 = -23.6611$ (p.u.) and $\lambda_3 = -15.2825$ (p.u.), respectively. Meanwhile, the final power, heat and gas outputs/demands of each participant, shown in Table 9.1, are within their operational ranges, which implies all the set of inequality constraints are satisfied. Therefore, the optimization goal is fulfilled. According to the results above, we can also get that the cooperative power, heat and gas market clearing prices are 32.6887 (p.u.), 23.6611 (p.u.) and 15.2825 (p.u.), respectively. In addition, the final profit of each EB is also summarized in Table 9.1. Each EB may play different roles based on its and neighbors' state. Let us use EB4 as an illustration. Note that $\Delta p_{4,t_0} > 0$ and $\Delta g_{4,t_0} > 0$, so it serves as a power and gas supplier, and these parts of excess self-generated power and heat will be sold to its neighbors. Meanwhile, $\Delta h_{4,t_0} < 0$, so it also serves as a heat consumer and will buy deficit heat from its neighbors.

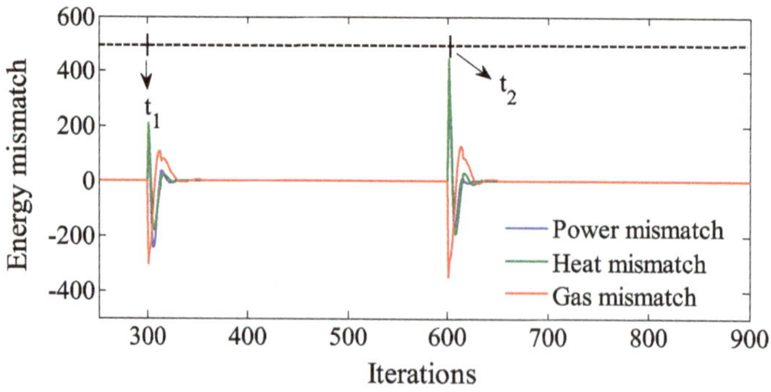

Fig. 9.7 Energy mismatch in case 2

9.4.2 Plug and Play Test

In this case study, the focus is on testing the plug and play performance of the proposed algorithm. All the initial conditions are the same as in case 1. On the basis, at t_1, the DFG of EB1, DCHP of EB2 and EL of EB5 are removed from the system at the same time, and the variables related to them are set to zero. From Figs. 9.7, 9.8 and 9.9, it can be seen that the total power, heat and gas mismatches converge to zero; meanwhile, the estimated power, heat and gas dual variables converge to their

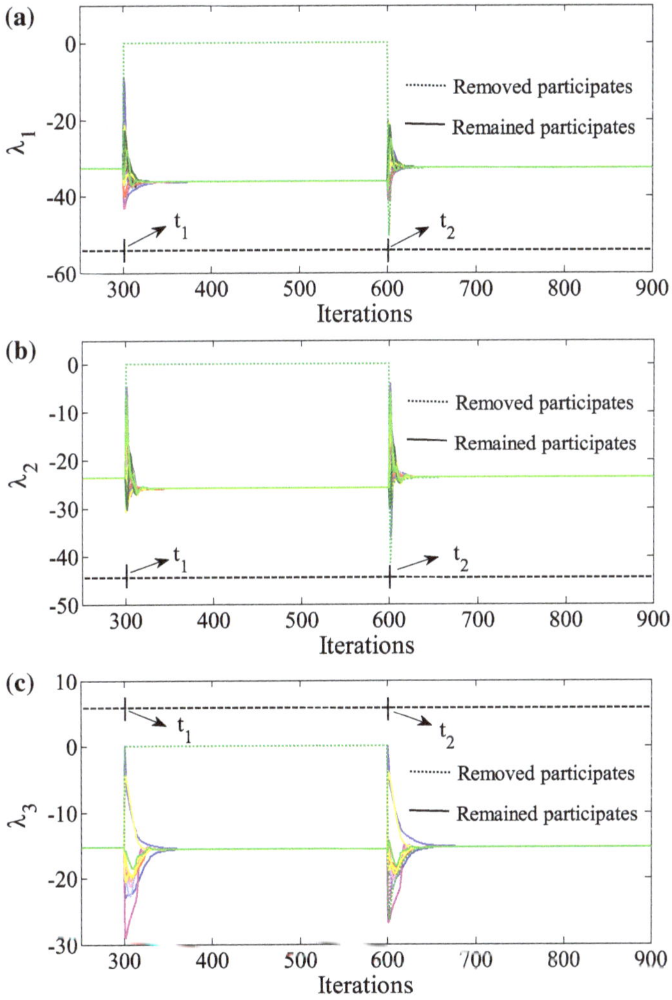

Fig. 9.8 Lagrange dual variable in case 2: **a** power dual variable, **b** heat dual variable, **c** gas dual variable

corresponding optimal values. Of course, the remaining participates have to adjust their power, heat and gas outputs/demands to compensate for the amount of power, heat and gas previously supplied or consumed by the disconnected DFG, DCHP and EL, and finally converge to a new solution. Further, at t_2, they are all plugged again back to the system. The simulation results are also shown in Figs. 9.7, 9.8 and 9.9. It can be seen that the system again converges to the new solutions responding to the new topological change. In addition, the final convergence solutions after t_2 are the same as the one prior to the disconnection. This implies that the proposed algorithm provides good plug and play capability.

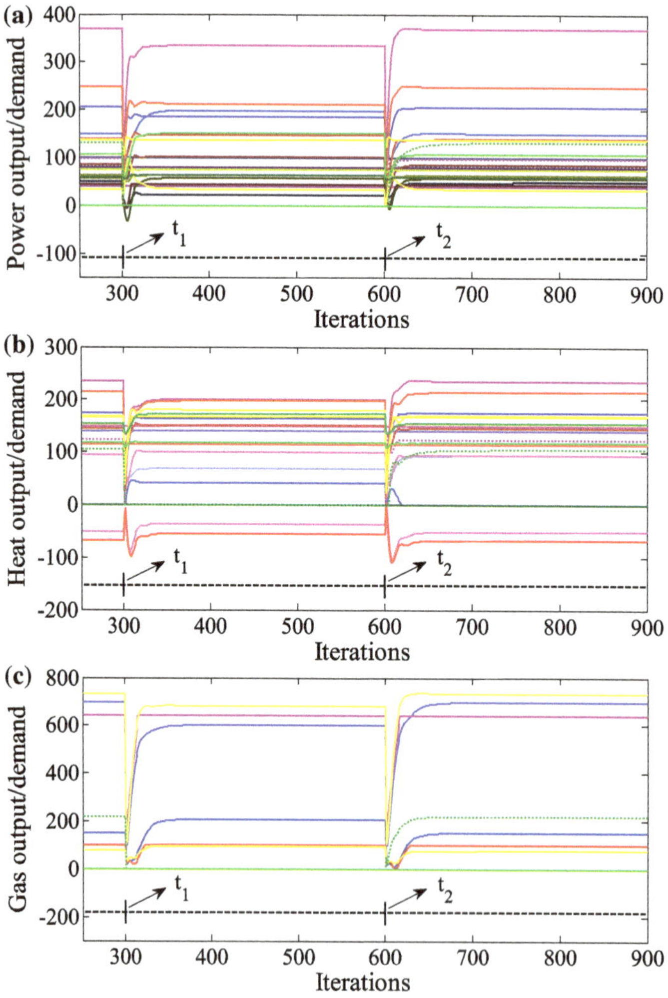

Fig. 9.9 Energy outputs/demands in case 2: **a** power outputs/demands, **b** heat outputs/demands, **c** gas outputs/demands

9.4.3 Scalability Analysis

This case study focuses on analyzing the computational scalability of the proposed distributed-consensus-ADMM algorithm with respect to the number of nodes (or participants). The number of nodes is set as $[10, 30, 50, \cdots, 970]$ with total 50 groups. Meanwhile the percentage of DRG, DFG, DPSD, DRHD, DFHD, DHSD, DCHP, DGP, and EL are set as 10, 10, 10, 10, 10, 10, 10, 10 and 20% in each example. The stopping criterion of the proposed algorithm is based on calculating the solution accuracy of the following equation

$$MSE = \frac{1}{N} \sum_{i=1}^{n} \sum_{j=1}^{m_i} \parallel \boldsymbol{e}_{ij,t}(k) - \boldsymbol{e}_{ij}^{*} \parallel^2 / \parallel \boldsymbol{e}_{ij}^{*} \parallel^2 \tag{9.54}$$

where N is the number of nodes. The algorithm is set to stop once MSE below the preset accuracy denoted as acc.

Next, we will analyze the trend of the average number of iterations, expressing the average computational complexity, with the growth of the number of nodes under the same acc. For each group of nodes, it will be randomly tested with 100 times. By employing the same data fitting method in [36], the fitting curve and functions with $acc = 10^{-4}$ are shown in Fig. 9.10. Therein, $R^2 \leq 1$ represents the coefficient of determination, and the higher value of R^2 is, the better fitting effect becomes. From Fig. 9.10, it can be observed that the average number of iterations for the proposed algorithm is approximately logarithmic (but not exponential) growth with the increase of the number of nodes. Thus, we can verify that the proposed algorithm can provide good scalability on the part of computations burden. In addition, for each participant, the number of communication times directly comes from the number of computation iterations. Therefore, it can be concluded that the proposed algorithm also exhibits better scalable behavior in terms of communications burden.

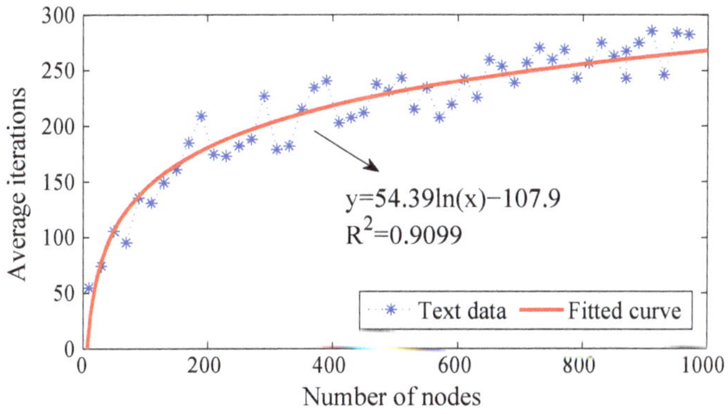

Fig. 9.10 Average number of iterations required for convergence

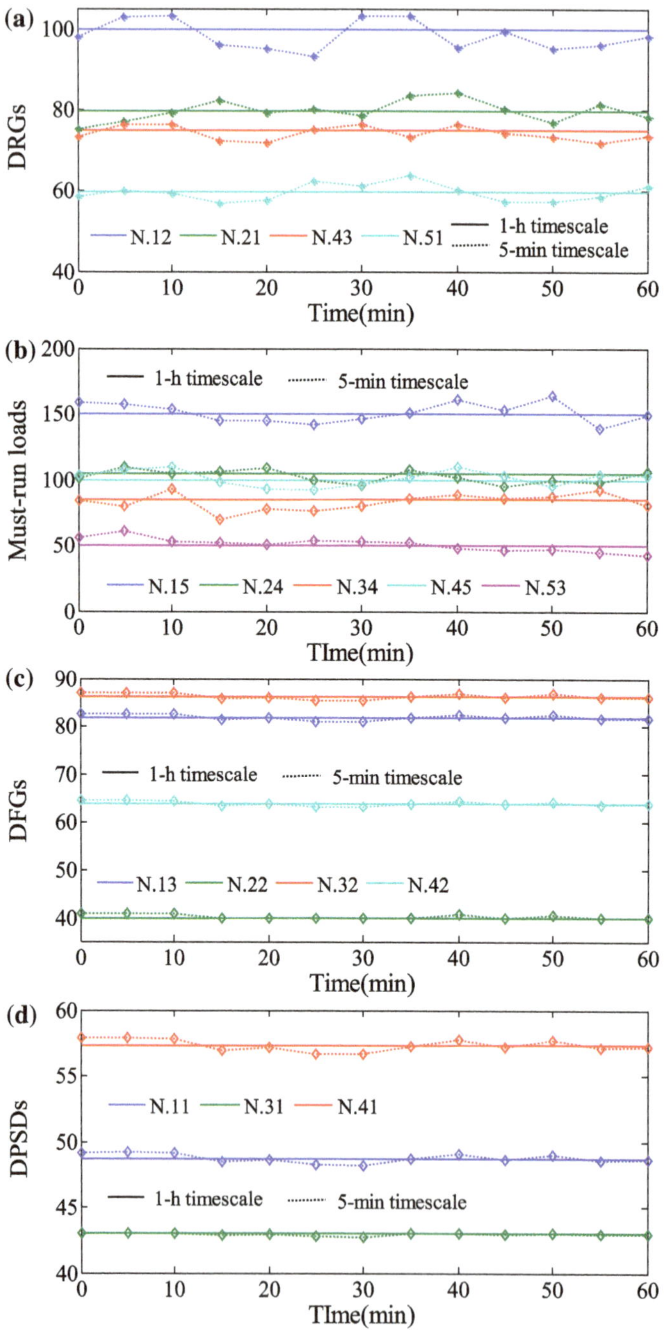

Fig. 9.11 5-min power dispatching results

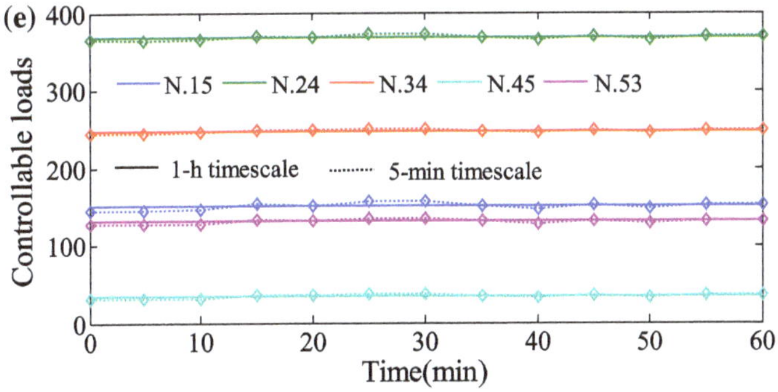

Fig. 9.11 (continued)

9.4.4 Short-Term Power Scheduling Adjustment

This case study focuses on analyzing the 5-min power dispatching problem on the basis of 1-h dispatching results in case 1. The 5-min power must-run loads variations and fluctuations of DRGs are shown in Figs. 9.11a–b, while the corresponding dispatching results of DFGs, DPSDs and controllable loads during each scheduling horizon are shown in Figs. 9.11c–e. It can be seen that we can not only adjust the power generations of DFGs and DPSDs but also the power demand of controllable loads to smooth out the renewable energy fluctuations and loads variations. Note that, for each DFG, DPSD or controllable power load, the 5-min dispatching results can track its hourly dispatching result in a better way, to further ensure the effectiveness of maximum social welfare in hourly scheduling. In addition, the participant, which holds larger capacity, is able to undertake task for accommodating the power variations and fluctuations. For example, the changes of the 5-min dispatching curve of participant N.15 with $p_{15,t}^{l,\max} = 1000$ (p.u.) is larger than that of participant N.53 with $p_{53,t}^{l,\max} = 750$ (p.u.) as seen in Fig. 9.11e. From the discussion mentioned above, the designed 5-min power scheduling adjustment strategy can not only meet the requirement of hourly scheduling goal but also smooth out 5-min loads variations and fluctuations of DRGs.

9.5 Conclusion

In this paper, an innovative energy management framework has been proposed for the future EI. Along with the EB, the system model has been established, which can better reflect the features and requirements of EI in a more precise way. Furthermore, the EMP of EI has been finally formulated as a distributed non-linear coupling optimization problem, which can be effectively solved by the proposed distributed-consensus-ADMM algorithm. In addition, the proposed algorithm is a fully distributed algorithm, where each participant requires only the information from its neighbors to estimate the optimal energy market clearing price and calculate the optimal energy outputs/demands. The effectiveness of the proposed algorithm has been demonstrated by several simulation results.

Solutions

Problems of this chapter

The solution is revealed here.
Problem Heading
(a) The solution of first part is revealed here.
(b) The solution of second part is revealed here.

9.6 Energy Management Framework of EI

References

1. A.Q. Huang, M.L. Crow, G.T. Heydt et al., The future renewable electric energy delivery and management (FREEDM) system: the energy internet. Proc. IEEE **99**(1), 133–148 (2011)
2. Q. Sun, Y. Zhang, H. He et al., A novel energy function-based stability evaluation and nonlinear control approach for energy internet. IEEE Trans. Smart Grid **8**(3), 1195–1210 (2017)
3. Q. Sun, R. Han, H. Zhang et al., A multi-agent-based consensus algorithm for distributed coordinated control of distributed generators in the energy internet. IEEE Trans. Smart Grid **6**(6), 3006–3019 (2015)
4. T. Niknam, R. Azizipanah-Abarghooee, A. Roosta et al., A new multi-objective reserve constrained combined heat and power dynamic economic emission dispatch. Energy **41**(1), 530–545 (2012)
5. E. Abdollahi, H. Wang, R. Lahdelma, An optimization method for multi-area combined heat and power production with power transmission network. Appl. Energy **168**, 248–256 (2016)
6. M.R. Benam, S.S. Madani, S.M. Alavi et al., Optimal configuration of the CHP system using stochastic programming. IEEE Trans. Power Deliv. **30**(3), 1048–1056 (2015)
7. T. Fang, R. Lahdelma, Optimization of combined heat and power production with heat storage based on sliding time window method. Appl. Energy **162**, 723–732 (2016)
8. D. Papadaskalopoulos, G. Strbac, Nonlinear and randomized pricing for distributed management of flexible loads. IEEE Trans. Smart Grid **7**(2), 1137–1146 (2016)
9. N. Rahbari-Asr, M.Y. Chow, J. Chen et al., Distributed real-time pricing control for large scale unidirectional V2G with multiple energy suppliers. IEEE Trans. Ind. Inf. **12**(5), 1953–1962 (2016)
10. W. Zeng, M.Y. Chow, A reputation-based secure distributed control methodology in D-NCS. IEEE Trans. Ind. Electron. **61**(11), 6294–6303 (2014)
11. S. Yang, S. Tan, J. Xu, Consensus based approach for economic dispatch problem in a smart grid. IEEE Trans. Powers Syst. **28**(4), 4416–4426 (2013)

12. G. Chen, F.L. Lewis, E.N. Feng et al., Distributed optimal active power control of multiple generation systems. IEEE Trans. Ind. Electron. **62**(11), 7079–7090 (2015)
13. H. Xing, Y. Mou, M. Fu et al., Distributed bisection method for economic power dispatch in smart grid. IEEE Trans. Powers Syst. **30**(6), 3024–3035 (2015)
14. F. Guo, C. Wen, J. Mao et al., Distributed economic dispatch for smart grids with random wind power. IEEE Trans. Smart Grid **7**(3), 1572–1583 (2016)
15. Y. Xu, W. Zhang, W. Liu, Distributed dynamic programming-based approach for economic dispatch in smart grids. IEEE Trans. Ind. Inf. **11**(1), 166–175 (2015)
16. G. Binetti, A. Davoudi, F.L. Lewis et al., Distributed consensus-based economic dispatch with transmission losses. IEEE Trans. Powers Syst. **29**(4), 1711–1720 (2014)
17. G. Chen, J. Ren, E.N. Feng, Distributed finite-time economic dispatch of a network of energy resources. IEEE Trans. Smart Grid (2017). https://doi.org/10.1109/TSG.2016.2516017
18. G. Binetti, A. Davoudi, D. Naso et al., A distributed auction-based algorithm for the nonconvex economic dispatch problem. IEEE Trans. Ind. Inf. **10**(2), 1124–1132 (2014)
19. Z. Wang, W. Wu, B. Zhang, A fully distributed power dispatch method for fast frequency recovery and minimal generation cost in autonomous microgrids. IEEE Trans. Smart Grid **7**(1), 19–31 (2016)
20. N. Rahbari-Asr, U. Ojha, Z. Zhang et al., Incremental welfare consensus algorithm for cooperative distributed generation/demand response in smart grid. IEEE Trans. Smart Grid **5**(6), 2836–2845 (2014)
21. Y. Xu, Z. Li, Distributed optimal resource management based on the consensus algorithm in a microgrid. IEEE Trans. Ind. Electron. **62**(4), 2584–2592 (2015)
22. W. Zhang, Y. Xu, W. Liu et al., Distributed online optimal energy management for smart grids. IEEE Trans. Ind. Inf. **11**(3), 717–727 (2015)
23. J. Wu, B. Zhang, K. Wang et al., Optimal economic dispatch model based on risk management for wind-integrated power system. IET Gener. Transm. Distrib. **9**(15), 2152–2158 (2015)
24. C. Wan, Z. Xu, P. Pinson et al., Probabilistic forecasting of wind power generation using extreme learning machine. IEEE Trans. Powers Syst. **29**(29), 1033–1044 (2014)
25. N. Nikmehr, S.N. Ravadanegh, Optimal power dispatch of multi-microgrids at future smart distribution grids. IEEE Trans. Smart Grid **6**(4), 1648–1657 (2015)
26. Z. Yang, R. Wu, J. Yang et al., Economical operation of microgrid with various devices via distributed optimization. IEEE Trans. Smart Grid **7**(2), 857–867 (2016)
27. Y. Zhang, N. Rahbari-Asr, J. Duan et al., Day-ahead smart grid cooperative distributed energy scheduling with renewable and storage integration. IEEE Trans. Sustain. Energy **7**(4), 1739–1747 (2016)
28. J. Qiu, Z.Y. Dong, J.H. Zhao et al., Multi-stage flexible expansion co-planning under uncertainties in a combined electricity and gas market. IEEE Trans. Power Syst. **30**(4), 2119–2129 (2015)
29. Z. Bao, Q. Zhou, Z. Yang et al., A multi time-scale and multi energy-type coordinated microgrid scheduling solution - part I: model and methodology. IEEE Trans. Powers Syst. **30**(5), 2257–2266 (2015)
30. Z. Bao, Q. Zhou, Z. Yang et al., A multi time-scale and multi energy-type coordinated microgrid scheduling solution - part II: optimization algorithm and case studies. IEEE Trans. Powers Syst. **30**(5), 2267–2277 (2015)
31. S. Boyd, L. Vandenberghe, *Convex Optimization* (Cambridge University Press, Cambridge, 2004)
32. N. Rahbari-Asr, Y. Zhang, M.Y. Chow, Consensus-based distributed scheduling for cooperative operation of distributed energy resources and storage devices in smart grids. IET Gener. Transm. Distrib. **10**(5), 1268–1277 (2016)
33. A. Teixeira, E. Ghadimi, I. Shames et al., The ADMM algorithm for distributed quadratic problems: parameter selection and constraint preconditioning. IEEE Trans. Signal Process. **64**(2), 290–305 (2016)

34. T.H. Chang, M. Hong, X. Wang, Multi-agent distributed optimization via inexact consensus ADMM. IEEE Trans. Signal Process. **63**(2), 482–497 (2015)
35. A. Shabanpour-Haghighi, A.R. Seifi, An integrated steady-state operation assessment of electrical, natural gas, and district heating networks. IEEE Trans. Power Syst. https://doi.org/10.1109/TPWRS.2015.2486819
36. N. Rahbari-Asr, M.Y. Chow, Cooperative distributed demand management for community charging of PHEV/PEVs based on KKT conditions and consensus networks. IEEE Trans. Ind. Inf. **10**(3), 1907–1916 (2014)

Chapter 10
Model-Free Energy Optimization for Energy Internet

Abstract With the high penetration of distributed energy, the scale of current energy network becomes larger. At the same time, it also has the problems of complex computation and slow convergence. In order to realize the rational planning and utilize various energy resources and improve the reliability and economy of the overall system, the problems of system stability, economic operation and power flow calculation must be considered comprehensively. As a kind of machine learning, reinforcement learning has strong intelligence and rapidity, which can realize the optimal control of the system. Aiming at the energy management model of regional energy Internet, this chapter studies how to transform energy management into Q learning model, and uses Q learning algorithm to verify the validity of the model. In the meantime, for the optimization scheduling problem of large-scale system, this chapter expands the optimal power flow model of energy internet into the optimal operation structure composed of multiple We-Energies based on the previous one, and uses the distributed reinforcement learning algorithm to optimize the large-scale energy internet scheduling which considers the average consistent information search to achieve the optimization process for cooperating and communicating multiple We-Energies.

10.1 Introduction

Optimal energy flow (OEF) is widely used to realize the interconnected multicarrier system (IMS) of economic and security operation to reduce the network loss [1]. The IMS is conventionally modeled as a multitude of subsystems in which the operation of energy subsystems was scheduled and optimized individually [2]. "We-Energy" (WE), as a novel energy interacting area for energy internet, exchanges energy (electricity, district heat and natural gas) with others by the advanced communication, electronic conversion and automatic control technology [3]. In order to enhance the performance of environment, economic and security, many papers have a comprehensive research on it. In [4–9], the optimal power flow of electricity and natural gas combined system is discussed. Paper [10] discussed the scheduling model of

Q. Sun, *Energy Internet and We-Energy*, Renewable Energy Sources & Energy Storage, https://doi.org/10.1007/978-981-13-0523-8_10

multiple energy system which is based on distributed CHP device. In this model, the city network of electricity, heating and natural gas is coupled. Energy production and consumption matching problem is summarized at CHP system level in [11]. The optimal operation of electricity and heating combined system is studied in [12], which considered the heating network constrains and proved that CHP system can promote to dispose the wind power.

In this chapter, two issues concerning optimal performance of energy internet are considered. The first research is conducted on energy management of We-Energy. The combined of multiple types of energy is one of the specific characteristics of Energy Internet. The Energy Internet can be assumed as a cluster of distributed energy resources and loads, which contains various types of energy resources such as electricity, gas, heat and so on [13]. The use of different kind of energy brings great benefit to Energy Internet, which allows multiple end users to make options according to their own power demands, hence increasing the flexibility of the power system and weakening the impact of traditional energy supplier. However, using distributed generations indiscriminately may also impose undesirable effects on power system. Therefore, issues on optimal energy management come into play. A lot of researches concerning control and operation of power system have been done in recent years. Several common optimization objectives including lower cost of carbon and minimum operating cost have been discussed in [14]. In [15], authors proposed a smart energy management system in order to minimize the operating cost of the micro-grid. Only electricity is discussed during optimization process while other types of energy resources are not considered in the paper. The authors in [16] proposed a micro-grid scenario consists of combined heat and power generation, as well as power and thermal energy storage devices. And an online algorithm has been put forward to optimize the cost of whole system.

However, the optimized economic dispatch does not always satisfy the demands when taking pollutants emission into account. So, multi-objective energy management has drawn attention from researchers so as to realize optimization both economically and environmentally. The authors in [17] proposed an intelligent energy management system (IEMS) for a CHP-based micro-grid, and minimized the operation cost and the net emission simultaneously. An efficient modified bacterial foraging optimization algorithm was used to find the optimal set points of the system. Reference [18] proposed a Stackelberg game-based optimization model, and a differential evolution-based heuristic algorithm was designed to reach the Stackelberg equilibrium. But in the previous studies, there is a lack of consideration of specific characteristics of Energy Internet, such as openness, sharing and peer-to-peer integration.

Another research is conducted on optimal power flow. Paper [19] proposed the generalized heuristic algorithm to study the optimal power flow of multiple energy system. While with the increasing utilization of co-generation plants such as electricity, natural gas and local district heating systems that make a strong coupling in IMS [20], the structure of network becomes more complex. There is a challenge to find the optimal strategy in such a way for this class of complex nonlinear multi objective optimization problems. The traditional optimal power flow algorithm such as linear programming [21], interior-point method [22], is unable to obtain the global optimal solution with these problems and conventional artificial intelligence such as algorithms genetic algorithm, particle swarm optimization have the disadvantages of slow computation speed for the large-scale network. The domestic and foreign scholars made in-depth study in this question for more efficient algorithm. In addition to improve the basic heuristic algorithm for OEF [23], distributed algorithm has become a research focus [24]. Meanwhile, some hybrid algorithm has been discussed. Paper [25] presents the reinforcement learning combined with simulated annealing (SA) algorithm to solve the optimal reactive power dispatch.

Recently, reinforcement learning algorithm (RL) as a kind of machine learning algorithms attracts people's attention. Some learning strategies on the basis of RL to solve deterministic optimal control problems in continuous state spaces can be found in some studies such as [26–35]. The distributed reinforcement learning (DRL) which is a new branch of reinforcement learning algorithm has been developed rapidly in various areas including distributed control, robotic teams, collaborative decision support systems, and economics [36]. DRL is defined to be composed of multiple agents, the whole system will achieve the learning goals through each agent executing part of reinforcement mission independently. All performance of DRL exhibits its advantage on the col problems, and the features aim at strategic decision make DRL widely used [37]. Each Pareto optimal solution is also a Nash equilibrium for a fully cooperative game [38], which means that if one agent is provided with greatest possible reward in a combined action, the reward received by the other agents must also be maximized.

In this chapter, the energy management of We-Energy is discussed. An Energy Internet model consisting of combined heat and power unit (CHP), photovoltaic unit, heating only unit and storage device is constructed. To construct an environmental-friendly and low-operating cost energy consumption structure, a multi-objective optimization model is proposed. Furthermore, in order to satisfy the power and heat demands of the We-Energy simultaneously as well as realize minimum operating cost and pollutant emission, an intelligent energy management system (IEMS) is presented. In particular, reinforcement learning method has been implemented to formulate the optimal operating strategy. Eligibility trace theory is also been introduced to accelerate the computational process.

What's more, the optimal energy flow in interconnected multicarrier systems where electric, heat and natural gas systems are coordinated. We propose a double-deck optimal model to improve the performance of security and environment for IMS. The proposed formulation for large-scale systems can be solved by machine learning algorithms which could find the optimization strategy intelligently. We present a hybrid reinforcement learning algorithm (HRL) for distributed multicarrier energy network, which computes a global optimal policy in cooperative subsystems on the basis of the implementation of independent optimization for subsystems. A policy is defined as a set of actions deriving from the reward function connecting the environment.

10.2 Reinforcement Learning Applied to Energy Management

10.2.1 Reinforcement Learning on Markov Decision Processes

As a main class of machine learning methods, reinforcement learning (RL) is an effective means for making sequential decision under uncertainties. In a reinforcement learning system, a reinforcement learning agent aims to find an optimal action policy by trial-and-error interaction with its uncertain environment. At each time step, the learning agent perceives the state of the environment but it is not provided with explicit information of the action to take. The agent autonomously selects a random action with certain probability and the current state of the environment therewith transits into its successive state. After that, the learning agent can receive a reward signal that evaluates the effect of this action.

A Markov decision process can be characterized as the formulation of a sequential decision-making problem. Therefore, reinforcement learning can be described by a finite *Markov* decision process.

A finite Markov decision process can be characterized as a 4-tuple $\{S, A, R, P\}$, where S is the state space of a finite set of states, A is the action space of a finite set of actions, R is the reward function, P is the matrix of state transition probability. $r(s, a, s')$ and $p(s, a, s')$ represents the instant reward and probability of the state transition from state s to state s' taking action a. An action policy of the MDP can be defined as $\pi : S \rightarrow A$.

The objective function of reinforcement learning is to receive the largest discounted reward. Therefore the state value function can be defined as

$$V^{\pi}(s) = \sum_{t=0}^{\infty} \gamma^t r(s_t, a_t)|s_0 = s, a_t = \pi(s_t), \tag{10.1}$$

where $\gamma \in (0, 1)$ is the discount factor, s_0 is the initial environment state. According to the equation above, the optimal action policy can be described as

$$\pi^* = \arg\max_{\pi} V^{\pi}(s), \tag{10.2}$$

The action-state value function can be defined as

$$Q^{\pi}(s_t, a_t) = r(s_t, a_t) + \gamma V^{\pi}(s_{t+1}), \tag{10.3}$$

According to the equation above, the optimal action policy can be described as

$$\pi^* = \arg\max_{a \in A} Q^{\pi}(s, a), \tag{10.4}$$

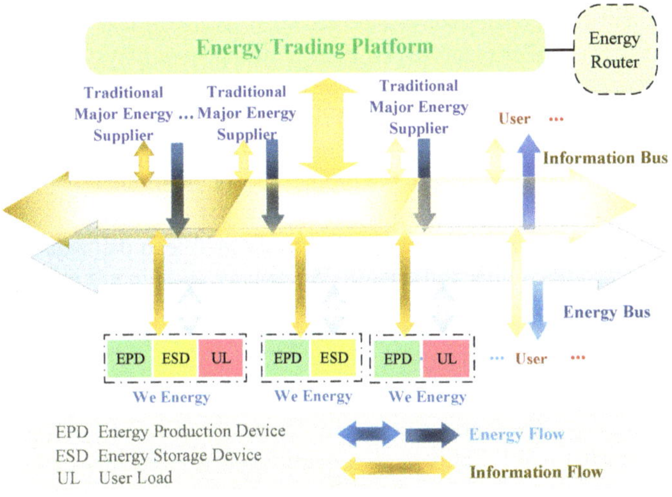

Fig. 10.1 Operation mode of We-Energy in the energy internet

10.2.2 The Q Learning Applied to Energy Management

10.2.2.1 Modeling of Energy Management and Multi-objective Optimization

In the Energy Internet, different structures of energy prosumers bring about various energy demands. Consequently, as a novel energy interaction area of Energy Internet, We-Energy no longer follows the track of traditional energy network where different types of energy are supplied independently. WE is capable to transform various types of energy such as electricity, district heat and natural gas into desired energy and exchange with others using advanced communication, electronic conversion and automatic control technology. That is to say, WE is no longer a passive consumer, but also a potential energy supplier during the energy interaction process.

As shown in Fig. 10.1, WEs are connected to information bus and energy bus simultaneously which can realize bi-directional interaction of information and bi-directional transmission of energy. Information of each WE can be submitted to the information bus and useful data of other WEs located in the Energy Internet can be abstracted from information bus as well, which achieves bi-directional information interaction. WE is able to supply excess energy to others through energy bus and can get compensatory energy when needed. Various types of energy can be transmitted through energy bus. In this section, we take electricity and heat into account simultaneously.

Basically, a WE is comprised of three units including energy production unit, energy storage unit and user load. Therefore, WE can be classified into seven types in which traditional pattern of energy network is embraced as well. WE only consists of user load and energy supplier can be regarded as traditional energy users and traditional energy supplier respectively. Energy storage unit plays an important role in energy management of We-Energy and can improve its performance and economic efficiency. In this section, we adopt the WE structure consists of three units at the same time.

In this research, we consider a WE composed of both electrical and thermal producers, users and storage devices.

The goal of IEMS is to satisfy electrical and thermal load demands considering both economic and environmental criteria. We-Energy participates in the open market, buying and selling power to the Energy Internet via energy bus and information bus.

A. Objective function

In expectation of a more sustainable and environmental friendly dispatch of We-Energy, we consider a WE holds more renewable energy resources and less fossil fuels. Thus, the objective function should be formulated as

$$f = \min(f_1, f_2), \tag{10.5}$$

where f_1 is the operating cost function of energy dispatch, f_2 is the pollutant emission function of energy dispatch.

B. Operating cost

As We-Energy is a comprehensive combination of energy producers in the Energy Internet, multi types of primary energy resources including fossil fuels, natural gas and renewable energy are used to satisfy the demand of users. The optimization of economic benefit is to minimize the operating cost of energy suppliers. The operating cost of We-Energy includes the cost of photovoltaic, gas-fired CHP and heat only unit. The objective function is introduced as

$$f_1 = \min \sum_{t=1}^{T} \left(\sum_{i=1}^{N_{CHP}} F_{CHPi,t} + \sum_{i=1}^{N_{PV}} F_{PVi,t} + \sum_{i=1}^{N_{Heat}} F_{Heati,t} + P_{Grid,t} \times bid_t \right), \tag{10.6}$$

where N_{CHP}, N_{pv}, N_{Heat} is the total number of natural gas-fired CHP, photovoltaic and heat only unit respectively. $F_{PV_{i,t}}$, is the cost of the ith photovoltaic and the cost is related to the of active power. $F_{CHP_{i,t}}$,, $F_{Heati,t}$ is the cost of the ith CHP and heating only unit at time step t respectively. $P_{Grid,t}$ is the active power abstracted from the Energy Internet, while bid_t is the corresponding electrical price.

The nonlinear cost function of a CHP unit can be explained as below

$$\begin{aligned} F_{CHPi,t} = {} & a_i \times P^2_{CHPi,t} + b_i \times P_{CHPi,t} + c_i + d_i \times H^2_{CHPi,t} \\ & + e_i \times H_{CHPi,t} + f_i \times H_{CHPi,t} \times p_{CHPi,t}, \end{aligned} \tag{10.7}$$

where a_i,b_i, c_i, d_i, e_i,f_i are the generation parameters of the ith natural gas-fired CHP, $P_{CHP_{i,t}}$ is the active power of the ith natural gas-fired CHP at time t.

The cost function of a heat only unit is expressed as a quadratic function

$$F_{Heatj,t} = a_j \times H_{Heatj,t}^2 + b_j \times H_{Heatj,t} + c_j, \tag{10.8}$$

where a_j, b_j, c_j are the generation parameters of the jth heating only unit, $H_{Heatj,t}$ is the active power of the jth natural gas-fired CHP at time step t.

C. Pollutant emission

Due to the aggravation of environmental pollution and energy shortage, there is a rising trend of reducing consumption of coal, natural gas and other traditional fossil fuels. In order to minimize the pollutant emission, the use of clean renewable energy should be maximized, and the objective function is expressed as

$$f_2 = \min \sum_{t=1}^{T} \left[\begin{array}{l} \left(\sum_{i=1}^{N_{CHP}} \alpha_i H_{CHPi,t}^2 + \beta_i H_{CHPi,t} + \gamma_i \right) \\ + \left(\sum_{j=1}^{N_{Heat}} \alpha_j H_{Heatj,t}^2 + \beta_j H_{Heatj,t} + \gamma_j \right) + E_{Grid,t} \end{array} \right], \tag{10.9}$$

where α_i, β_i, γ_i are the emission parameters of the ith gas-fired CHP, and α_j, β_j, γ_j are the emission parameters of the jth heating only unit. As can be seen in (10.9), the emission function of CHP unit and heat only unit is quadratic function.

In this section, the emission during the production of electricity is also taken into consideration. The emission mainly caused by burning fossil fuels in thermal power plants, considering coverage fraction of these plants, E_{Grid} can be defined as follows

$$E_{Grid,t} = \alpha P_{Grid,t}^2 + \beta P_{Grid,t} + \gamma', \tag{10.10}$$

where α, β, γ' are emission parameters of thermal power plants.

D. Constraints

There are several constraints should be taken into consideration. Power balance between electrical demand and electrical supply, and balance between thermal demand and thermal supply are expressed as follows

$$\sum_{i=1}^{N_{CHP}} P_{CHPi,t} + \sum_{i=1}^{N_{PV}} P_{PVi,t} + P_{Grid,t} + P_{ch,t} - P_{dis,t} = P_{load,t}, \tag{10.11}$$

$$\sum_{i=1}^{N_{CHP}} H_{CHPi,t} + \sum_{i=1}^{N_{Heat}} H_{Heati,t} = H_{load,t}, \tag{10.12}$$

where $P_{ch,t}$, $P_{dis,t}$ are the charging and discharging rate of the storage unit at time step t respectively. $P_{load,t}$, $H_{load,t}$ are the electrical load demand and heat demand respectively.

Equations (10.13) and (10.14) can be expressed as follows as well

$$\sum_{i=1}^{N_{CHP}} P_{gasi,t} \times \eta_e^C + \sum_{i=1}^{N_{PV}} P_{PVi,t} + P_{Grid,t} + P_{ch,t} - P_{dis,t} = P_{load,t}, \tag{10.13}$$

$$\sum_{i=1}^{N_{CHP}} P_{gasi,t} \times \eta_h^C + \sum_{i=1}^{N_{Heat}} H_{Heati,t} = H_{load,t}, \tag{10.14}$$

where $P_{gasi,t}$ is the gas input of the natural gas-fired CHP at time step t, η_e^C, η_h^C are the output ratio of electric power and heat respectively.

In addition, the output power of all the units should satisfy its upper and lower bound, which can be expressed as

$$P_{PVi,t} \le P_{PV}^{\max}, \tag{10.15}$$

$$P_{ch,t} \le P_{ch}^{\max}, \tag{10.16}$$

$$P_{dis,t} \le P_{dis}^{\max}, \tag{10.17}$$

$$P_{CHP}^{\min} \le P_{CHP,t} \le P_{CHP}^{\max}, \tag{10.18}$$

$$H_{CHP}^{\min} \le H_{CHP,t} \le H_{CHP}^{\max}, \tag{10.19}$$

$$H_{Heat,t} \le H_{Heat}^{\max}, \tag{10.20}$$

$$E_{stor} \le E_{stor}^{\max}, \tag{10.21}$$

where $P_{PV}^{\max}$, $P_{ch}^{\max}$, $P_{dis}^{\max}$, $P_{CHP}^{\max}$, $H_{CHP}^{\max}$, $H_{Heat}^{\max}$, $E_{stor}^{\max}$ are the upper limit of each device respectively. $P_{CHP}^{\min}$, $H_{CHP}^{\min}$ are the lower bound of output electrical power and heat of CHP respectively.

10.2.2.2 Reinforcement Learning Method

A. Q-learning with eligibility trace

The aim of Q-learning is to learn the value of each action taken from the action space at each state, which is defined to be the predicted total discounted reward received by the agent over the future as a result of taking that action from the action space. The one-step Q-learning is defined as follows [11]:

$$\begin{aligned} Q_{t+1}(s_t, a_t) = Q_t(s_t, a_t) + \eta[r_{t+1} + \gamma \max_{a \in A} Q_{t+1}(s_{t+1}, a) \\ - Q_t(s_t, a_t)], \end{aligned} \tag{10.22}$$

At each time step t, the action value $Q_t(s_t, a_t)$ is recorded. After selecting a subsequent state s_{t+1}, an immediate reward r_{t+1} is obtained and $\max_{a \in A} Q_{t+1}(s_{t+1}, a)$ is picked out by searching a lookup table that stores the action values for each state.

The parameter $\gamma \in [0, 1)$ is the discount factor. If the discount factor is small, the agent tends to care more about the immediate reward rather than the rewards received in the future. Thus, in order to make the agent more "farseeing", a large discount factor is chosen in this section. The parameter $\eta \in [0, 1)$ is the learning rate which determines how far the agent is adjusted towards the estimated value. A large factor allows the agent to learn faster, and vice versa. In this section, a large learning rate is selected to shorten the learning process.

In order to accelerate the training process, the eligibility trace theory is introduced to the Q-learning algorithm. The updating rule of the eligibility trace is expressed as follows:

$$e_{t+1}(s,a) = \begin{cases} \gamma\lambda e_t(s,a) + 1 & s = s_t, a = a^* \\ 0 & s = s_t, a \neq a^* \\ \gamma\lambda e_t(s,a) & s \neq s_t \end{cases}, \tag{10.23}$$

where $e_t(s, a)$ is the eligibility trace of the state-action pair at time step t, λ is the trace decay parameter, a^* is the optimal action at time step t. A larger decay parameter makes the algorithm converge faster, so a large λ is adopted in this section.

Therefore, in consideration of eligibility, the updating rule of Q-learning can be rewritten as

$$\begin{aligned} Q_{t+1}(s_t, a_t) = Q_t(s_t, a_t) + \eta[r_{t+1} + \gamma \max_{a \in A} Q_{t+1}(s_{t+1}, a) \\ - Q_t(s_t, a_t)]e_t(s, a). \end{aligned} \tag{10.24}$$

B. State space and action space

The state space of the intelligent energy management system is as follows:

$$S = \{s \mid bid_t\}. \tag{10.25}$$

The action space of the intelligent energy management system can be described as

$$A = \left\{a \mid P_{gas,t}\right\}. \tag{10.26}$$

C. Reward function

After taking an action, the IEMS receives an immediate reward to evaluate the selected action. Since our goal is to optimize the operation of We-Energy overall rather than merely optimize a single objective, the reward function should take the aforementioned two objectives into consideration simultaneously. In addition, constrains should be satisfied as well. Therefore, the reward function is defined as

$$r = -((1 - \omega)f_1 + \omega f_2 + KN), \tag{10.27}$$

where parameter $\omega \in (0, 1)$ is weight, f_1 is the value of the operating cost function, f_2 is the value of the pollutant emission function, K is a positive number, N is the number of the inequality constraints that not be satisfied.

D. Action selection policy

The action selection policy allows the agent to select an action a_i at state s with a probability of $p(s, a_i)$ according to the action values. The policy can be described as

$$p(s, a_i) = \frac{e^{Q(s,a_i)/\tau}}{\sum_{a_i} e^{Q(s,a_i)/\tau}}, \tag{10.28}$$

where τ is a parameter called temperature, which determines the randomness of the exploration. If a lower temperature is selected, the agent tends to select the action with higher action value, while a higher temperature makes the agent act more randomly.

10.2.3 Simulation and Results

In order to verify the effectiveness of the proposed energy management strategy for the We-Energy, the following simulation model is built as shown in Fig. 10.2. We consider a We-Energy model consists of a combined heat and power unit, a photovoltaic unit, a heating only unit and a storage unit.

Table 10.1 explains the limits of output power of each device in the proposed system. Cost coefficients and emission coefficients are presented in Tables 10.2 and 10.3.

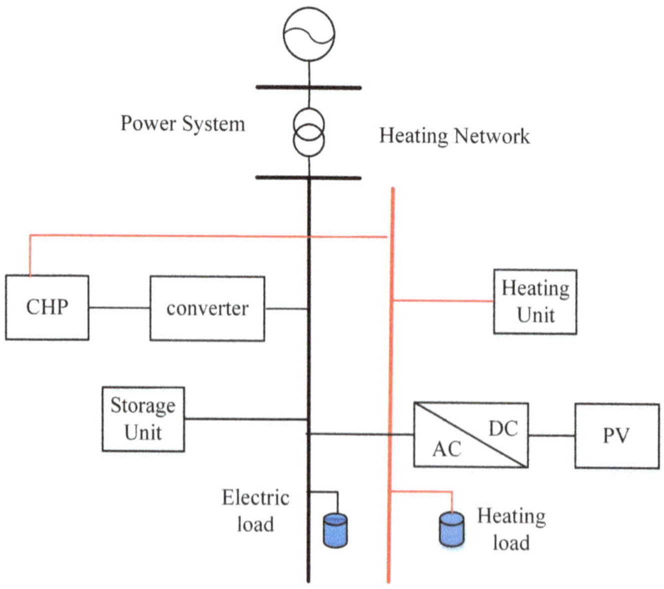

Fig. 10.2 Simulation model of We-Energy

Table 10.1 Installed device

ID	Type	Min power (kW)	Max power (kW)
1	PV	0	25
2	CHP (electrical)	0	60
3	CHP (heat)	0	80
4	Heat	0	50

Table 10.2 Cost fuction coeffecients

Devices	Cost parameters					
	a_i	b_i	c_i	d_i	e_i	f_i
CHP	0.0065	1.21	2	0.003	4	0.61
Heat	0.038	2.011	65	–	–	–

Table 10.3 Emission function coeffecients

Devices	Emission parameters		
	α_i	β_i	γ_i
CHP	0.08	– 2	11
Heat	0.7	– 2	5
Power plants	0.46	– 1.3	3.27

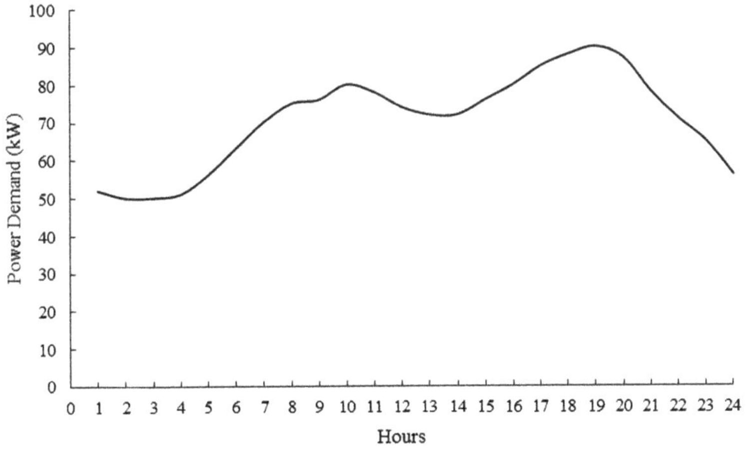

Fig. 10.3 Electrical power demands of a day

Power demand and heat demand are shown in Figs. 10.3 and 10.4 respectively. The output power of photovoltaic is shown in Fig. 10.5. The day ahead market price of electricity is proposed in Fig. 10.6. The operating cost of photovoltaic is shown in Fig. 10.7.

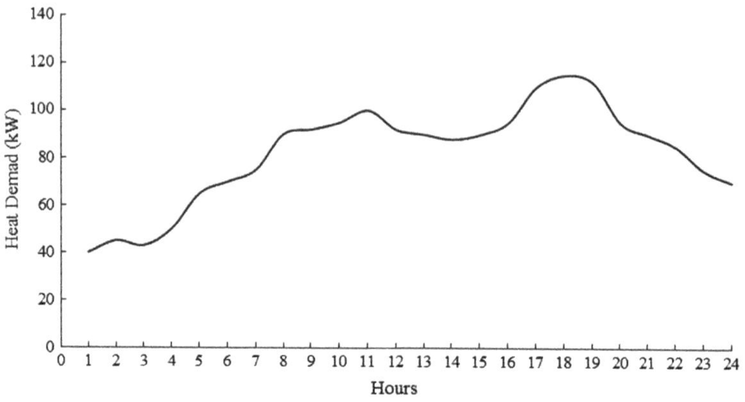

Fig. 10.4 Heat demands of a day

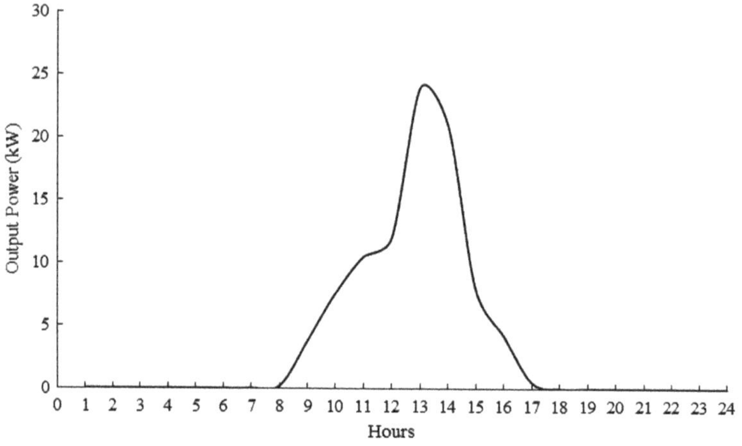

Fig. 10.5 Output power of a photovoltaic

In this section, we consider CHP produces electrical power and heat at a fixed ratio of 1:1.3. Therefore, electrical power and heat generation scheduling for proposed WE is shown in Figs. 10.8 and 10.9.

As can be seen in Figs. 10.8 and 10.9, reinforcement learning algorithm is incorporated into the energy management system and realizes optimal power scheduling. Considering the benefits of renewable sources on reducing pollutant emission, power generated by photovoltaic is consumed at first. In the peak hours, when the market prices are high (9:00–21:00), storage device transfers power to load while less power is obtained from Energy Internet. When the market price is low, power is bought from Energy Internet to fully charge the storage device. Therefore, optimization is realized both environmentally and economically.

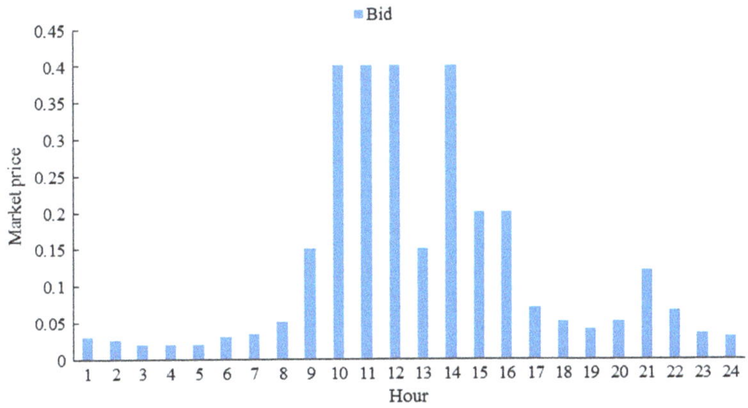

Fig. 10.6 Day ahead market price

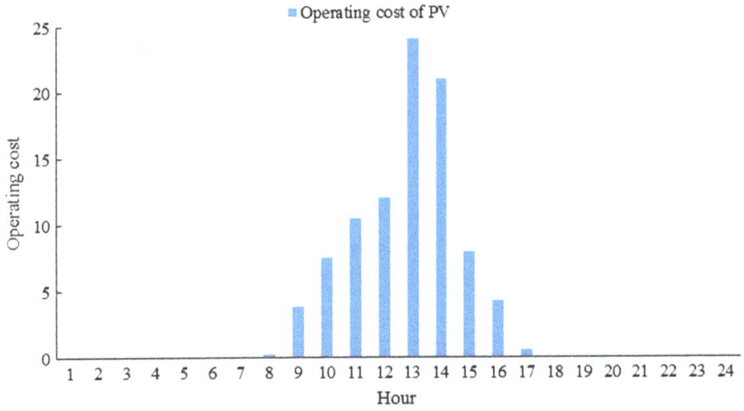

Fig. 10.7 Operating cost of photovoltaic

10.3 MFEO Energy Internet Scheme with Distributed Reinforcement Learning

10.3.1 Distributed Reinforcement Learning Algorithm

As a branch of reinforcement learning, distributed reinforcement learning is defined as a global structure composed of a plurality of agents. Each agent independently performs some or all of the reinforcement learning tasks while allowing the entire system to achieve the set learning goals. In the current research field, the theory and method of agent research is an important method to solve large-scale and complex information interconnection system, while multi-agents with different structures, distributed performance, dynamic characteristics and large-scale autonomy multi-

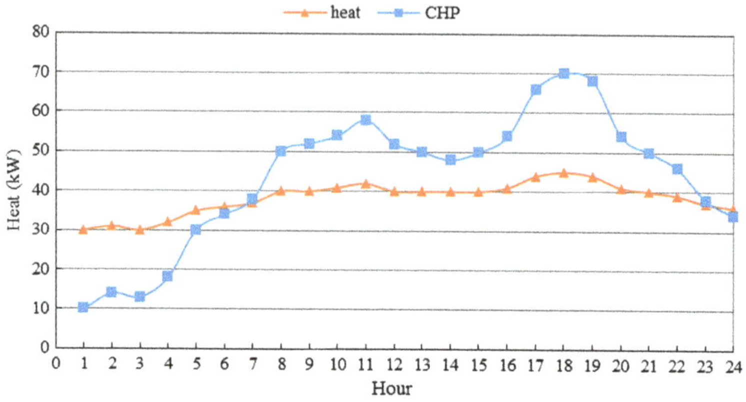

Fig. 10.8 Heat scheduling of We-Energy

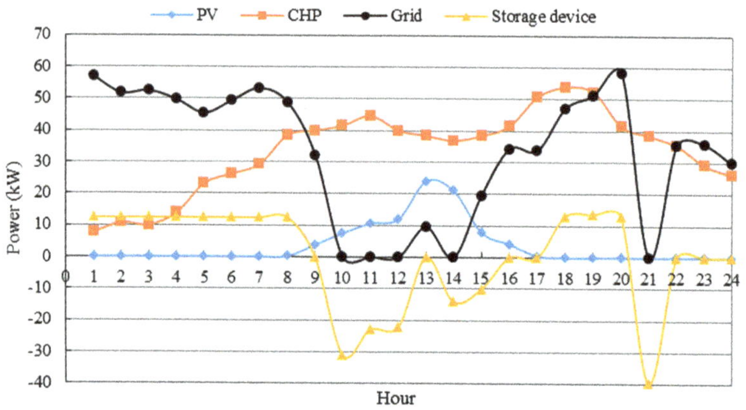

Fig. 10.9 Power scheduling of We-Energy

agent systems can not only achieve individual intelligent optimization, but also make the whole system more responsive, intelligent and social.

At present, the intelligence and fast speed of the system operation are higher and higher, and the information technology and distributed processing power have been developed rapidly. The current study of distributed reinforcement learning model is divided into four types considering the feature of the established system structural differences, including the central reinforcement learning (RLC), independent reinforcement learning (RLI), group reinforcement learning (RLG) and social reinforcement learning (RLS). A multi-agent network based on distributed reinforcement learning is established, and the information of rewards and punishments received by adjacent agent can be obtained by only a small amount of transmission information between each agent. At the same time, in the learning process, according to the overall performance of the system, the application of iterative methods to influence the characteristics of the non-adjacent agent has reached the global optimization objective of the system.

The main feature of the central reinforcement learning model is the learning objective set by the collaborative feature of multi-agent. Based on the classical reinforcement learning algorithm, the optimal interconnection strategy is obtained. In essence, the central reinforcement learning is to make the system distributed optimization problem as a learning objective and to study centralized learning.

The feature of independent reinforcement learning model is that each agent is an independent study module in the whole system, in the process of reinforcement learning, each agent takes the action strategy with the best reward value according to the interaction feedback between itself and the environment. Each agent as an independent agent has no contact with other agents and the agents receive signals from the system's distribution mechanism. This kind of learning mechanism to retain the agent independence of the individual, the optimal target for the whole system is not easy to balance, but also to change the agent number given high degree of freedom, number of convergence problem in the process of reinforcement learning agent. It is suitable for complex system with more agents.

The characteristic of group reinforcement Learning is to set the state of all agents as a state set and define all the actions as interconnected actions, the Q-table of each agent's reinforcement learning is formed by the correspondence between the interconnected state, interconnected actions and the Q value. Therefore, group reinforcement learning is essentially a group behavior in the learning process. Each agent needs to consider the status and actions of other agents which leads to the state set and action set of the whole system are relatively large and the learning speed is slow. Group reinforcement learning is more suitable for the case of a smaller number of agents. The most essential difference of the mathematical model between the group reinforcement learning and the independent reinforcement learning lies in the definition of multi-agent state sets.

The social reinforcement learning model is an extension of independent reinforcement learning, which combines the independent reinforcement learning model with the social structural system and economic model. The combination of social attributes and economic structure enhances the intelligence of multi-agents to conduct more compatible collaborative work and competition at a time to promote the optimization of the system.

10.3.2 Related Issues of Model-Free Energy Optimization

With the high proportion of new energy terminals such as distributed power, electric vehicles and distributed energy storage components which have diversified features including energy production, storage and consumption in power grids, energy internet is characterized by complex nonlinearity, strong uncertainty and strong coupling. As a result, there may be situations where a complete and accurate model of energy internet could not be established in solving practical problems. The reinforcement learning method is based on the autonomy-based evaluation of the learning process considering the action that feedback from the environment without accurate model. This method of replacing rapid mathematical feedback with fast autonomic feedback has a good application effect for solving complex, uncertain and unstructured environmental system problems. Therefore, this chapter adopts the reinforcement learning to realize the model-free optimization of energy internet.

A collaborative optimization control structure composed of multiple We-Energies is established. The optimal scheduling of energy internet is realized through the distributed parallel optimal calculation of multiple We-Energies.

According to the basic description of distributed reinforcement learning model and the establishment of the structure of the collaborative optimization, the analysis to solve the problem of optimal operation for energy internets is discussed as follows.

(1) For a large-scale energy internet with distributed power, the architecture of the energy internet is more complex, and the energy subject with the form of We-Energy will increase. On this basis, the classical centralized optimization method for solving the optimal operation is complicated to deal with the larger system.
(2) In solving complex target function, the traditional planning method, such as Newton method is unable to solve the limitations which are short of flexibility for dealing with such network characteristics and robustness is weak, and it cannot achieve ideal results.

To sum up, the method of distributed parallel coordination optimization is applied to solve the problem of large-scale hybrid optimal operation. This chapter presents a new approach combining with distributed reinforcement learning algorithm to solve hybrid optimal operation problem. According to the Q-learning algorithm which is used to solve the energy management problem has certain feasibility, therefore, the integration of reinforcement learning with the energy Internet structure based on multiple We-Energies interconnection constitutes a distributed reinforcement learning optimization model, but the following key issues need to be considered in model establishment:

(1) For each We-Energy form the independent optimization individual, we make the energy action space not exceed 50,000 according to the actual situation to avoid the influence of the large action space for the learning process.
(2) In the process of reinforcement learning, the objective function of each We-Energy and the interconnection action of We-Energy must be considered at the same time. The actions taken from the We-Energy will affect adjacent We-Energy, thus affecting the overall performance.
(3) The objective function is calculated objectively with the method of parallel computation and serial computation. In the actual energy internet model, the objective function considers the energy loss of each We-Energy while considering the energy loss of the whole energy internet.

10.3.3 Distributed Reinforcement Learning Model for Hybrid Optimal Energy Flow

10.3.3.1 Mathematical Model

The energy internet is a large-scale energy coupling network with electricity, gas and heat. Compared with the power systems, optimal power flow problem of the IMS system becomes more complicated caused by the complex structure. In order to improve efficiency of the method to solve optimization planning problems in energy system, a double-deck multicarrier energy network model is proposed in this chapter.

With interconnections of the IMS system, it is well known that the large-scale systems can be changed as the presence of different political parties. Each party which is defined as "We-Energy" plays the double role of producers and consumers (prosumers) in energy internet. The main body of WE will be the individual, company or community that consists of energy production or storage devices such as distributed generation, energy storage, CCHP and so on. WEs coordinate with each other to guarantee multi energy to reliable transport. In addition, each WE is connected to be considered as the nodes of interconnected multicarrier systems.

Several kinds of WE are presented in this model which consists of energy production device, energy storage device, and user load. Some of them will be connected to district heating plant and gas source. Meanwhile some WE will contain distributed generation such as wind power plant electricity, gas and heat using the coupling way to transmit, while energy hub is defined as an energy carrier to coupling link electricity/gas/heat. Multiple energy inputs will be transformed to other forms of energy as the output of the system. The model of energy hub can be described as

$$\underbrace{\begin{bmatrix} L_e \\ L_g \\ L_h \end{bmatrix}}_{L} = \underbrace{\begin{bmatrix} C_{11} & C_{12} & C_{13} \\ C_{21} & C_{22} & C_{23} \\ C_{31} & C_{32} & C_{33} \end{bmatrix}}_{C} \underbrace{\begin{bmatrix} P_e \\ P_g \\ P_h \end{bmatrix}}_{P}, \tag{10.29}$$

where matrix L stands for electricity, gas and heating output, matrix P are input of electricity, gas and heating from the corresponding grid respectively. Matrix C is the conversion coupling matrix which is the mapping from energy input to the energy output. Then, using energy hub to analyze multi energy condition will implement the collaborative energy flow optimization calculation in IMS.

In order to highlight the importance of the environmental economic and secure benefits of IMS, a double nested optimization model is constructed for both minimizing the WE consumption and optimizing the voltage stability of IMS.

The first layer is considered to minimize the energy cost of system operating for each WE at different energy types so as to make the best use of renewable energy. The total cost of operation is sum of the multi-fuel type consumed by the IMS multiplied by the fuel cost. The objective of the total cost study is as follows:

$$F_1 = \min\left(\sum_{t=1}^{T}\sum_{i=1}^{N^G}(a_i^G(P_i^G)^2 + b_i^G P_i^G + c_i^G) + \sum_{t=1}^{T}\sum_{i=1}^{N^G} Q_i^G + \sum_{t=1}^{T}\sum_{i=1}^{N^F} Q_i^F\right), \tag{10.30}$$

T stands for the number of scheduling time. a_i^G, b_i^G and c_i^G are the coefficients of CHP generators, P_i^G is the active power of generator, N_i^G is the number of CHP generator sets. Also N_i^H is the total number of coal-fired boiler sets. Q_i^G is the heat output from CHP, while Q_i^F is the heat output from traditional coal-fired boiler.

The aim of second layer is to maintain voltage stability of the global system based on voltage constrains. It can be presented as follows:

$$F_2 = \min\frac{1}{n}\sum_{j=1}^{n}\left|V_j - V^*\right|, \tag{10.31}$$

where V_j means the voltage of node j. V^* is the node voltage rating and n is the number of node. Through considering the voltage deviation of each WE, the security of IMS will be improved.

With double-deck optimization structure, these objectives are to be met with such thoroughness and confidence as to be embedded into planning or operation problems in multicarrier energy systems.

In order to achieve this objective, an OEF model must meet the following requirements with the consideration of electrical network, heating network and natural gas network operation constraints. The corresponding constraints are presented in the following.

In electrical network, the active power balance equation and reactive power balance equation connected to the ith bus can be calculated as follows:

$$P_{E,i}^G = P_{E,i}^D + \sum_{j=1}^{N} V_i V_j (G_{E,ij}\cos\theta_{ij} + B_{E,ij}\sin\theta_{ij}), \tag{10.32}$$

$$Q_{E,i}^G = Q_{E,i}^D + \sum_{j=1}^{N} V_i V_j (G_{E,ij}\sin\theta_{ij} - B_{E,ij}\cos\theta_{ij}), \tag{10.33}$$

It can be seen that power balance condition is met with that generator power injection is equal to load demand plus losses in IMS.

In heating network, the energy balance in IMS is expressed by

$$P_{H,i}^S = P_{H,i}^D + \sum_{j=1}^{N_H} P_{H,ij}^L, \tag{10.34}$$

It should be noted that the equation satisfies the equality of heating power producers with the load demand plus heat energy users losses.

While in natural gas network, the gas power balance is expressed by

$$P^S_{G,i} = P^D_{G,i} + \sum_{j=1}^{N_G} P^{comp}_{G,ij} + \sum_{j=1}^{N_G} P^L_{G,ij}, \tag{10.35}$$

In this equation, gas power injection is commensurate with the load demand plus the compressor power and natural gas power.

Meanwhile, it also should be subjected to some inequality constraints of the whole network.

10.3.3.2 Proposed RL Method

Considering the energy network structure as well as the optimal energy flow model, the hybrid reinforcement learning algorithm was applied to solve the problem innovatively.

According to the distributed model of IMS, each WE coordinates and interacts with each other to solve the complex problems. Unfortunately, the centralized approach does not fit well with this narrowly defined double-deck model. They tend to enhance the calculation difficulty and require consideration of multiple aspects. Hybrid Reinforcement Learning (HRL) is an effective way to improve the learning efficiency and solve the problem of "dimension disaster". For the characteristics of the model, the first layer can use the distributed RL for each WE while centralized RL is put into use for the second layer.

A. Implementation of DRL for IMS

DRL is a method which expands the single-agent RL. In DRL, each agent can obtain rewards from adjacent agent with a little information. The global system use iteration to influence non-adjacent agent so as to optimize the performance of the whole system based on reinforcement learning.

Combining the IMS and the distributed reinforcement learning, the implementation of optimal energy flow is the generalization of the Markov decision process

Definition 1 A OEF of IMS is a tuple $(S, A_1, \ldots, A_n, P, R_1, \ldots, R_n)$ where n is the number of WEs, S is the discrete set of environment states, $A_i(i = 1, \ldots, n)$ are the sets of actions available to the WEs, yielding the joint action set $A = A_1 + \cdots + A_n$ that every WEs parallel compute for reinforcement learning which is different from the centralized algorithm. $P : S \times A \times S \to [0, 1]$ means the state transition probability. $R_i(i = 1, \ldots, n)$ are the direct reward functions of each WE.

In the OEF algorithm, through taking into consideration the objects and constrains, control variables in IMS are output power of each WE by adjusting the energy storage equipment and so on. The vector of state variables can be defined as follows:

$$u = [\theta, |V|, \dot{m}, v_g, T_{s,load}, T_{r,load}]^T, \tag{10.36}$$

where π is the vector of pressure. θ, $|V|$ are vectors of unknown angles and magnitudes of voltage. $\dot{m}$ stands for the vector of pipeline mass flows and T^S, T^r are vectors of supply and return temperatures.

For DRL, in order to achieve the goal of OEF, rewards in reinforcement learning should be combined with the objective function and constraint conditions.

The local immediate reward value R of each WE need satisfy the constraint conditions to ensure the validity of the calculation results for each subsystem. Each WE will obtain the optimal strategy by maximizing reward function values.

Definition 2 The local reward for WE is defined as

$$R_{i,0}^K = \begin{cases} 0, & \textit{if constraints are violated} \\ \frac{1}{F_2^K(X)}, & \textit{otherwise} \end{cases}, \tag{10.37}$$

Every WE will check control variables through connected transmission lines to see whether they meet the corresponding boundary conditions. If all of constraints are satisfied, the local reward signal will be set to the negative objective function. Otherwise, it will be zero. The local rewards are applied to each WE to guide action strategy.

The aim of OEF is to seek a best strategy from the action space, so that the global reward is presented as an average value of summation of local rewards from each WE.

$$R^K = \frac{1}{n}\sum_{i=1}^{n} R_{i,0}^K \tag{10.38}$$

The structure of DRL is shown below as Fig. 10.10. Through the multi-energy flow calculation for IMS, the running status of each WE will be acquired. Afterwards, the local reward of WE will be obtained from the information interaction with environment according to Definition 2. Then, the global reward will be updated with the local reward if all the information is available in information fusion unit. The Q-learning unit will operate based on RL iteration rule to find the optimal strategy. Meanwhile, combined with the prior knowledge for initial action set, the learning state and learning efficiency could be improved.

B. Implementation of RL for each WE

In order to minimum energy cost of each WE, RL algorithm is utilized to search for the optimal strategy which considers the operating state of internal equipment from WE.

RL is a method for single-agent which can be achieved by Markov decision process (MDP) modeling. A four-tuple (S, A, P, r) is defined to express the approach, where S

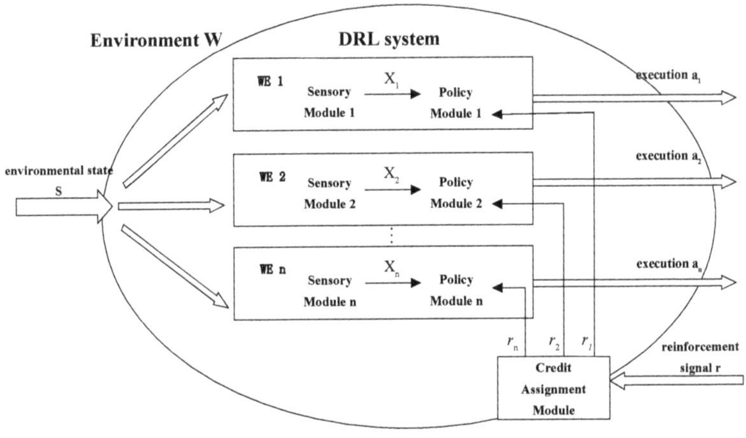

Fig. 10.10 Distributed reinforcement learning optimal energy flow structure

is the limited environmental state space, A stands for the limited action sets. $P(s, a, s')$ stands for the probability that state S transfers to the state s' under control action a and $r(s, a, s')$ is the immediate reinforcement signal given by environment when the state s transfers to the state s' after action a. In an optimal energy flow algorithm of IMS, Q-learning is used to evaluate state of system after an action without an environmental model.

The function Q refers to the optimal reward discount value of WE with action a at state s, denoted by

$$Q(s, a) = r(s, a, s') + \gamma \max_{a \in A} Q(s', a), \tag{10.39}$$

where s, s' are the current state and the state of the next moment. γ is the discount factor which the value of it often is $0 < \gamma < 1$.

Definition 3 Given the RL iteration rule for WE, the Q-learning operation is defined as

$$Q_0(s, a) = 0 \quad \textit{for} \text{ all } s \in S, u \in U, \tag{10.40}$$

$$Q_t(s, a) = \begin{cases} Q_t(s, a) & \textit{if } s \neq s_t \textit{ or } a \neq a_t \\ (1 - \alpha_t)Q_{t-1}(s, a) + \alpha_t[r_t(s, a) + \gamma \max Q_{t-1}(s', a)] & \textit{if } s = s_t \textit{ and } a = a_t \end{cases}, \tag{10.41}$$

where α_t is the learning factor with $0 < \alpha_t < 1$,α_t indicates the proportion of update part in Q value. If its corresponding state s or the action a_t has no samples in state space, the value of this pair will not update. Otherwise, the update rule will be used to approach its value.

Definition 4 The reward for WE is defined as

$$r_{i,0}^{k} = \begin{cases} 0, & if\ constraints\ are\ violated \\ \frac{1}{F_1^k(X)}, & otherwise \end{cases}, \tag{10.42}$$

The action sets are made up of equipment actions in WE and the reward is designed as the reciprocal of reward function. It can be seen that the less energy consumption, the more rewarding under the condition of satisfying the constraints. Greedy strategy is adopted for always choosing the highest Q-value movement in the current state.

In the second layer OEF algorithm, each WE uses two types of equipment to realize operation control strategy, conventional energy components and the energy coupling unit. The first one is power source and the second one is conversion units of each We-Energy such as electrical transformers, power electronic devices, gas compressors, heat exchangers or boilers and others.

RL method only needs to respond to the control effect of assessment information on the basis of the above equipment adjustment. Using RL for each WE to implement energy optimization of energy cost respectively has a higher robustness.

10.3.4 Simulation and Results

In this section, we apply the proposed HRL algorithm to the OEF problem for IMS with the structure of nine interconnection WE which is shown in Fig. 10.11. All the parameters are expressed in per-unit value. In power system, the apparent power per-unit value is 100 MVA, voltage basic value is defined in 1 kV and the scope of bus voltage is [0.9, 1.1]. Power basic value in natural gas network is 100 MW and the pressure is set for 10 bar. Thermal power is 100 MW and the temperature basic value is 100 °C, while time delay of heating pipe network is set for 1 h. Table 10.4 shows the efficiencies of devices for each WE (Fig. 10.12).

In layer 1, the curve of the heat output power given by WEs is shown in Fig. 10.14. In order to minimum voltage deviation in IMS, we should adjust the output of each WE under the system constrains. According to maximum limits of each WE, the power output values of each device is divided into 10 grades on the basis of their maximum limits with 20% fluctuating value.

The action set size of DRL in IMS is $10 \times 6 = 60$ which is smaller than the size of centralized RL in 10^6. Figure 10.13 shows the optimal power flow result of the WE in the first layer. Figure 10.14 shows the learning step of layer 1 for IMS. The reward is the reciprocal of the average voltage deviation. The learning factor α of iteration rule is set for 0.85 and the discount factor γ is 0.2. It can be seen in the chapter that the reward converges to the reward of 25 after 1680 steps. Meanwhile, if we modify the learning factor α to be 0.6, the results are presented in Fig. 10.15. Compared with Fig. 10.14, the learning steps rise to 1960 steps. It shows that learning factor affects

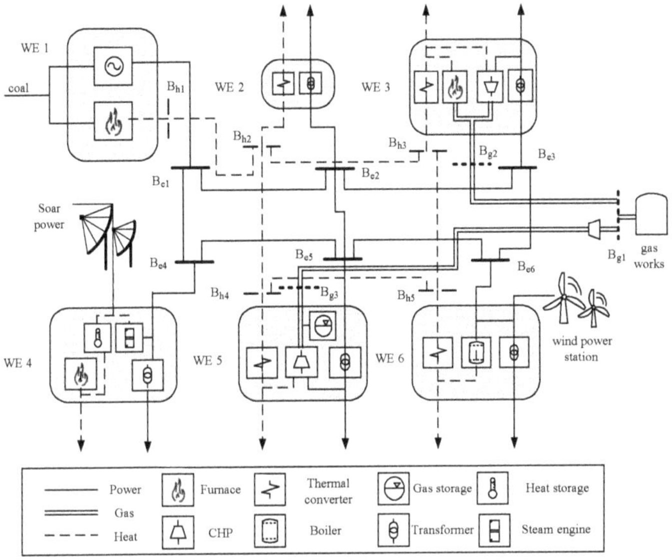

Fig. 10.11 Energy internet simulation diagram

Table 10.4 Test network conversion efficiency of energy equipment

Devices	Efficiency	Capacity
The transformer in WE 1	$\eta_{ge} = 0.3$	2000
The furnace in WE 1	$\eta_{gh} = 0.85$	700
The furnace in WE 3	$\eta_{gh} = 0.7$	700
The CHP in WE 3,5	$\eta_{ge} = 0.35, \eta_{gh} = 0.45$	700
The furnace in WE 4	$\eta_{gh} = 0.9$	300
The heat storage in WE 4	$\eta_h^{ch} = 0.9, \eta_h^{dis} = 0.9$	700
The boiler in WE 4	$\eta_{gh} = 0.9$	150
The gas storage in WE 4	$\eta_g^{ch} = 0.9, \eta_g^{dis} = 0.9$	300

the convergence speed that the higher the learning factor, the better the convergence speed.

Then, compared with voltage deviation before optimization, the indicator of DRL shows that voltage deviation has reduced from 0.045 to 0.038.

In layer 2, According to the devices in each WE, all kinds of control equipment are considered as control variable including generator, boil, CHP, electric boiler, thermal storage devices and gas storage device. In order to set each WE action space with unified standard, the action variables will be discretized in this chapter.

The output of generator is divided into 5 grades according to its limits. The power input values of boil will be in 10 grades based on its capacity. The same as boil,

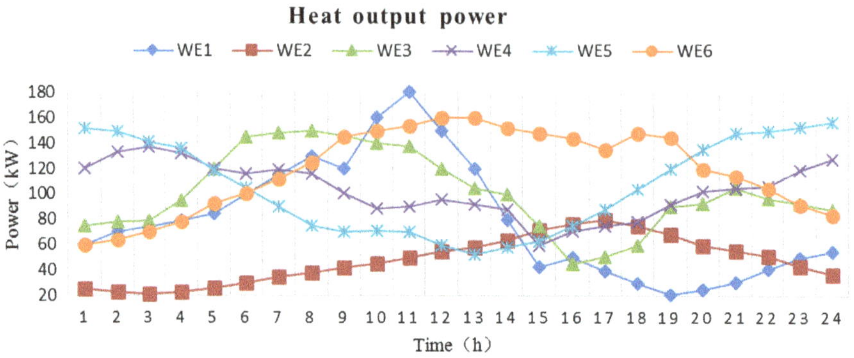

Fig. 10.12 The heat output power of We-Energies

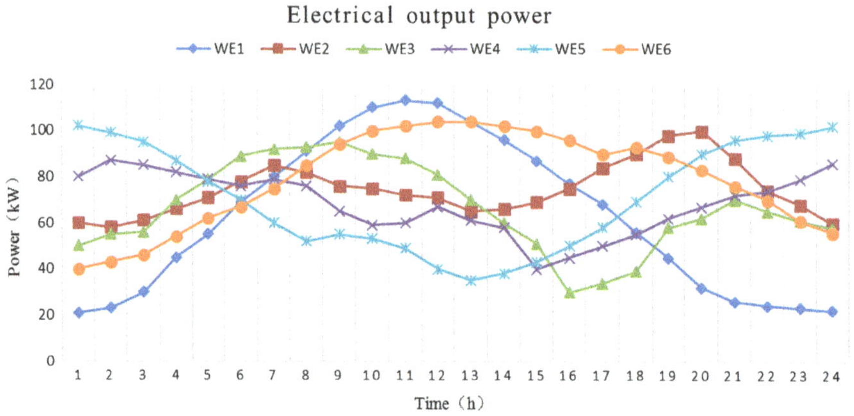

Fig. 10.13 The electrical output power of We-Energies

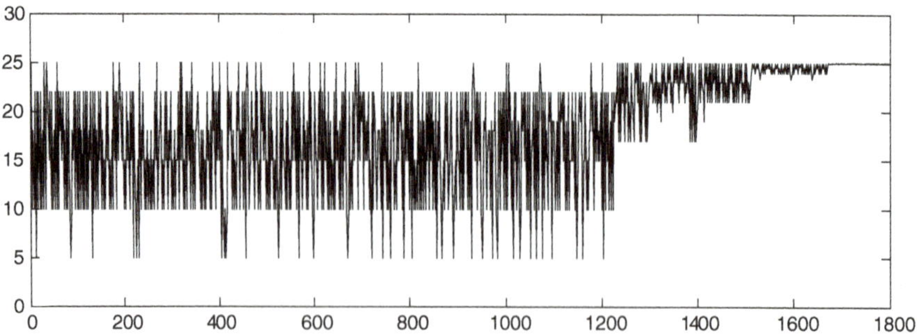

Fig. 10.14 Distributed RL process at $\alpha = 0.85, \gamma = 0.2$

the power input of CHP is divided into 5 grades. The power input of electric boil is in 10 grades on average based on their maximum limits. The power input values

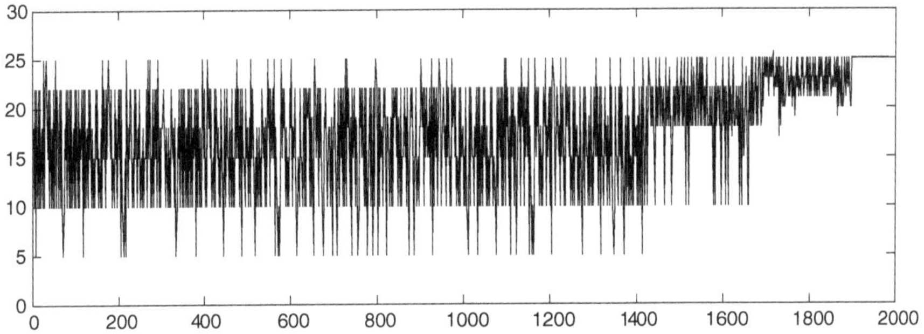

Fig. 10.15 Distributed RL process at $\alpha = 0.6, \gamma = 0.2$

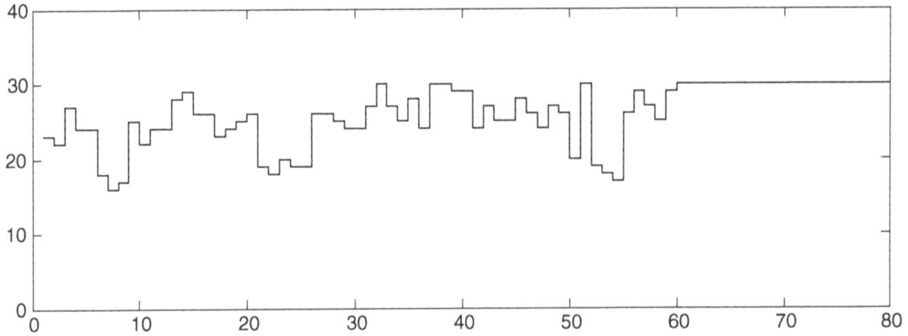

Fig. 10.16 DRL optimization process

of thermal storage devices are in 3 grades and the power input values of gas storage device is in 5 grades

As we can see that there are 5 control variables in WE1. With the grades defined above, the number of actions for WE1 can be calculated as $5 \times 10 \times 5 \times 10 \times 3 = 7500$. WE2 can be calculated as $3 \times 10 = 30$. Other WE follows a similar pattern to WE1.

Through the RL for each WE, the optimal energy cost has reduced 21%-26% from the objective function (Fig. 10.16).

Based on the system simulation result for Modified 6-bus, hybrid reinforcement learning is verified to be satisfied with the optimal energy flow in interconnected multicarrier systems.

10.4 Conclusion

This chapter proposes a model free optimal model and presents the RL algorithm that can drive these agents to parallel learn behaviors. The model free optimal model not only reduces the loss of the system but also improves the safety and reliability, especially considering the accuracy of information. In addition, using the RL algorithm for this optimal energy model has some advantages. The method of utilizing each WE to undertake the task independently to reach the coordinated system is suitable for hierarchical control mode of energy internet. It has the ability to solve the Inadaptability of multi-objective function and real-time performance of large-scale network computing.

References

1. A. Shabanpour-Haghighi, A.R. Seifi, Energy flow optimization in multicarrier systems. IEEE Trans. Ind. Inf. **11**(5), 1067–1077 (2015)
2. J. Hu, Q. Sun, T. Fei, A game-theoretic pricing model for energy internet in day-ahead trading market considering distributed generations uncertainty. in *IEEE Symposium Series on Computational Intelligence* (2016)
3. R.Z. R´ıos-Mercado, C. Borraz-S´anchez, Optimization problems in natural gas transportation systems: a state-of-the-art review. Appl. Energy **147**, 536–555 (2015)
4. C.M. Correa-Posada.; P. Sanchez-Martin, Integrated power and natural gas model for energy adequacy in short-term operation. IEEE Trans. Power Syst., **30**(6), 3347–3355 (2015)
5. A. Alabdulwahab, A. Abusorrah, X. Zhang, X, Coordination of interdependent natural gas and electricity infrastructures for firming the variability of wind energy in stochastic day-ahead scheduling. IEEE Trans. Sustain. Energy **6**(2), 606–615 (2015)
6. M. Chaudry, N. Jenkins, M. Qadrdan et al., Combined gas and electricity network expansion planning. Appl. Energy **113**(6), 1171–1187 (2014)
7. X. Xu, H. Jia, H.D. Chiang et al., Dynamic modeling and interaction of hybrid natural gas and electricity supply system in microgrid. IEEE Trans. Power Syst. **30**(3), 1212–1221 (2015)
8. X. Zhang, M. Shahidehpour, A. Alabdulwahab et al., Optimal expansion planning of energy hub with multiple energy infrastructures. IEEE Trans. Smart Grid **6**(5), 2302–2311 (2015)
9. C. Liu, M. Shahidehpour, J. Wang, Coordinated scheduling of electricity and natural gas infrastructures with a transient model for natural gas flow. *Chaos* **21**(21) (2011)
10. X. Zhang, G.G. Karady, S.T. Ariaratnam, Optimal allocation of CHP-based distributed generation on urban energy distribution networks. IEEE Trans. Sustain. Energy **5**(5), 246–253 (2014)
11. X. Fei, M. Yong, Combined electricity-heat operation system containing large capacity thermal energy storage. Proc. CSEE **34**(29), 5063–5072 (2014)
12. G. Zepeng, K. Chongqing, Operation optimization of integrated power and heat energy systems and the benefit on wind power accommodation considering heating network constraints. Proc. CSEE **35**(14), 3596–3604 (2015)
13. Z. Pan, H. Sun, Q. Guo, Energy Internet oriented security analysis method for multi-energy flow. Power Syst. Technol. **40**, 1627–1634 (2016)
14. Q. Huang, M.L. Crow, G.T. Heydt, J.P. Zheng, S.J. Dale, The future renewable electric energy delivery and management (FREEDM) system: the energy internet. Proc. IEEE **99**(1), 133–148 (2011)
15. C. Chen, S. Duan, T. Cai, B. Liu, G. Hu, Smart energy management system for optimal microgrid economic operation. IET Renew. Power Gener. **5**(3), 258–267, (2011)
16. X. Ou, Y. Shen, Z. Zeng, G. Zhang, L. Wang, Cost minimization online energy management for microgrids with power and thermal storages. in *24th International Conference on Computer Communication and Networks (ICCCN), Las Vegas, NV* (2015), pp. 1–6
17. M. Motevasel, A.R. Seifi, T. Niknam, Multi-objective energy management of CHP (combined heat and power)-based micro-grid. Energy **51**, 123–136 (2013)
18. L. Ma, N. Liu, J. Zhang, W. Tushar, C. Yuen, Energy management for joint operation of CHP and PV prosumers inside a grid-connected microgrid: a game theoretic approach. IEEE Trans. Ind. Inf. **12**(5), 1930–1942 (2016)

19. A. Shabanpour-Haghighi, A. R. Seifi, Energy flow optimization in multicarrier systems. *IEEE Trans. Ind. Inf.* **11**(5) (2015)
20. T. Krause, G. Andersson, K. Frohlich, A. Vaccaro, Multiple-energy carriers:modeling of production, delivery, and consumption. *Proc IEEE,* **99,** 15–27 (2011)
21. B. Stott, J.L. Marinho, OAlsac, Review of linear programming applied to power system rescheduling. in *Power Industry Computer Applications Conference* (1979), pp. 142–154
22. K.A. Clements, P.W. Davis., K.D. Frey, An interior point algorithm for weighted least absolute value power system state Eestimation (1991)
23. R.Z. Ríos-Mercado, C. Borraz-Sánchez, Optimization problems in natural gas transportation systems: a state-of-the-art review. Appl. Energy **147**, 536–555 (2015)
24. Q. Sun, R. Han, H. Zhang et al., A multiagent-based consensus algorithm for distributed coordinated control of distributed generators in the energy internet. IEEE Trans. Smart Grid **6**(6), 3006–3019 (2015)
25. Y. Xu, W. Zhang, W. Liu et al., Multiagent-based reinforcement learning for optimal reactive power dispatch. IEEE Trans. Syst. Man Cybern. Part C **42**(6), 1742–1751 (2012)
26. K. Doya, Reinforcement learning in continuous time and space. Neural Comput. **12**(1), 219–245 (2000)
27. Z. Chen, S. Jagannathan, Generalized Jamilton–Jacobi–Bellman formulation -based neural network control of affine nonlinear discrete-time systems. *IEEE Trans. Neural Netw.* **19**(1), 90–106 (2008)
28. T. Dierks, B. TThumati, S. Jagannathan, Optimal control of unknown affine nonlinear discrete-time systems using offline-trained neural networks with proof of convergence. Neural Netw. **22**(5–6), 851–860 (2009)
29. D. Liu, Q. Wei, Policy iteration adaptive dynamic programming algorithm for discrete-time nonlinear systems. IEEE Trans. Neural Netw. Learn. Syst. **25**(3), 621–634 (2014)
30. H. Modares, F.L. Lewis, Naghibi-Sistani M B. Integral reinforcement learning and experience replay for adaptive optimal control of partially-unknown constrained-input continuous-time systems. Automatica **50**(1), 193–202 (2014)
31. H. Zhang, L. Cui, X. Zhang et al., Data-driven robust approximate optimal tracking control for unknown general nonlinear systems using adaptive dynamic programming method. IEEE Trans. Neural Netw. **22**(12), 2226–2236 (2011)
32. R. Kamalapurkar, H. Dinh, S. Bhasin et al., Approximate optimal trajectory tracking for continuous-time nonlinear systems. Automatica **51**, 40–48 (2015)
33. B. Kiumarsi, F.L. Lewis, H. Modares et al., Reinforcement [formula omitted]-learning for optimal tracking control of linear discrete-time systems with unknown dynamics. Automatica **50**(14), 1167–1175 (2014)
34. H. Zhang, Q. Wei, Y. Luo, A novel infinite-time optimal tracking control scheme for a class of discrete-time nonlinear systems via the Greedy HDP iteration algorithm. IEEE Trans. Syst. Man Cybern. Part B Cybern. **38**(4), 937–942 (2008)
35. R. Kamalapurkar, L. Andrews, P. Walters et al., Model-based reinforcement learning for infinite-horizon approximate optimal tracking. in *Decision and Control.* (IEEE, 2015), pp 5083–5088
36. C. Yu, M. Zhang, F. Ren et al., Emotional multiagent reinforcement learning in spatial social dilemmas. IEEE Trans. Neural Netw. Learn. Syst. **26**(12), 3083–3096 (2015)
37. P. Plamondon, B. Chaib-Draa, A.R. Benaskeur, A Q-decomposition and bounded RTDP approach to resource allocation. in *Autonomous Agents & Multiagent Systems/Agent Theories, Architectures, and Languages* (2007), pp 1–8
38. L. Matignon, G.J. Laurent, N.L. Fort-Piat, Coordination of independent learners in cooperative Markov games. Piat (2009)

Lightning Source UK Ltd.
Milton Keynes UK
UKHW02n1137170718
325839UK00001B/45/P

9 789811 305221